Operations Research: Concepts and Applications

Operations Research: Concepts and Applications

Edited by **Courtney Hoover**

NY RESEARCH PRESS

New York

Published by NY Research Press,
23 West, 55th Street, Suite 816,
New York, NY 10019, USA
www.nyresearchpress.com

Operations Research: Concepts and Applications
Edited by Courtney Hoover

International Standard Book Number: 978-1-63238-498-0 (Hardback)

Printed in the United States of America.

Contents

Permissions

List of Contributors

Preface

This book has been an outcome of determined endeavour from a group of educationists in the field. The primary objective was to involve a broad spectrum of professionals from diverse cultural background involved in the field for developing new researches. The book not only targets students but also scholars pursuing higher research for further enhancement of the theoretical and practical applications of the subject.

Operations research is mainly concerned with the application and creation of decision making tools. This discipline employs mathematical science, statistical analysis, and mathematical optimization. It is often applied to real world problems to determine the maximum and minimum of quantities such as profit, output and inventory. This book covers some topics of crucial importance related to this field such as simulation, discrete systems, time series analysis, decision support systems, product management, quality management, logistics, warehousing, etc. It elucidates new techniques and their applications in a multidisciplinary approach. It attempts to understand the multiple branches that fall under this discipline and how such concepts have practical application in the current scenario. The topics covered in this book offer the readers new insights in the field of operations research.

It was an honour to edit such a profound book and also a challenging task to compile and examine all the relevant data for accuracy and originality. I wish to acknowledge the efforts of the contributors for submitting such brilliant and diverse chapters in the field and for endlessly working for the completion of the book. Last, but not the least; I thank my family for being a constant source of support in all my research endeavours.

Editor

Statistical Data Analyses on Aircraft Accidents in Japan: Occurrences, Causes and Countermeasures

Kunimitsu Iwadare, Tatsuo Oyama

National Graduate Institute for Policy Studies (GRIPS) 7-22-1 Roppongi Minato-ku, Tokyo, Japan
Email: oyamat@grips.ac.jp

Abstract

We investigate the major characteristics of the occurrences, causes of and counter measures for aircraft accidents in Japan. We apply statistical data analysis and mathematical modeling techniques to determine the relations among economic growth, aviation demand, the frequency of aircraft/helicopter accidents, the major characteristics of the occurrence intervals of accidents, and the number of fatalities due to accidents. The statistical model analysis suggests that the occurrence intervals of accidents and the number of fatalities can be explained by probability distributions such as the exponential distribution and the negative binomial distribution, respectively. We show that countermeasures for preventing accidents have been developed in every aircraft model, and thus they have contributed to a significant decrease in the number of accidents in the last three decades. We find that the major cause of accidents involving large airplanes has been weather, while accidents involving small airplanes and helicopters are mainly due to the pilot error. We also discover that, with respect to accidents mainly due to pilot error, there is a significant decrease in the number of accidents due to the aging of airplanes, whereas the number of accidents due to weather has barely declined. We further determine that accidents involving small and large airplanes mostly occur during takeoff and landing, whereas those involving helicopters are most likely to happen during flight. In order to decrease the number of accidents, i) enhancing safety and security by further developing technologies for aircraft, airports and air control radars, ii) establishing and improving training methods for crew including pilots, mechanics and traffic controllers, iii) tightening public rules, and iv) strengthening efforts made by individual aviation-related companies are absolutely necessary.

Keywords

Statistical Data Analysis, Aircraft Accidents, Causes of Aircraft Accidents, Accident-Prevention

Measures, Mathematical Model Analysis, Exponential Distribution, Negative Binomial Distribution

---

1. Introduction

Plane crashes, excluding those related to the Self-Defence Forces, have occurred more than 10 times a year on average during the period from 1974 to 2010 in Japan. With the growth of the economy, e.g. gross domestic product (GDP), aircrafts have been becoming a more common andpopular means of transportation for ordinary citizens. Now, most people's international travel is conducted by airplane. Although the demand growth for aircraft has been very high, the number of airplane accidents has been decreasing over the last 30 to 50 years. Airplane accidents have been efficiently prevented by modern technological innovations. In this paper, we aim to investigatethe major characteristics of the occurrences and causes of and countermeasures for aircraft accidents in Japan. We apply statistical data analysis and mathematical modeling techniques to determine the relationships among economic growth, aviation demand, the frequency of aircraft/helicopter accidents, the intervals of accidents, and the number of fatalities.

Regarding aircraft accident studies, Bazargan and Vitaly [1] investigated the data obtained from 40,000 accidents that occurred in the United States between 1983 and 2002 in order to determine the level of influence on the accidents of various factors such as the gender, age and length of experience of pilots who had been involved in accidents.They concluded that inexperienced pilots had a high probability of causing accidents due to various types of human errors. Chang and Yang [2] classified survival factors for in-flight safety into the following categories: design and equipment of the aircraft, education for pilots and crew and cooperation among them, passengers' behavior and safety education for them, and the ability to cope with emergencies. They selected 47 items from these categories and ranked them in terms of the level of importance. Based on their ranking, they stated that proper instructions by pilots and crew in case of emergency and evacuation and training for managing a crowd of people were of particular importance.

In Japan, Rinoie [3] reported various efforts made for the future, discussing the methods of ensuring safety from an operational viewpoint, which included the improvement of cockpit functions, the development of a system that would enable safe and solid flight operation, and also reforms to the pilot licensing system. Shiroyama et al. [4] discussed establishing an aircraft accident investigation system adopted in the United States, introducing a research method based on the collection of experts' knowledge, support for victims and their families from third-party organizations, and procedures for pursuing legal responsibility. Shiroyama et al. [4] emphasized that Japan should review the existing situation in terms of adequate manpower, the establishment of an investigation framework and proper treatment of bereaved families. Nakasuga [5] suggested the need for psychologial training and fail-safes in order to cope with "panic" from the aspect of human factors, explaining the relationship between aircraft and people in the process of automation in light of aspects of human nature such as miscommunication and habituation. Kamiyama [6] investigated carefully various types of safety measures based on data of plane crashes he statistically compiled. As far as we know, there is scant literature where plane accident data have been statistically analyzed.

This paper aimed to investigate the historical trend and transitional characteristics of aircraft accidents in Japan to determine the properties and characteristics related with their occurrences and causes, identify how accident-prevention policies and measures were taken in addition to indicating that the accident data regarding the occurrence of accidents and the number of fatalities were in accordance with various mathematical models. Based on the results of these attempts, we try to forecast for the future and discuss safety and related measures in order to contribute to air traffic safety. There have been various studies on aircraft accidents at home and abroad as mentioned above, most of which have focused on contributing factors in a particular accident and have analyzed measures to deal with them in detail. Our paper aims to look at aircraft accidents from a panoramic viewpoint in terms of descriptive statistics. We believe that there are very few arguments over what the priority of an effective measure will be when we are required to deal with a certain accident. Also, we try to evaluate whether past measures are appropriate or not. In Japan, the Japan Transport Safety Board, a branch of the Ministry of Land, Infrastructure, Transport and Tourism (MLIT), publishes comprehensive information regarding accident data. By investigating the highly reliable accident data, this paper aims to contribute to the formulation of effective measures against aircraft accidents in Japan.

In the next chapter, we utilize aircraft accident data regarding Japan's major means of air transportation such as small and large airplanes and helicopters over the past 37 years. We analyze accidents due to the aging of airplanes, changes in the number of fatalities, causes of the accidents, and other factors. In Chapter 3, we identify the history of the accidents and measures to prevent them as well as the characteristics of their trends in terms of accident factors. In Chapter 4, we build mathematical models by investigating the economic growth, aviation demand, the relationship between the aviation demand and the accident figures, the occurrence intervals of accidents, and the number of fatalities during a certain period of time. Finally, in the Summary and Conclusions we discuss the major findings, and highlight the necessary issues that should be included in countermeasures for preventing accidents for small and large airplanes and helicopters.

2. Aircraft Accidents in Japan

2.1. Airplane and Helicopter Accidents

In Article 76, Paragraph 1 of the Civil Aeronautics Act established in 1952, aircraft accidents are defined as follows: i) crash, collision and fire on an aircraft, ii) casualties and damage to properties due to the aircraft accident, iii) missing and dead among those who were inside the aircraft, iv) minor collision with other aircraft, and v) other aircraft-related accidents specified in the ministerial ordinance of the MLIT. Government regulations require the reporting of the above accidents. The so-called aircraft as a means of transportation range from small airplanes accommodating a few passengers to large craft that are able to carry hundreds of people. There are also helicopters that fly on various missions, including lifesaving, news gathering activities, observation and monitoring. In this study, when analyzing accidents involving airplanes and helicopters, we classify aircraft weighing more than 5700 kg as large airplanes, and those weighing less than 5700 kg as small airplanes. In our data analysis, we use accident data made publicly available by the Japan Transport Safety Board on their website [7].

Table 1 shows the numbers of airplane and helicopter landings and accidents with small airplanes figure and their respective fatalities from 1974 to 2010 in parentheses. Thus the number of large airplane accidents and fatalities can be calculated by deducting the data in parentheses from the corresponding total in **Table 1**. Missing data in **Table 1** are not available. The number of small airplane accidents marked a record high of 28 in 1976 while the lowest is 2 in 2006 and 2009. The coefficient of variation is 0.54 for small airplanes accidents data. The highest number of fatalities was 17 in 1986, whereas there were none in 1993, 1999, 2006 and 2009. The coefficient of variation is 0.87 for fatalities data.

Regarding the number of accidents involving large airplanes, the highest is 7 in 1976, and the lowest is 0 in 2010. The coefficient of variation is 0.51, smaller than the above small airplanes case. The number of fatalities in 1985 is 520 people, the highest on record. The number of years that there are no fatalities is 22. The coefficient of variation is 4.22, the largest among all types of aircraft accidents data. In terms of aircraft in general, the highest number of accidents is 35 in 1976 while the lowest is 5 in 2006 and 2009. The coefficient of variation is 0.47.

Figure 1 is a piecewise linear graph showing the number of such accidents as well as the number of aircraft landings. From 1974 to 2010, the total number of accidents involving small airplanes was 351, which accounts for 75.1% of the total number of 467 aircraft accidents for the entire period. **Figure 2** shows the changes over the period in the number of helicopter accidents as well as the number of landings. The highest number of accidents is 24 in 1978, whereas years 2003 and 2006 both marked 1, the lowest. The coefficient of variation is as small as 0.50. The number of fatalities marked a record high of 31 in 1990, whereas it was 0 in 1995, 1999 and 2006. The coefficient of variation is 1.03, indicating a considerable variation.

1) Landings and accidents of airplanes/helicopters

Figure 1 indicates that, for the last 37 years, the growth in aviation demand in Japan shows an increase in the number of landings of about 15,000 annually. On the other hand, the number of accidents decreases by about 0.39 annually. **Figure 2** suggests that the number of landings and accidents involving helicopters has gradually decreased over the last 20 to 30 years. As for helicopters, the number of landings decreases annually by about 435, and the annual accident figure decreases by about 0.36. The relationship between the aviation demand and accident figures will be discussed in Section 4.1.

The number of accidents per 100,000 landings per year varies.However, they are generally on the decrease, dropping to 0.5 in 2006 after peaking at 6.7 in 1976. Helicopter figures have also been decreasing after peaking

Table 1. Number of airplane and helicopter landings, accidents and fatalities.

Year	Airplane				Helicopter		
	Landing (10^3 times)	Number of accidents		Number of fatalities		Landing (10^3 times)	Number of accidents	Number of fatalities

Year	Landing (10^3 times)	Number of accidents		Number of fatalities		Landing (10^3 times)	Number of accidents	Number of fatalities
1974	536	21	(16)	3	(2)		17	6
1975	521	17	(15)	5	(5)		15	1
1976	521	35	(28)	16	(14)		17	5
1977	551	14	(10)	2	(2)		12	3
1978	579	15	(10)	7	(6)		24	8
1979	610	20	(15)	5	(5)		16	1
1980	610	15	(12)	2	(2)		23	2
1981	622	15	(11)	33	(7)		15	10
1982	631	17	(13)	6	(4)		11	5
1983	629	14	(13)	10	(9)		14	2
1984	634	7	(3)	3	(2)		10	4
1985	637	15	(12)	524	(4)		18	6
1986	653	19	(14)	17	(17)		11	3
1987	681	21	(16)	14	(14)		8	3
1988	704	7	(5)	7	(6)		13	3
1989	734	11	(8)	7	(5)		11	7
1990	747	13	(12)	5	(5)		15	31
1991	778	11	(9)	1	(1)	31	17	18
1992	800	6	(4)	5	(4)	29	9	3
1993	806	8	(5)	0	(0)	27	16	2
1994	856	9	(6)	266	(2)	29	15	6
1995	896	8	(6)	5	(4)	26	4	0
1996	925	18	(12)	12	(7)	26	10	13
1997	949	14	(11)	7	(7)	26	6	8
1998	982	16	(14)	15	(15)	25	5	2
1999	982	9	(7)	0	(0)	28	8	0
2000	996	7	(6)	3	(3)	30	12	5
2001	996	7	(6)	6	(6)	31	11	3
2002	1020	8	(5)	5	(5)	27	11	4
2003	1027	12	(9)	11	(11)	27	1	4
2004	1043	13	(11)	3	(3)	28	6	5
2005	1063	9	(7)	1	(1)	27	7	8
2006	1097	5	(2)	0	(0)	27	1	0
2007	1100	9	(3)	2	(2)	27	7	6
2008	1077	10	(6)	3	(1)	23	4	2
2009	1060	5	(2)	0	(0)	23	6	5
2010	1060	7	(7)	4	(4)	23	4	14
Total	30113	467	(351)	1015	(185)	538	410	208
Average	813.9	12.6	(9.5)	27.4	(5.0)	26.9	11.1	5.6
Standard deviation	197.8	6.0	(5.2)	94.3	(4.4)	2.4	5.6	5.8
Maximum	1100	35	(28)	524	(17)	31	24	31
Minimum	521	5	(2)	0	(0)	23	1	0

Values in parentheses are the number of accidents and fatalities of small airplane. Source: The Japan Transport Safety Board [7], Aviation Statistics Catalogue, Japan Aeronautic Association, 1984-85, 1998-99, 2001, 2011 [8].

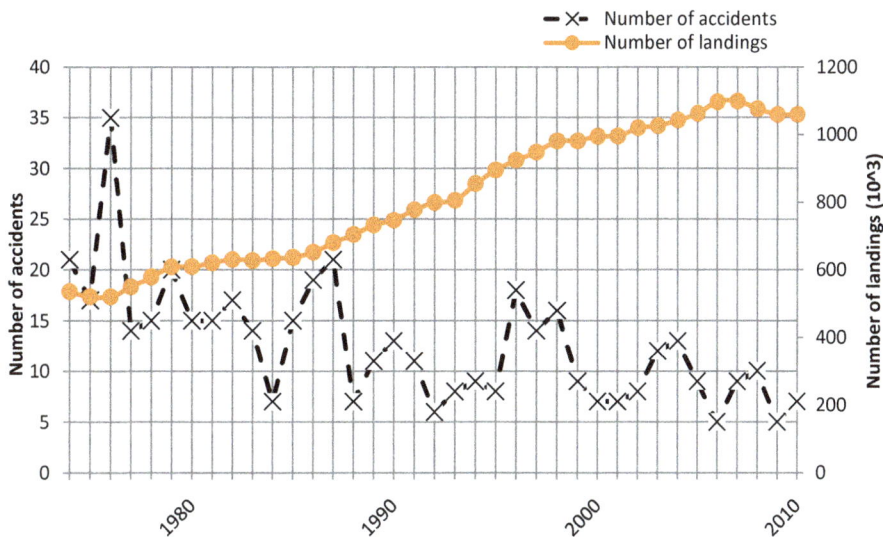

Figure 1. Landings and accidents for aircraft (1974-2010).

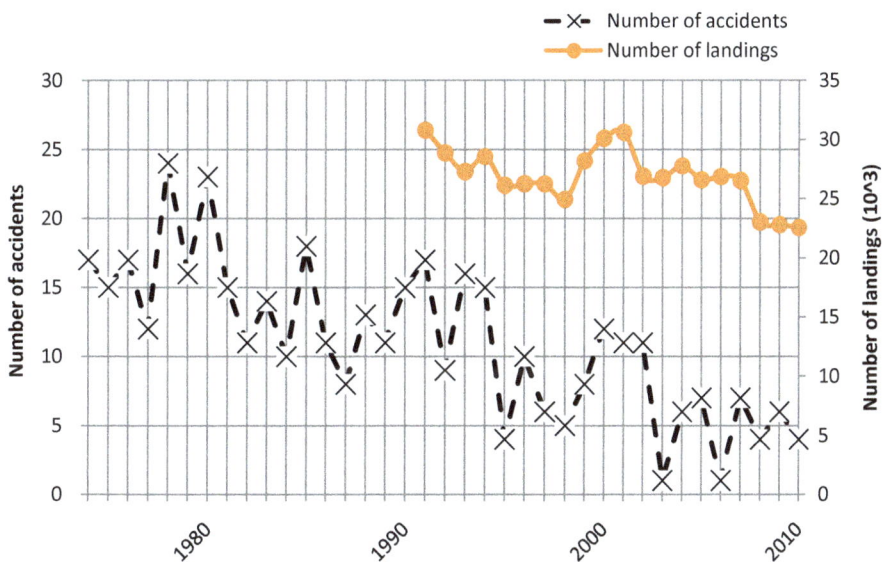

Figure 2. Landings and accidents for helicopter (1974-2010).

at 58.7 in 1993. As the number of landings is unavailable separately for both small and large airplanes, we could not make a simple comparison. However, as a means of air transportation, we can say there are numerous helicopter accidents. The accident figures vary with the season. For the entire period from 1974 to 2010, summer (June to August) has the highest accident figure of 191, followed by autumn (September to November) of 80, and 82 in spring (March to May). The lowest is 57 in winter (December to February). These seasonal trends are due to helicopter use. For comparison, we can see no particular seasonal variation in accidents of other types of aircraft. As described above, the number of helicopter accidents per landing is 16 times greater than for other types of aircraft. In order to reduce the number of accidents per landing, the first step is to implement preventive measures against helicopter accidents.

2) Aircraft accidents by type

Figure 3 shows the 3-year and 5-year moving average values respectively for large airplanes. The number of accidents for both small and large aircraft types has been decreasing every year. However, large airplanes display more variations and a smaller degree of decrease, compared to small airplanes. Furthermore, we can also see the possibility that the number of accidents increases or decreases about every 10 years. Especially, the crash

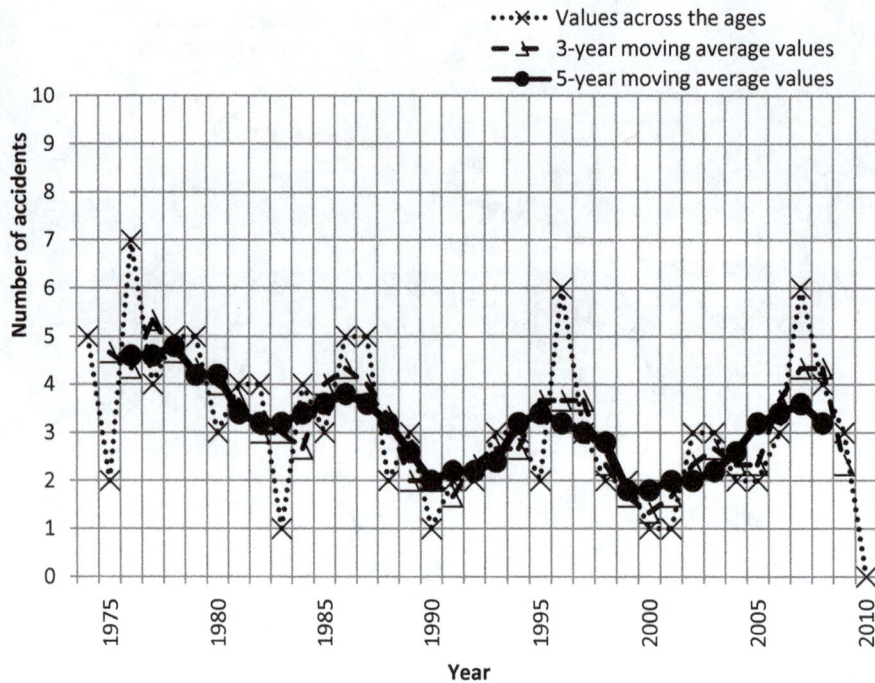

Figure 3. 3-year and 5-year moving averages for the number of accidents involving large airplanes (1974-2010).

of a JAL plane killing 520 people occurred in 1985, and the China Airlines accident killing 264 people happened in 1994. There were many other accidents around this period. In 2007, about 10 years after the China Airlines accident, the number of accidents was relatively high. Therefore, we can see that the number of accidents involving large airplanes increases approximately on a 10-year cycle.

2.2. Fatalities Due to Airplane and Helicopter Accidents

Regarding fatalities from aircraft accidents during the period from1974 to 2010, the highest was 524 people in 1985 whereas there was no fatalities in 1993, 1999, 2006 and 2009. It is still vivid in our memory that, on August 12, 1985, a Japan Airlines Jumbo Jet Flight 123 from Tokyo to Osaka crashed into a ridge near Mount Osutaka in Gunma Prefecture, taking the lives of 520 people. There is no doubt that this plane crash is the worst in Japan's avaiation history. However, also in the last 30 years, China Airlines Flight 140 crashed at Nagoya Airport, killing 264 people on board [7]. Such large airplane accidents become worse as the sizes of aircraft become larger, taking numerous lives once an accident occurs.This definitely has the potential of creating a serious social issue. Regarding helicopter accidents, the highest number of fatalities during our study period was 31 in 1990. There were no fatalities in 1995, 1999 and 2006. The high number of fatalities in 1990 is due to crashes involving large helicopters carrying 10 passengers and 6 passengers.

Calculating the standard deviation (SD) corresponding to the annual fatalities data during our study period, we found that those for small airplanes and helicopters were relatively low:4.4 and 5.8, respectively. On the other hand, the SD for large airplanes is as high as 94.6. Meanwhile, the SDs of accident figures for small airplanes and helicopters were 5.1 and 5.6, respectively. The SD of the number of large airplane accidents is relatively small at1.6. This means that the annual data of accident figures for large airplanes varies only a little; however, the number of fatalities is very likely to increase drastically once a large accident occurs.

Figure 4 and **Figure 5** show the past trend of the number of fatalities from accidents involving small and large airplanes, and helicopters, respectively. Note that the JAL Jumbo Jet crash in 1985 and the China Airlines crash in 1994 are both excluded from the graph due to their enormous fatalities, 520 and 264, respectively. In **Figure 4**, the high number of fatalities in 1982 is due to an accident at Haneda Airport on February 9, killing 24 people. In **Figure 5**, the high number of fatalities in 1990 is due to crashes involving 2 helicopters having a large number of passengers, 10 and 6, respectively.

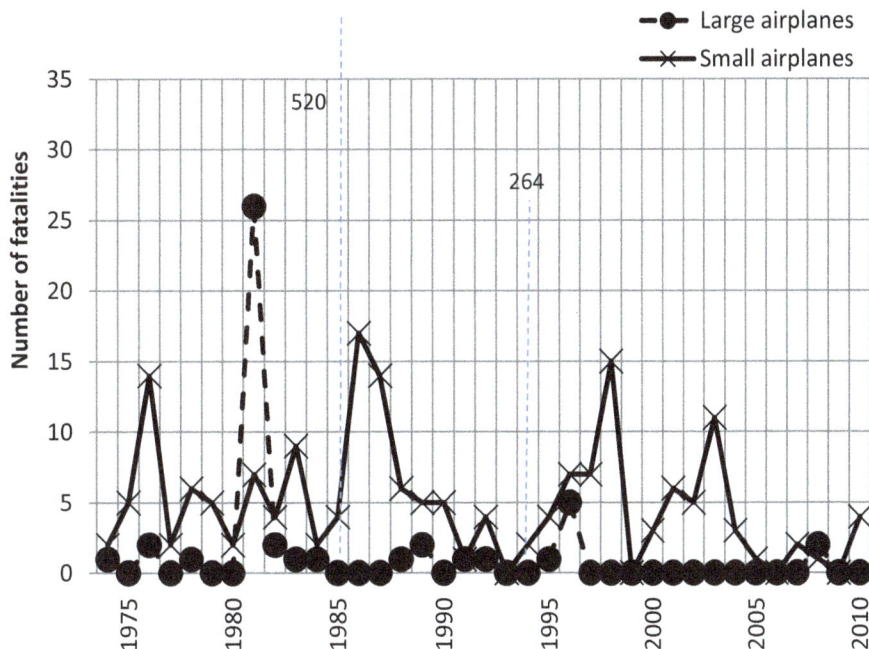

Figure 4. Fatalities from aircraft accidents (1974-2010).

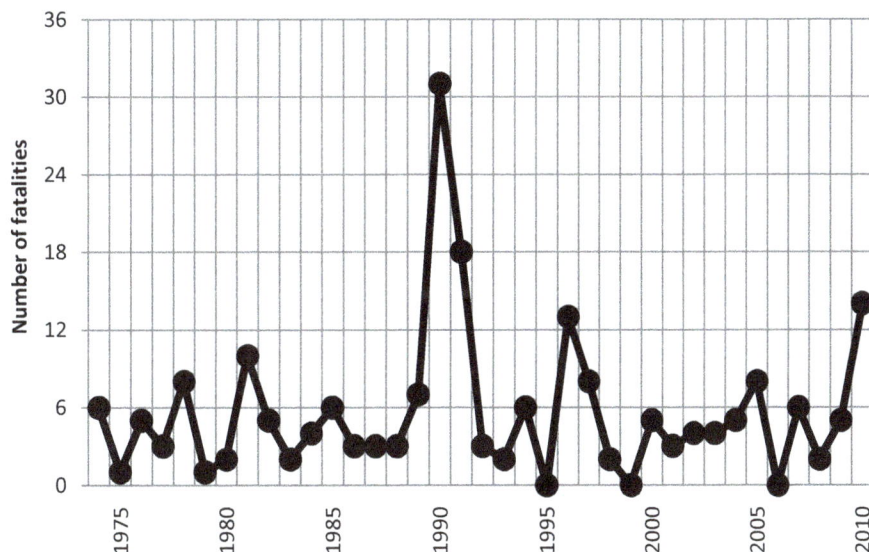

Figure 5. Numbers of fatalities from helicopter accidents (1974-2010).

Table 2 shows the number of fatalities from accidents for each period when we divide the entire study period between 1974 and 2010 into three periods: Period I: 1974-1985, Period II: 1986-1997, and Period III: 1998-2010. From **Table 2**, we can see that small airplanes and helicopters have a large number of fatalities in Period II. The increased number of fatalities in Period II, despite the fact that the number of accidents declined, is due to an increase in the number of accidents involving small airplanes and helicopters accommodating a relatively large number of passengers.

3. Causes of Airplane/Helicopter Accidents and Accident-Prevention Measures

3.1. Causes of Airplane and Helicopter Accidents

Using the data during the period from 1974 to 2007 in order to investigate the causes of airplane/helicopter ac-

Table 2. Number of accidents and fatalities.

Period		Large airplanes		Small airplanes		Helicopters	
I	Accidents	47	(3.9)	158	(13.2)	192	(16.0)
	Fatalities	554	(46.2)	62	(5.2)	52	(4.3)
II	Accidents	37	(3.1)	108	(9.0)	135	(11.3)
	Fatalities	274	(22.8)	72	(6.0)	97	(8.1)
III	Accidents	32	(2.5)	85	(6.5)	83	(6.4)
	Fatalities	2	(0.2)	51	(3.9)	58	(4.5)
Total	Accidents	116	(3.1)	351	(9.5)	410	(11.1)
	Fatalities	830	(22.4)	185	(5.0)	207	(5.6)

Values in Parentheses are annual averages.

cidents, we define three periods in this chapter slightly differently from in the previous chapter as Period I: 1974-1985, Period II: 1986-1996, and Period III: 1997-2007, respectively, because causes of accidents are under investigation from 2008.

1) Causes of small airplane accidents

Figure 6 shows the percentages of individual causes for small airplane accidents. Since one accident may occur as a result of a combination of multiple causes, the cumulative total number of causes may not equal the total number of accidents. As shown in **Figure 6**, 234 accidents (56.7%) of all the 336 accidents are due to pilot error. In terms of other causes having a high percentage, equipment failure caused 64 accidents (15.5%), adverse weather conditions accounted for 56 accidents (13.6%), and improper maintenance resulted in 24 accidents (5.8%). Thus, pilot error, equipment failure and weather factors account for more than 80% of all the causes. Especially, small airplanes are mostly controlled manually by the pilot during the entire flight including takeoff and landing rather than staying on autopilot, which resulted in the highest percentage of pilot error. Pilot error during takeoff and landing is classified into 4 major factors: operation, judgment, procedure and guidance. Operational errors include insufficient deceleration and premature veering. Judgment errors are typically a delay in making a decision to abort takeoff and a misreading of altitude. Procedural errors usually indicate mistakes in regular procedures such as failing to lower the landing gear. Misguidance occurs during training. The breakdown of the major error factors is as follows: 109 operational errors (46.6%), 102 judgment errors (43.6%), 45 procedural errors (19.2%) and 18 misguidance (7.7%). The percentages provided for individual error factors represent their ratios to the total number of accidents. Since one accident may occur as a result of a combination of multiple errors, the total number does not amount to 100%. Equipment failures consist of 42 engine failures (66.7%), 14 malfunctions of the landing gear (22.2%) and 7 defects in other parts of the aircraft (11.1%). Accidents due to adverse weather conditions are mostly caused by clouds and fog, resulting in 27 accidents (46.4%) and 5 accidents (8.9%), respectively. These two factors alone account for more than half of the entire weather-triggered accidents. Most such accidents occur due to poor visibility. Other factors include wind and air turbulence, accounting for 13 accidents (21.4%) and 9 accidents (16.1%), respectively, from which we can see that aircraft are very susceptible to wind as they fly in the air.

In **Figure 7**, we show the causes of small airplane accidents based on classifying into two situations: during takeoff/landing and other situations(mostly during flight). From **Figure 7**, we can see the overall decrease in the frequency of major accident causes such as pilot error, bad weather, equipment failures and improper maintenance, in each period. The number of accidents due to pilot error factors during takeoff/landing dropped drastically from 77 in Period I to 38 in Period II.

Table 3 shows the changes in the number of small airplane accidents in the breakdown of pilot error factors foreach period. All types of errors remarkably decreasedin frequency from Period I to Period II. That is, there are various improvement efforts including enhanced operability, and new attempts at error proofing in order to avoid pilot error including judgment errors and procedural errors, such as warning tones and improved layout within the cockpit. Furthermore, the number of errors during training is also reduced, from which we can assume that there have also been some improvements in education. All of these contributed to the decrease in

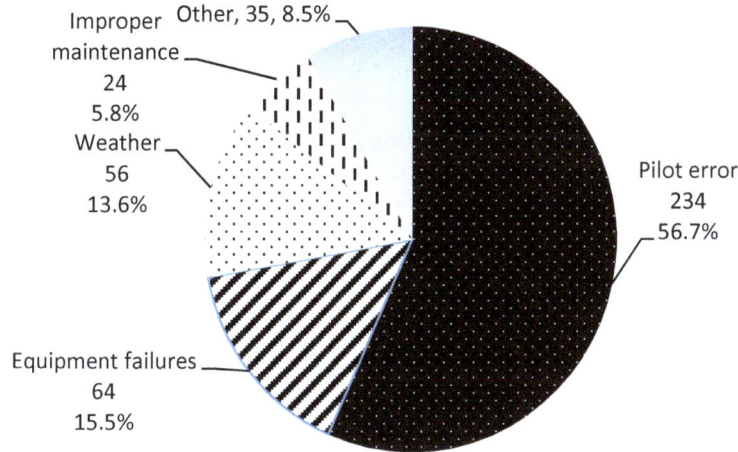

Figure 6. Accident causes for small airplanes (1974-2007).

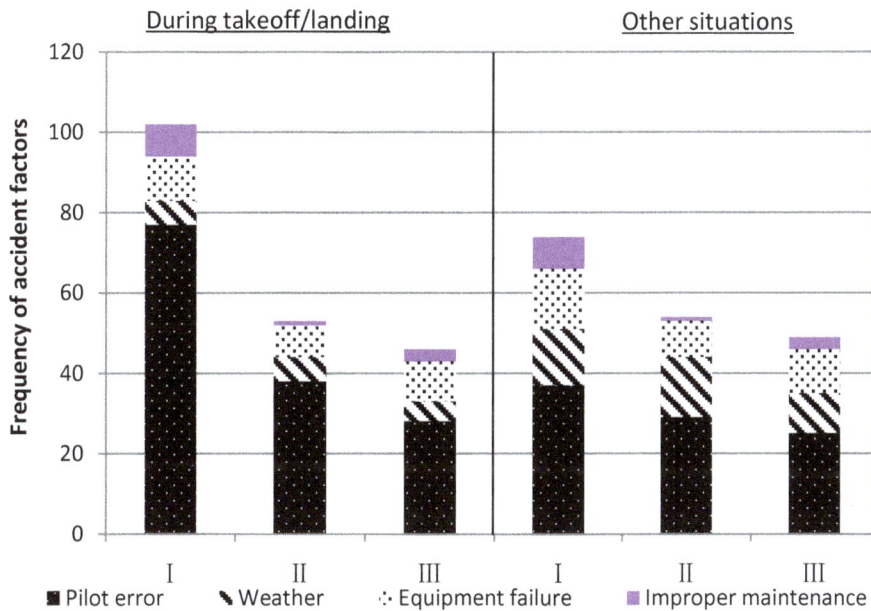

Figure 7. Accident factors for small airplanes (1974-2007).

Table 3. Types of pilot error for small airplanes accidents.

Period	During TL					During FL				
	OE	JE	PE	MG	Total	OE	JE	PE	MG	Total
I	40	19	20	10	89	18	23	3	1	45
II	23	10	8	3	44	11	25	0	0	36
III	12	6	11	2	31	5	19	3	2	29
Total (%)	75 (45.7)	35 (21.3)	39 (23.8)	15 (9.1)	164 (100.0)	34 (30.9)	67 (60.9)	6 (5.5)	3 (2.7)	110 (100.0)

TL: Take Off and Landing; FL: Flight; OE: Operational Errors; JE: Judgment Errors; PE: Procedural Errors; MG: Misguidance.

the number of small airplane accidents. There are still old models of aircrafts in service. As they are replaced by new models, we expect that the number of accidents due to pilot error will further decrease.Accidents during flight are mostly due to judgment errors in pilot errors, the number of which is 23, 25 and 19 in Periods I, II and

III, respectively. Though judgement errors during flight are very likely to result in fatalities, we can see that technical improvements are far from enough to help reduce such errors. A solution to this issue is expected to contribute greatly to a reduced number of accidents and fatalities.

2) Causes of large airplane accidents

There were 109 large airplane accidents in total during the period from 1974 to 2007. Accident causes consist of weather, pilot error, equipment failure and improper maintenance, which account for 38 accidents (32.2%), 37 (31.4%), 12 (10.2%) and 4 (3.4%), respectively, as shown in **Figure 8**. The other 27 accidents are due to birds, collisions with obstacle at the airport, deaths by disease, and some other unknown reasons. While pilot error is less than that of small airplanes in terms of the percentage, they still account for a major part of the causes of large airplane accidents. As shown in **Figure 9**, about 90% of accidents are due to air turbulence and wind, from which we can see that large airplanes are very susceptible to strong wind. Comparing accident factors during takeoff and landing with those that occurred in other situations, the biggest factor is pilot error, which accounts for 28 accidents (51.9%) that occurred during takeoff and landing. On the other hand, accidents that happened in other situations, in other words, those that happened mainly during flight, are mostly due to weather factors, accounting for 26 accidents (40.6%). In **Figure 10**, the period is divided into three periods I, II, and III. Further, large airplane accidents are classified into two situations: during takeoff and landing and other

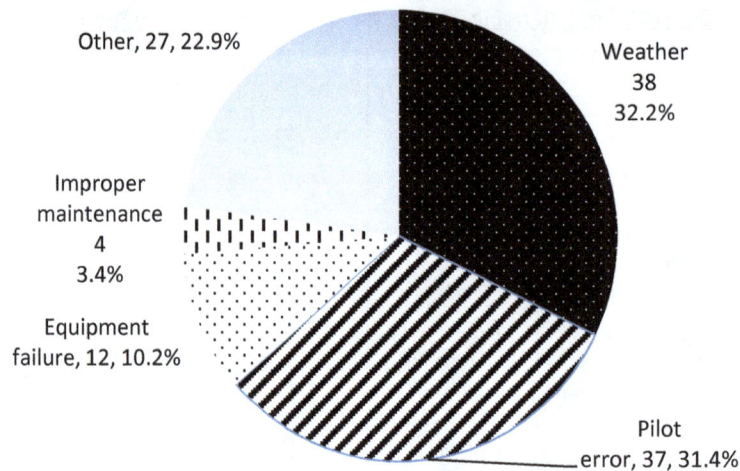

Figure 8. Accident factors for large airplanes (1974-2007).

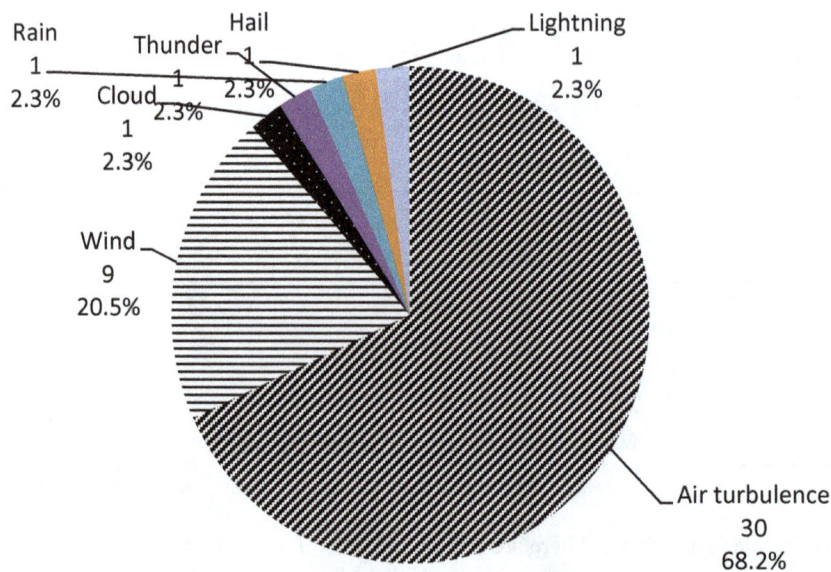

Figure 9. Weather factors for large airplane accidents (1974-2007).

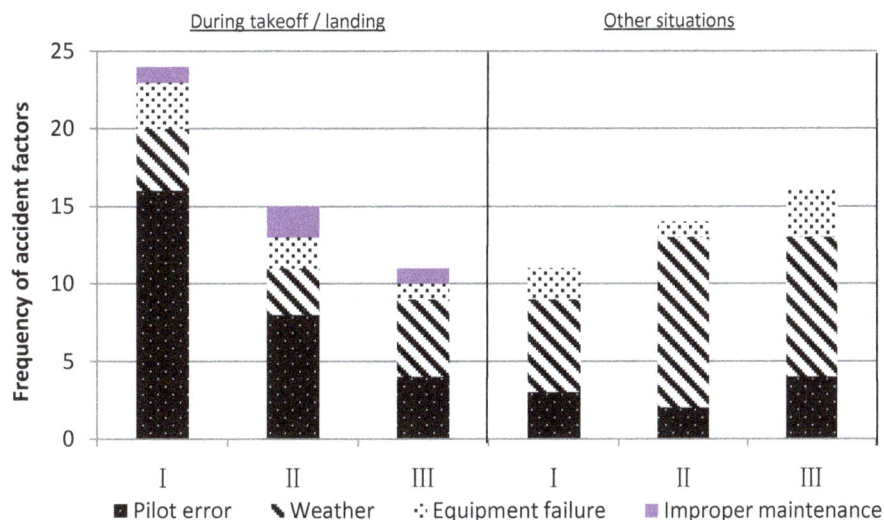

Figure 10. Accident factors for large airplanes (1974-2007).

Situations (mostly during flight). We can see that the number of accidents decreases substantially from Period I to Period II. This is mostly due to technological developments in aviation policy concerning air traffic services, aircraft and airport control, all of which were rapidly upgraded after two big accidents in 1971 before Period I. Such developments are likely to be the result of the implementation of various measures, including a ban on flying in clouds, upgrading of traffic lanes for private airlines and the Self Defense Forces, installation of radars, and securing a sufficient number of air traffic controllers. The number of accidents due to weather in Period I, II and III are 10, 14 and 14, respectively. This implies that reducing aircraft accidents due to weather is not so easy.

Most accident causes during takeoff and landing are due to pilot error, whereas those in other situations are attributed to weather conditions. **Table 4** shows the changes in frequency for each type of pilot error during takeoff and landing, and flight for each period from 1974 to 2007.Pilot errors due to operational errors and judgment errors are on a decreasing trend. This is presumably attributed to the improvement in aircraft performance and efforts associated with Crew Resource Management (CRM). The increased number of operational errors during flight in Period III is due to excess maneuvering by the pilot. The number of accidents due to adverse weather conditions was 6 (54.5%) in Period I, which increased to 11 (78.6%) in Period II. This is possibly because of the recent shifting weather patterns as a result of global warming. However, it decreased to 9 (56.3%) in Period III, 2 accidents less than in Period II. This decrease is probably due to an increase in pilots' knowledge and technological development. In 1982, Delta Air Lines Flight 191 was hit by air turbulence and crashed during final landing in the United States, killing 135 people. After this accident, the U.S. Federal Aviation Administration (FAA) encouraged improvements in weather radar technology and revisions in technical training (see Rodriguesand Cusick [9]).

The number of pilot errors in other situations was 2 (14.3%) in Period II and 4 (25.0%) in Period III. The frequency of equipment failures also increased slightly from 1 (7.1%) in Period II to 3 (18.8%) in Period III. We assume that this is due to the drastic increase in aviation demand. However, the increase in the number of pilot errors is presumably due to the following factors. In Period I, the factors include mere operational errors by the pilot, insufficient communication with the air traffic controller and incorrect assumptions. None of them are external factors such as natural phenomena, which implies there is still room for improvement. On the other hand, a large number of accidents in Periods II and III are due to natural phenomena and incorrect responses thereto. Accidents due to the factors in Period I are likely to be reduced relatively easily by technological developments, intensive training and improved communication methods. On the contrary, factors associated with natural phenomena have not yet been eliminated, as air turbulence occurs quite frequently and appropriate responses thereto are quite problematic.

3) Causes of helicopter accidents

Figure 11 shows the causes of helicopter accidents. Through the entire period, there were 396 helicopter accidents, the causes of which are mainly pilot error, accounting for 276 accidents, nearly 70% of all accidents.

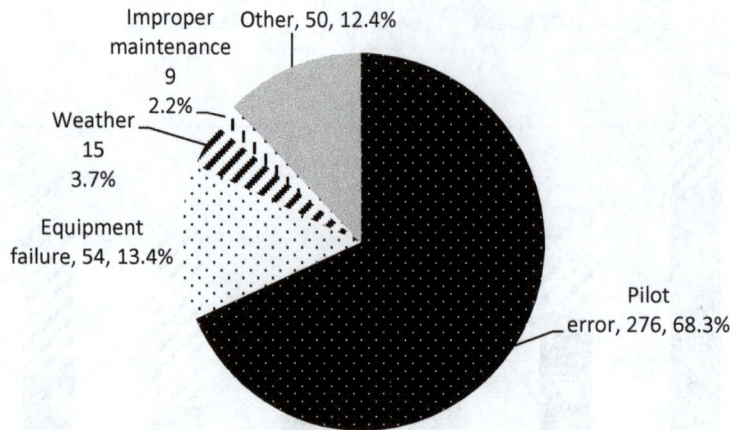

Figure 11. Accident causes for helicopters (1974-2007).

Table 4. Types of pilot error for large airplanes accidents.

Period	During TL					During FL				
	OE	JE	PE	MG	Total	OE	JE	PE	MG	Total
I	11	4	2	0	17	1	2	1	0	4
II	6	4	2	0	12	1	1	0	1	3
III	4	1	0	1	6	3	1	0	1	5
Total (%)	21 (60.0)	9 (25.7)	4 (11.4)	1 (2.9)	35 (100.0)	5 (14.3)	4 (11.4)	1 (2.9)	2 (5.7)	12 (100.0)

TL: Take Off and Landing; FL: Flight; OE: Operational Errors; JE:Judgment Errors; PE: Procedural Errors; MG: Misguidance.

The number of major accident factors including pilot error (276 accidents, 68.3%), equipment failure (54 accidents, 13.4%), weather (15, 3.7%), and improper maintenance (9, 2.2%) accounts for almost 90% of all accidents. Thus, we know that almost 70% of helicopter accidents are due to various types of pilot error. Weather factors consist of fog accounting for 5 accidents, clouds and air turbulence, each of which accounts for 3 accidents, and wind and snow, each of which accounts for 2 accidents. Here, we can see that low visibility due to fog and clouds, air turbulence and wind are all closely related to the adverse impact on the operation of aircraft.

Figure 12 shows the changes in the frequencies of major helicopter accident factors in individual periods from Period I to Period III. The major accident factors are pilot error, weather, equipment failure and improper maintenance. From **Figure 12**, we find that helicopter accidents occur more often during flight than during takeoff and landing. This is in contrast with small and large airplanes, which have more accidents during takeoff and landing than during flight. There is an increase in the percentage of accidents due to pilot error that occurred mainly during flight, rather than during takeoff and landing, whereas those due to other factors decreased.This indicates that it is possibly more difficult to prevent pilot error during flight than to reduce other accident factors.

The most common use for helicopters at the time of an accident is crop-dusting or other similar operations which accounted for 145 accidents (36.6%) followed by the transfer of personnel which accounted for 113 accidents (28.5%), load carrying which accounted for 63 accidents (15.9%), 37 accidents (9.3%) during training, transporting the body of a helicopter which accounted for 20 accidents (5.1%), and 18 accidents during maintenance flights (4.5%). As agricultural chemicals for rice are intensively applied from June to August, we can see the seasonal feature that many helicopter accidents occur in summer. As we the investigate purposes of the use of helicopters involved in accidents for each period, we note that the frequency of crop-dusting purposes has decreased. One of the factors for such a change is possibly that crop-dusting is no longer performed by manned helicopters, which have been replaced by unmanned helicopters.

The number of accidents decreased from 192 in Period I to 129 in Period II, dropping by about 30%, which further decreased by 40% to 75 in Period III.The number of accidents due to judgment errors in pilot error is 150, accounting for more than 70%. **Table 5** shows the breakdown of pilot error for helicopter accidents. Accidents

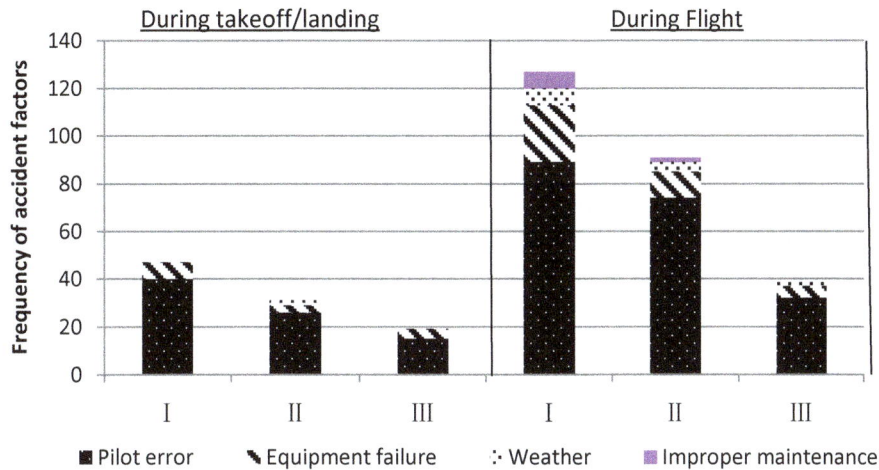

Figure 12. Frequency of accident factors for helicopters (1974-2007).

Table 5. Types of pilot error for helicopters accidents.

Period	During TL					During FL				
	OE	JE	PE	MG	Total	OE	JE	PE	MG	Total
I	28	14	0	7	49	20	73	0	0	93
II	21	7	0	2	30	21	55	1	1	78
III	14	4	2	2	22	6	22	0	0	28
Total (%)	63 (62.4)	25 (24.8)	2 (2.0)	11 (10.9)	101 (100.0)	47 (23.6)	150 (75.4)	1 (0.5)	1 (0.5)	199 (100.0)

TL: Take Off and Landing; FL: Flight; OE: Operational Errors; JE: Judgment Errors; PE: Procedural Errors; MG: Misguidance.

due to judgment errors by the pilot, collisions in particular, seem to be very difficult to eliminate, even if helicopter technologies are mechanically and electronically improved. These days, however, various new systems have been established, such as one that produces a warning tone or activates an automatic braking function when recognizing an obstacle, which has already been utilized in cars. When they are further upgraded and linked to GPS so that they can be applied to helicopters, the number of accidents due to collisions may possibly decrease.

3.2. Accident Prevention Measures and Their Effects

In order to prevent aircraft accidents, we consider the following three elements to be important: i) technology innovation for accident prevention and aircraft design; ii) training of cabin attendants, airport traffic controllers and mechanics; iii) air traffic public administration and policy making.

Regarding technology innovation for accident prevention and aircraft design, prevention systems for collision with mutual aircraft or with the ground, avoidance of human error by introducing autopilot systems, safety designs preventing fatal break down in case of some functional defects and inspection methods for old aircraft can possibly prevent aircraft accidents. Additionally, designs for considering survival factors are also important in order to increase the survival rate in case of accidents. In the past, the Minimum Safe Altitude Warning (MSAW) was developed and introduced in ground control in 1975, the Ground Proximity Warning System (GPWS) was required in commercial aircraft in 1977. In the same year, the Radar Data Processing system (RDP) and the Automated Radar Terminal System (ARTS) were installed and a radar network system covering all the airspace of Japan was completed, resulting in the enhancement of the safety of flight and the efficiency of air traffic control as described by Sugie [10], Yanagida [11] and Tsumajika [12]. Shimada [13] mentioned that a glass cockpit effectively preventing pilot error was installed in 1981. In 1978, the adoption of a new design principle called "Damage Tolerance Design" was announced in Japan. Damage tolerance design includes a structure that prevents a fatal breakdown even if there is a minor scratch less than a few millimeters in size that may not be detected during a regular checkup.

The second important element for preventing accidents is training of all concerned personnel. Nakasuga [5] suggested the necessity of psychological training for pilots especially in case of "panic". Also we believe that improving communication with other aircrafts as well as with controllers, the copilot and cabin attendants in an aircraft, improving inspectors' skill and enhancing general health management are very important for reducing aircraft accidents. After the accidents in the 1970s, Japan Airlines (JAL) tightened its standards for the employment examination for pilots and the promotional examination for captains, and established the Special Committee on Safety Measures, in which an intensive reanalysis of past accidents was conducted in order to take preventive measures against common pilot errors (see Yanagida [11]). In the 1980s, JAL was the first to adopt cockpit resource management, which was created from the viewpoint of human behavior, focusing on how to further enhance performance as a team and to create a related training method in order to eliminate human error. Sincethen, Crew Resource Management (CRM) for all crew in aircraft has been adopted in each airline (Sugie [10], Yanagida [11]). In response to an accident due to mental illness in 1982, the Council for Civil Aviation conducted deliberations where measures were undertaken to improve the health management of and operational control for pilots. The establishment of the Japan Aeromedical Research Center was part of such measures (see Yamamoto [14]).

Finally, we emphasize the importance of air traffic public administration and policy making. According to Masui and Takahashi [15] and Tazaki [16], on July 3, 1971, Toa Domestic Airlines Flight 533, a YS-11 called "Bandai" en route from Sapporo to Hakodate Airport crashed into the mountainside of Yokotsudake due to inclement weather, killing all 68 people on board. After the Bandai accident, while the Transport Ministry was still reviewing the enhanced aviation security measures for aviation administrative agencies on July 30, 1971, All Nippon Airways Flight 58, a Boeing 727 airliner, collided with a Japan Air Self-Defense Force F-86F fighter jet over Shizukuishi, Iwate Prefecture, killing all 162 people on board. In response to these two accidents, the government drew up the Outline of Aviation Security Measures, which included accelerated improvements in aviation security facilities such as the Instrument Landing System (ILS), the VHP Omni directional Range(VOR) and the Distance Measuring Equipment (DME), the prompt introduction of a long-range radar surveillance network covering all parts of Japan, other safety measures such as securing sufficient numbers of aviation security personnel, including air traffic controllers, as well as reviewing the establishment of related administrative agencies in order to improve the current aviation security framework. Additionally, they established "the outline of air traffic safety emergency measure". Thus, the government decided to separate training and testing areas for military and civilian use. Further, as recommended by the ICAO, flights in Visual Meteorological Condition (VMC) were prohibited, eliminating rules that allowed flights at varying altitudes. "The memorandum regarding adjustments between the aviation administration by the transport ministry and operations of the self-defense forces" was revised and incorporated to serve as a report focusing on air traffic safety, rather than for the old operational coordination purposes. Accelerated modernization of the air traffic control system was strongly agreed upon within the Japan Civil Aviation Bureau (JCAB). The JCAB established the flight control investigation headquarters, in which emergency security measures were formulated and the air control system was reviewed.

We can say that pilot, aircraft, airports, cabin attendants, airport traffic controllers as well as technology and techniques comprise construct ageneral technical system. Therefore, comprehensive policy making and enforcement from the safety perspective of the technical system are important, and subsequently, we can reduce the number of accidents and mitigate the damage. In addition, Shiroyama et al. [4] emphasized that surveying system and approach to the bereaved are also important.

In order to reduce the number of accidents, we consider that measures against adverse weather conditions for large airplanes are effective, whereas those against pilot error are necessary for small airplanes and helicopters. As for large airplanes, measures against air turbulence are essential. Regarding small airplanes, as described in Chapter 3.1, the number of judgment errors during flight has hardly decreased. Therefore, technological innovations and improved education that could help reduce judgment errors would be effective. It would also be necessary to adopt error proofing that prevents procedural errors during takeoff and landing and to enhance training in order to avoid operational errors. Although error proofing has already been adopted in the latest model of small airplanes, airplanes manufactured 30 years ago may be involved in accidents today. Accordingly, it would probably be a good idea to develop a device for error proofing that can also be installed in older aircraft, and to create regulations that require a mandatory application of such a device. As for helicopters, there are a large number of collisions with obstacles during flight. Therefore, it would be helpful to have a system in which enhanced preliminary observation is required. Also the authorities can provide advance warning of the danger of obstacles upon receipt of flight notification.

Other systems have also been developed such that warning tones are emitted or a control function such as automatic braking is activated when obstacles are detected like the ones utilized in cars. They may possibly be upgraded so they are connected with the GPS and are thus able to issue a warning based on the collision prediction. Their utilization in small airplanes and helicopters may possibly contribute to the further decrease in the number of collision accidents.

4. Mathematical Model Analyses on Airplane/Helicopter Accidents

4.1. Aviation Demand and Airplane/Helicopter Accidents

Figure 13 shows the changes in the number of landings and the value of GDP (Gross Domestic Product). Up to the year 1991, aviation demand increased as the GDP grew. From 1992 to 1998, the aviation demand stayed on an upward trend despite the downturn growth in the GDP. After 1998, the aviation demand slowed as well, taking a downward turn in 2007. In the future, we expect the possibility of high growth in aviation demand due to a possible increase in the GDP and upgrading of airports, among other factors.

When we plot the number of aircraft landings on the horizontal axis, and the number of accidents on the vertical axis using the data from 1974 to 2010, we obtain the graph as shown in **Figure 14**. If we define Japan's economic growth as the explanatory variable (x) and the number of aircraft landings as the dependent variable (y)[1], the parameter estimate for the simple regression model using the data between 1974 and 1991 is givenas follows:

$$y = 0.772x + 4.13$$
$$(21.4) \quad (38.3)$$

(1)

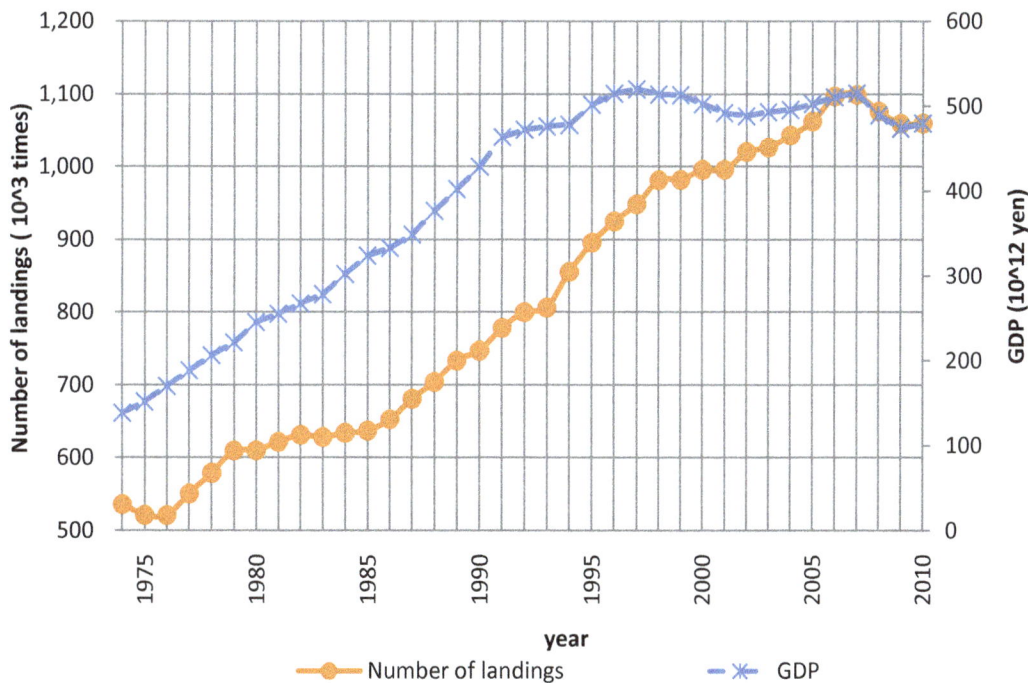

Figure 13. Aircraft landings and GDP (1974-2010). Source: Ministry of Internal Affairs and Communications Statistics Bureau, Japan Statistical Yearbook, 1981, 1985, 1989, 1993/94, 1997, 2002, 2007, 2010, 2014, Japan Statistical Association [17].

---

[1]Source: Ministry of Internal Affairs and Communications Statistics Bureau, Japan. Statistical Yearbook, 1981, 1985, 1989, 1993/94, 1997, 2002, 2007, 2010, 2014, Japan. Statistical Association [17].

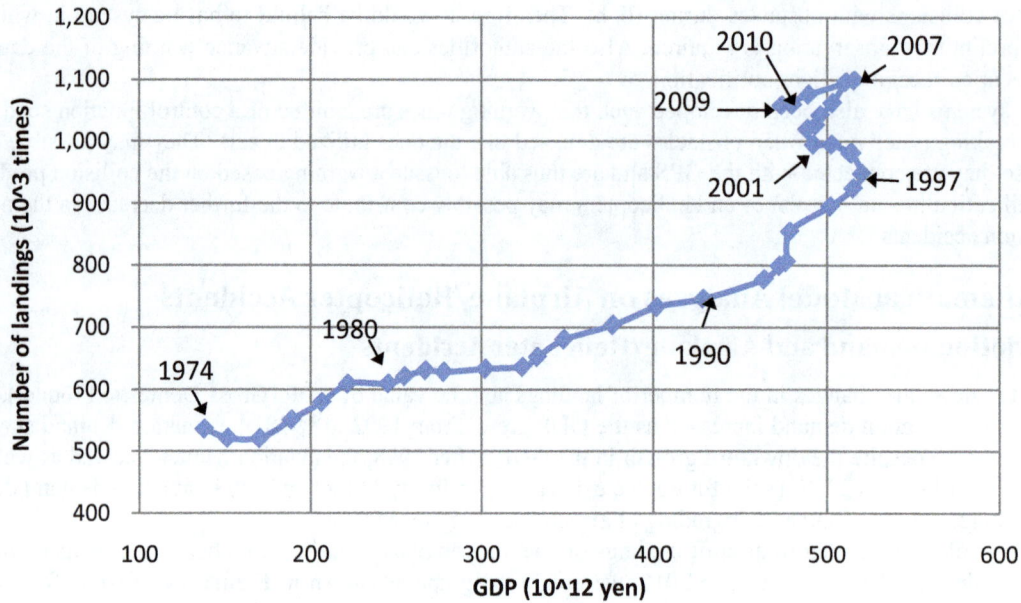

Figure 14. Changes in GDP and the number of aircraft landings (1974-2010).

Goodness of fit of the above model (1) is $R^2 = 0.976$, and the values in parentheses ( ) are t-values. From the above result we find that the number of aircraft landings grows around 772 corresponding to the unit increase (10^{12} yen) of GDP during the period from 1974 to 1991. The economy slowed down after 1991 and the GDP stayed aound 400 trillion yen, whereas the number of landings increased by approximately 20 thousand per year until 2005.

From the regression result in (1), we can see that aviation demand increases along with economic growth. Before 1991, GDP growth of 100 trillion yen corresponded to an increase in the number of landings by 77 thousand times.

We selectrailway transportation demand and automobile registration demand as the factors influencing aviation demand. Then we carry out a multiple linear regression analysis based on the data during the period from 1974 to 2004. If we define railway passengers' traveling distance demand (10^8passenger-kilometer) $(x_1)^2$, and the number of registered cars (100 units) $(x_2)^3$, as the explanatory variables and the number of aircraft landings (y) as the dependent variable, the parameter estimates of a multiple linear regression model areobtained as follows:

$$y = 0.194x_1 + 1.32x_2 + 240724.4$$
$$(1.68) \quad (40.4) \quad (7.32)$$

(2)

The goodness of fit of the above model (2) is $R^2 = 0.998$, and the values in parentheses ( ) are t-values. We know that the Japanese railway transportation demand has been increasing up to around 1990, then has been stagnant until 2004. On the other hand the automobile registration demand has been almost constantly increasing similar to the aviation demand which has been increasing throughout almost the whole period. Thus, we find that the automobile registration demand has been contributing more effectively to the increase of the aviation demand expressed by the number of aircraft landings.

When we plot the number of flights on the horizontal axis, and the number of accidents on the vertical axis from 1974 to 2010, we obtain the graph as shown in **Figure 15**. Now, in order to identify the relation between aviation demand and the number of aircraft accidents we define a quadratic function model between two variables: the number of aircraft landings (100 thousand times) (x) as the explanatory variable and the aircraft accident figure (y) as the dependent variable. The parameter estimates of a multiple linear model based on the data between 1974 and 1998 are given as follows:

---

[2]Source: Statistical Bureau, Ministry of Internal Affairs and Communication.
[3]Source: Automobile Inspection and Registration Information Association.

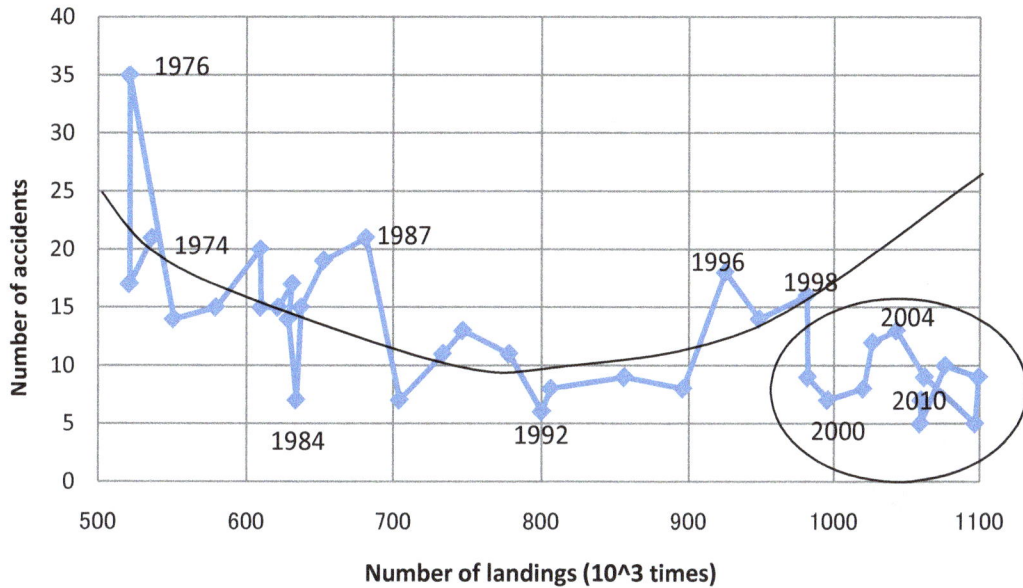

Figure 15. Aircraft landings and accidents (1974-2010).

$$y = 1.77x^2 - 28.2x + 122.3$$
$$(31.7) \quad (3.39) \quad (4.07)$$

(3)

The goodness of fit of estimate Equation (3) is $R^2 = 0.443$, and the values in parentheses ( ) are t-values. The minimum of y in the above quadratic function (3) is obtained at around $x = 7.97$. This implies that accidents decrease as the number of landings increases and when the number of landings is around 800 thousand, the number of accidents is at a minimum. In other words, the minimum value is obtained between 1992 when there were about 800,000 landings and 1995 when there were about 900,000 landings. According to the actual data, there were 6 accidents in 1992, 8 in 1993, 9 in 1994, and 8 in 1995, which is consistent with the period when the least number of accidents should occur in theory. Up to 1991, this is attributed to the progress in accident preventive measures, which was greater than the increase in the aviation demand. Specifically, as described below, while the air traffic services developed, the related technologies including aircraft and the air traffic control also advanced. There were also apparent effects of human resource development through education and training until 1991. However, after 1991, we note that accidents that are very hard to deal with cannot be prevented completely as old and decrepit airplanes are still in service, and it is very difficult to supervise privately owned airplanes. In the future, aviation demand and the number of accidents may increase slightly due to the improvement of airports and an increased number of LCCs (Low Cost Carriers). There is also a fear that unexpected airplane accidents may occur as new technologies are introduced and the wave of computerization prevails. However, we expect no major variations in the number of accidents.

4.2. Mathematical Model for Intervals of Accident Occurrence

Figure 16 and **Figure 17** show the distribution data of the number of days in the interval of accident occurrence in descending order in terms of airplanes (small and large) and helicopters, respectively. We divided our entire study period from 1974 to 2010 into three parts: Periods I, II and III. Then, we examined the goodness of fit of individual distribution data in terms of the exponential distribution, lognormal distribution, beta distribution and gamma distribution, utilizing "Statistically Fit Software version 2" provided by Green Mountain Software. The results of the goodness of fit of these distributions are indicated in **Table 6** and the estimated values of the parameters are shown in **Table 7**. The values in **Table 6** are the results of a chi-squared test. Those with an underline are not rejected in individual tests, and highlighted data shows the most fit data among all statistical models. Regarding the exponential distribution, 13 probability distributions of occurrence intervals out of 16 probability distributions of the number of occurrence days were not rejected in a chi-squared test. Regarding the airplanes

Mean value:29.0
Standard deviation:32.8
n=466

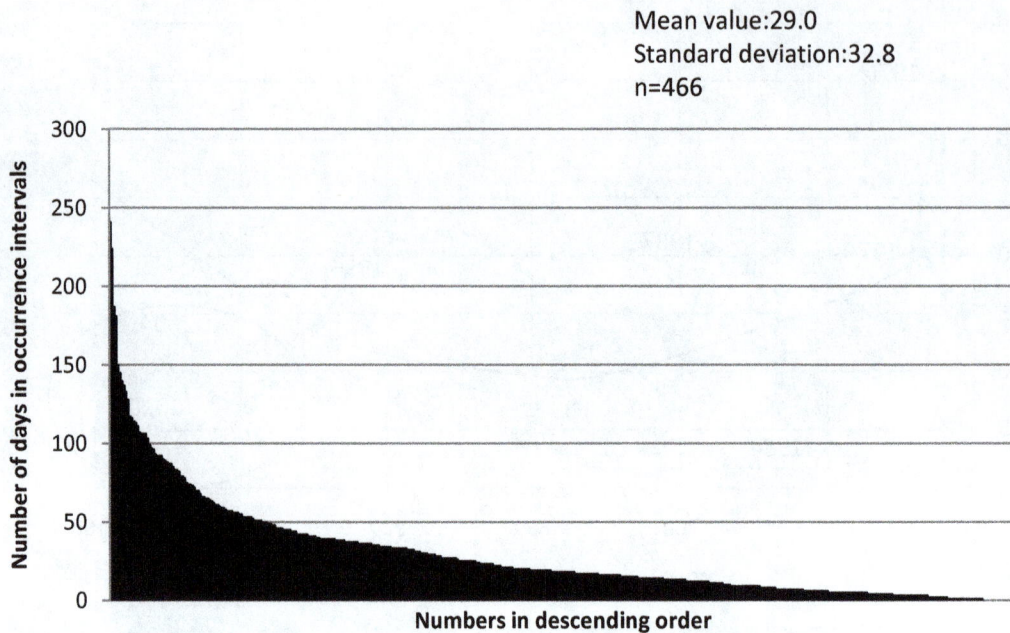

Figure 16. Number of days in occurrence intervals for airplane accidents (1974-2010).

Table 6. Results of the goodness of fit tests (1974-2010).

Type of aircraft	Period	Probability distribution			
		Exponential	Lognormal	Beta	Gamma
Airplane (large/small)	Entire period	29.90	34.50	40.80	33.80
	I	8.59	40.70	13.10	18.50
	II	10.30	10.60	8.21	8.43
	III	3.43	3.77	6.85	5.48
Small airplane	Entire period	18.10	38.10	28.10	29.60
	I	6.80	12.80	21.20	15.80
	II	11.90	10.70	17.00	20.70
	III	6.76	5.26	11.50	9.40
Large airplane	Entire period	5.83	13.70	2.66	10.60
	I	4.22	3.35	2.91	4.22
	II	0.97	1.78	3.14	1.51
	III	1.75	2.75	0.25	1.25
Helicopter	Entire period	39.30	35.10	60.70	39.60
	I	14.40	8.74	21.20	11.60
	II	19.40	15.70	15.20	17.80
	III	12.70	5.37	16.20	16.70

Shaded values in the table indicate the highest goodness of fit. Those with an underline were not rejected in individual tests.

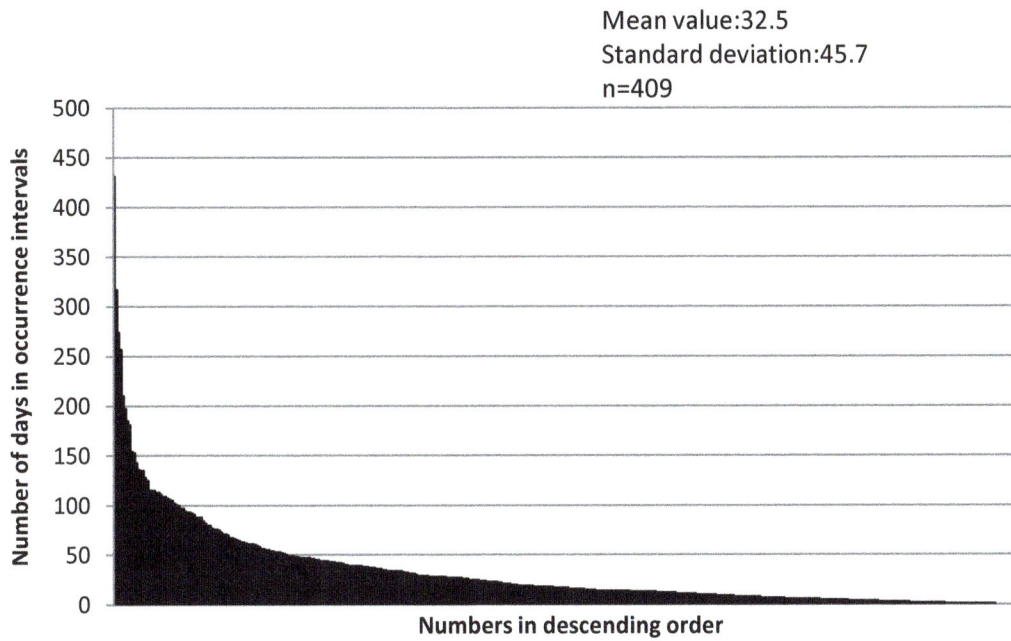

Mean value:32.5
Standard deviation:45.7
n=409

Figure 17. Number of days in occurrence intervals for helicopter accidents (1974-2010).

Table 7. Estimated parameters in individual distributions (1974-2010).

Types of aircraft	Period	Probability distribution							
		Mean value	Exponential	Lognormal		Beta		Gamma	
		(Actual measurement)	Beta (1/λ)	Mu (μ)	Sigma (σ)	p	q	Alpha	Beta
Airplane (large/small)	Entire period	29.0	29.0	2.8	1.2	0.8	27.6	1.1	26.4
	I	21.4	21.4	2.6	1.1	1.0	7.6	1.3	17.1
	II	30.2	30.2	2.9	1.2	0.8	3.9	1.1	27.5
	III	40.6	40.6	3.2	1.2	1.0	9.3	1.1	37.2
Small airplane	Entire period	38.6	38.6	3.1	1.2	0.9	10.7	1.0	36.8
	I	27.8	27.8	2.9	1.1	1.1	9.0	1.2	23.2
	II	40.6	40.6	3.2	1.2	0.9	10.7	1.1	37.2
	III	55.9	55.9	3.5	1.2	0.8	3.6	1.0	53.8
Large airplane	Entire period	113.0	113.0	4.2	1.2	0.8	2.3	1.1	100.8
	I	91.2	91.2	4.1	1.2	0.8	1.4	1.2	72.9
	II	118.1	117.1	4.3	1.1	0.8	1.7	1.2	98.2
	III	138.6	134.6	4.5	1.2	0.8	1.3	1.3	104.1
Helicopter	Entire period	32.5	32.5	2.8	1.3	0.8	8.9	0.8	38.7
	I	22.8	22.8	2.5	1.2	0.8	4.6	1.0	23.4
	II	32.2	32.2	2.8	1.3	0.8	17.5	0.9	37.0
	III	55.4	55.4	3.3	1.3	0.8	10.2	0.8	65.8

(small and large), 6 out of 12 distributions indicated the highest degree of goodness of fit. If those falling into the second highest are included, 10 distributions have a high degree of goodness of fit with the exponential distribution. As the mean value and the standard deviation are close, which is one of the properties of the exponential distribution, we can say that the probability of accident occurrence has the property of the exponential distribution. Accordingly, we herein define a formula by the exponential distribution. The probability density function of the exponential distribution is provided as follows:

$$f(t) = \lambda e^{-\lambda t} \tag{4}$$

When using the parameter estimate $\hat{\lambda}$, the expected value of the occurrence interval ($1/\hat{\lambda}$) in terms of airplanes (small and large) is 33 days during the entire period, while it is 23, 31 and 44 days in Period I, II and III, respectively. In the same manner, regarding small airplanes, it is 39 days (entire period), 28 (Period I), 41 (Period II), and 56 days (Period III). Regarding large airplanes, it is 113 days (entire period), 91 (Period I), 118 (Period II) and 139 days (Period III). Regarding helicopters, it is 33 days (entire period), 23 (Period I), 32 (Period II), and 55 days (Period III). Therefore, we can say that the process of accident occurrence in the air traffic services is in accordance with the Poisson process.

4.3. Mathematical Model of Fatalities in Accidents

The frequencies of fatalities from airplanes (large/small), small airplanes, large airplanes and helicopters for every 3 month interval are illustrated in **Table 8**. Fatalities exceeding 50 people are counted as 50 people. **Table 9** shows the mean value, standard deviation (SD) and coefficient of variation (CV) of fatalities in 3 months, 6 months and 1 year. The results of the goodness-of-fit tests of fatality counts in 3 months, 6 months and 1 year in

Table 8. Frequency distribution of number of fatalities in aircraft and helicopters.

Number of fatalities	Aircraft			Helicopters
	Total	Small	Large	
0	78	87	128	83
1	18	15	12	25
2	17	17	4	12
3	15	13	1	10
4	5	3	0	5
5	3	4	0	3
6	3	4	0	6
7	1	1	0	1
8	2	2	0	0
9	2	2	0	0
10	1	0	0	0
14	0	0	0	1
17	0	0	0	1
20	0	0	0	1
24	0	0	1	0
29	1	0	0	0
More than 50	2	0	2	0
Total	148	148	148	148

Table 9. Mean values and standard deviations in individual periods.

Types of aircraft	Period	Mean value		SD		CV	
All (large/small)	3 months	2.2	(6.9)	6.4	(47.8)	2.87	(6.96)
	6 months	4.4	(13.7)	8.8	(67.3)	1.98	(4.90)
	1 year	8.8	(27.5)	11.8	(94.3)	1.34	(3.44)
Small airplanes	3 months	1.3	(1.3)	2.0	(2.0)	1.62	(1.62)
	6 months	2.5	(2.5)	2.9	(2.9)	1.15	(1.15)
	1 year	5.0	(5.0)	4.4	(4.4)	0.87	(0.87)
Large airplanes	3 months	1.0	(5.6)	6.1	(47.8)	6.14	(8.52)
	6 months	2.0	(11.2)	8.5	(67.4)	4.30	(6.00)
	1 year	4.0	(22.5)	12.0	(94.6)	3.01	(4.21)
Helicopters	3 months	1.4	(1.4)	2.8	(2.8)	2.01	(2.01)
	6 months	2.8	(2.8)	4.2	(4.2)	1.48	(1.48)
	1 year	5.6	(5.6)	5.8	(5.8)	1.03	(1.03)

Values in parentheses are the calculated result based on original data, SD: Standard deviation; CV: Coefficient of Variation.

terms of individual distributions (beta, exponential, gamma, lognormal, negative binominal, normal and Poisson) are provided, utilizing "Statistically Fit Software version 2" provided by Green Mountain Software. The results of the goodness of fit of these distributions are indicated in **Table 10**, and the estimated values of the parameters are shown in **Table 11**. As shown in **Table 10**, the highest degree of the goodness of fit was observed in the negative binominal distribution. Regarding all airplanes (small and large), fatality distributions in 6 months and 1 year were not rejected in a chi-squared test. The 6-month distribution had the highest degree of the goodness of fit. Further, none of the fatality distributions in 6 months and 1 year regarding small airplanes and helicopters were rejected in a chi-squared test, in which the highest degree of the goodness of fit was observed. Fatality distributions regarding large airplanes were rejected in a chi-squared test. The probability density function of the negative binominal distribution is expressed as follows:

$$P(x) = \frac{(k+x-1)!}{x!} p^k (1-p)^x \tag{5}$$

In the above Formula (5) $P(x)$ represents the probability when there are deaths of x people within a certain period of time, and k is a constant. The negative binominal distribution demonstrates the probability distribution when there are x failures before succeeding k times. For instance, in **Table 11**, we have $k = 1$ and $p = 0.310$ in airplanes (large and small), which means that when there are 0 fatalities, the probability is 0.31. Similarly, we get $k = 3$ and $p = 0.375$ in small airplanes, which means that when there are 0, 1 or 2 fatalities, the probability is 0.375. In other words, when there is only a minor degree of variation in fatalities and the mean value is large, there is a high probability of fatalities in that time period, which means there is a low probability of 0 fatalities when $k = 1$. Therefore, in the negative binominal distribution in **Table 11**, as the time period for the fatality count increases, $i.e.$, 3 months, 6 months, and 1 year, the p value becomes smaller.

In the goodness-of-fit test in **Table 10**, the value for 1 year of large airplanes in the negative binominal distribution, and those for 1 year of airplanes (large and small) and 6 months and 1 year of large airplanes in the Poisson distribution were "not available". This result is due to the insufficient number of samples in some time periods. Regarding the occurrence intervals of air traffic accidents, the highest degree of the goodness of fit was observed in the exponential distribution. Further, we can say that the number of fatalities from air traffic accidents for a certain period of time is in accordance with the negative binominal distribution.

Table 10. Results of the goodness of fit tests (1974-2010).

Type of aircraft	Period	Probability distribution						
		Beta	Exponential	Gamma	Lognormal	Negative binominal distribution	Normal	Poisson
Airplane (large/small)	3 months	655	663	663	656	33.6	655	273
	6 months	58.6	64.2	62.9	64.2	11.7	122	77.2
	1 year	7.84	12	5.95	7.08	7.17	25.6	N.A.
Small airplane	3 months	360	395	395	519	19.5	369	86.7
	6 months	74.7	30.1	30.1	74.7	4.78	50.7	13.8
	1 year	9.62	2.32	9.62	9.62	2.05	7.19	18.9
Large airplane	3 months	1980	1980	1980	1980	78.3	1.98E+03	155
	6 months	480	481	481	481	56.6	674	N.A.
	1 year	63.5	63.5	63.5	63.5	N.A.	129	N.A.
Helicopter	3month	542	562	562	544	14.8	545	89.6
	6 month	54.1	61.9	61.9	132	2.16	59.5	25.7
	1 year	7.84	15	5.95	11.6	1.88	18.8	15.9

N.A.: not available. Values are results of a chi-squared test. Shaded Values are the results of the higher degree of goodness of fit. Those with an underline were not rejected in individual tests.

Table 11. Parameters of individual models.

Type of aircraft	Period	Probability distribution											
		Beta		Exponential	Gamma		Lognormal		Negative binominal		Normal		Poisson
		p	q	Beta $(1/\lambda)$	Alpha	Beta	Mu (μ)	Sigma (σ)	k	p	Mean	Sigma	Lamda
Airplane (large/small)	3 months	1.1	31188.0	2.2	1.0	2.2	1.0	0.9	1	0.310	2.2	6.4	2.2
	6 months	1.0	144431.0	4.4	1.7	2.7	1.2	0.9	1	0.184	4.4	8.8	4.4
	1 year	1.2	21278.6	8.8	1.6	5.7	1.8	0.9	1	0.102	8.8	11.7	8.8
Small airplane	3 months	2.3	1391.2	1.3	1.0	1.3	0.9	0.7	1	0.444	1.3	2.0	1.3
	6 months	1.9	215.1	2.5	1.0	2.5	1.0	0.7	1	0.286	2.5	2.9	2.5
	1 year	1.9	45.6	5.0	3.4	1.5	1.5	0.8	3	0.375	5.0	4.3	5.0
Large airplane	3 months	0.3	1.2	1.0	1.0	1.0	0.7	1.3	1	0.502	1.0	6.1	1.0
	6 months	0.3	1.2	2.0	1.0	2.0	0.8	1.3	1	0.335	2.0	8.5	2.0
	1 year	0.7	8.8	4.0	1.4	2.9	1.0	1.4	1	0.201	4.0	11.8	4.0
Helicopter	3 months	1.5	2558.0	1.4	1.0	1.4	0.8	0.8	1	0.416	1.4	2.8	1.4
	6 months	1.5	739.0	2.8	1.0	2.8	1.1	0.8	1	0.262	2.8	4.1	2.8
	1 year	1.8	68579.0	5.6	2.5	2.3	1.5	0.7	2	0.262	5.6	5.7	5.6

5. Summary and Conclusions

We attempted to investigate the past trend of accidents occurring in Japan for major air traffic services such as small and large airplanes and helicopters, respectively, and countermeasures taken to prevent aircraft accidents. Utilizing statistical data, we also conducted a mathematical modeling analysis. The findings we obtained were as follows.

1) Accident prevention measures have been advanced in all three types of air traffic services, *i.e.*, small airplanes, large airplanes and helicopters, contributing greatly to the decreased number of accidents. Especially, accidents mainly due to the pilot error show a significant decrease in small and large airplanes during takeoff and landing. On the other hand, we find that helicopter accidents occur more frequently during flight than during takeoff and landing, and the number of accidents caused by the major factor, pilot error, has also been decreasing. However, in terms of percentages of individual factors, accidents due to pilot error have increased only for helicopters. Although the decrease in the pilot error is significant, except for large airplanes, it still accounts for the highest percentage of accident factors.

2) It is clear that the number of accidents due to weather factors has not been reduced. Particularly for large airplanes, it has not been reduced at all, or rather, it has increased both in the total number and the percentage. This is an indication of the fact that measures to counter weather factors are extremely difficult to implement, which remains a major issue to be solved.

3) We found that the decrease in the number of accidents was achieved through various safety efforts made by every person involved in air traffic services, including strengthening of the rules in public policy, technological innovations in airplanes, helicopters and the surrounding facilities such as airports and air control radars, improvements in the training method for mechanics and traffic controllers and advancements in related techniques, the enhancement of public rules, and numerous measures taken by individual related companies. Further efforts to ensure the safety of old airplanes and helicopters are also necessary and required.

4) We conducted mathematical model analyses on economic growth, aviation demand, accident figures, intervals of accident occurrence, and fatalities. Results suggested that, in terms of economic growth, aviation demand and accident figures, our regression model could indicate a decrease in the accident figures together with the growing economy and aviation demand until 1991. We also found that, regarding the intervals of accident occurrence and the number of fatalities, a good fit for the probability distribution was observed for the exponential distribution and the negative binominal distribution, respectively.

As described above, we understand that small and large airplanes and helicopters have different characteristics, and therefore, it is necessary to develop separate accident prevention measures suitable for individual types of aircraft. Regarding small airplanes, as the airplane itself cannot be replaced easily, there are some accidents caused by using old airplanes. The number of accidents will not decrease even if warning tone systems for incorrect actions and the layout of the cockpit are improved in new models. Therefore, we assume that it will be effective to develop systems that will reduce the number of accidents due to the pilot error, such as a modularized warning tone system that can be installed in old models, and to make their installation mandatory.

Regarding large airplanes, the major issue is weather, for which individual airlines have been implementing various measures. The most effective way to reduce the number of large airplane accidents is to continue undertaking such measures. Regarding helicopters, there is the high possibility of accidents due to the pilot error, especially collisions, as helicopters often fly at low altitudes. We assume that it will help reduce the number of accidents through a thorough preliminary observation of the surrounding environment, enhanced pilot training, as well as the development of a warning system in which obstacles and GPS are closely linked to each other.

We consider that there may exist some other causes for aircraft accidents such as pilot's mental and psychological disorder, terrorist attacks and so on. How to deal with these troubles to prevent accidents will be the remaining and future problems. It is also very important to conduct a study and systematization from the viewpoint of how to survive an accident, to mitigate injury and to reduce the number of injuries.

References

[1]   Bazargan, M. and Vitaly, S.G. (2011) Impact of Gender, Age and Experience of Pilots on General Aviation Accidents. *Accident Analysis & Prevention*, **43**, 962-970. http://dx.doi.org/10.1016/j.aap.2010.11.023

[2]   Chang, Y.-H. and Yang, H.-H. (2010) Aviation Occupant Survival Factors: An Empirical Study of the SQ006 Accident. *Accident Analysis & Prevention*, **42**, 695-703. http://dx.doi.org/10.1016/j.aap.2009.10.018

[3]   Rinoie, K. (2008) Aviation Safety from Viewpoint of Aircraft Design and Operation. *Journal of Reliability Engineering Association of Japan*, **30**, 362-368.

[4]   Shiroyama, H., Murayama, A. and Kajimura, I. (2003) Legal System for Safety Concerning Aircraft Accidents in the United States: Suggestions for Japan. *Journal of the Sociotechnology Research Network*, **1**, 149-158.

[5] Nakasuga, S. (1995) A Study on Human Factors in Aircraft and Enhanced Safety. *Journal of Reliability Engineering Association of Japan*, **17**, 12-21.

[6] Kamiyama, T. (1976) Safety and Reliability of Aircraft. *Operations Research Society of Japan*, **21**, 424-431.

[7] The Japan Transport Safety Board, Transport Ministry's Aircraft Accident Investigation Commission, Aircraft Accidents Report. http://jtsb.mlit.go.jp/jtsb/aircraft/index.php

[8] Aviation Statistics Catalogue, Japan Aeronautic Association, 1984-85, 1998-99, 2001, 2011.

[9] Rodrigues, C.C. and Cusick, S.K. (2012) Commercial Aviation Safety. McGraw Hill, New York.

[10] Sugie, H. (2010) Professional Pilot. IKAROS Publications, Tokyo.

[11] Yanagida, K. (2010) Longing for Air Safety. 100 Years of Japanese Aviation. Japan Aeronautic Association.

[12] Tsumajika, E. (2010) Japan's Sky Returned to Japan after the War. 100 Years of Japanese Aviation. Japan Aeronautic Association.

[13] Shimada, H. (2011) 50 World Classic Passenger Aircraft. SB Creative, Tokyo.

[14] Yamamoto, T. (2010) Policy for the Development of Air Transport Businesses. 100 Years of Japanese Aviation. Japan Aeronautic Association.

[15] Masui, K. and Takahashi, H. (1987) Shift in Public Transport Policy. Nippon Hyoron Sha Co., Ltd., Tokyo.

[16] Tazaki, T. (2010) Shizukuishi Accident and Restructuring of the Aviation Security System. 100 Years of Japanese Aviation. Japan Aeronautic Association.

[17] Ministry of Internal Affairs and Communications Statistics Bureau, Japan Statistical Yearbook, 1981, 1985, 1989, 1993/94, 1997, 2002, 2007, 2010, 2014, Japan Statistical Association.

Using Sports Wagering Markets to Evaluate and Compare Team Winning Streaks in Sports

R. Alan Bowman[1], Thomas Ashman[2], James Lambrinos[1]

[1]School of Management, Union Graduate College, Schenectady, NY, USA
[2]Behavioral Sciences Collegium, Eckerd College, St. Petersburg, FL, USA
Email: lambrinj@uniongraduatecollege.edu

Abstract

Point spreads and money lines from sports wagering markets are used to evaluate the impressiveness of team streaks. Sports wagering data have previously been useful in assessing competitive balance in sports. Our approach was motivated by the amount of media scrutiny that accompanied the winning streak of the NBA's Miami Heat and the point streak of the NHL's Chicago Blackhawks which occurred simultaneously for the most part in 2013. The topic came to the forefront again with the 2014-2015 winning streak of the Atlanta Hawks. Three streaks are highlighted in our paper. The length of the streaks, the quality of the competition, injuries, and a variety of other factors were mentioned in the media but the discussion was limited to subjective opinions as no way of properly weighing relative influence of the factors was identified. Wagering markets provide an excellent source of information for making these judgments. Several complementary measures are described and the most impressive team streaks within and across professional baseball, basketball, football, and ice hockey are identified.

Keywords

Winning Streaks, Sports Wagering Markets, Baseball, Basketball, Football, Hockey

1. Introduction

In 2013, the Miami Heat in the NBA and Chicago Blackhawks in the NHL were simultaneously in the middle of long streaks (a 26-game winning streak for the Heat and a 24-game points streak for the Blackhawks). This fueled a very active (albeit brief) debate in the media about which streak was more impressive. Many factors were brought up in the discussions besides just the obvious length of the streak including the quality of competition, home vs. away games, travel, back-to-back games, injuries, and many others including the inherent difficulty of comparing streaks in one sport vs. another. In the 2014-15 NBA season, the Atlanta Hawks won 19 straight games. Unlike the Heat (with Lebron James, Dwyane Wade, and Chris Bosh), the Hawks were not perceived for the most part to be a dominating team—their streak was a bit more of a surprise. Should one consider this in judging the impressiveness of the Hawks' streak and possibly even conclude that the Hawks' streak was more impressive than the Heat's even though it was shorter? The NBA's Eastern Conference has recently been significantly weaker than the Western Conference. Did either the Heat or the Hawks simply beat up on Eastern patsies? Should this be taken into account and, if so, how? Is there any way to legitimately compare streaks in the same sport, much less across sports? Although this topic has generated considerable public interest and discussion, we are unaware of any serious treatment of it in the literature. We seek to provide an objective, unbiased, and rigorous quantitative approach for assessing the impressiveness of sports streaks that incorporates as many relevant factors as possible and to use our approach to evaluate and compare past winning streaks both within and across professional sports. We present our approach and results in this paper.

Since Mottley [1] suggested using OR techniques to analyze sports over 50 years ago, there has been a growing interest in doing so. Wright [2] provides an excellent summary, categorization, and analysis of articles that have addressed a wide variety of issues that fall under this broad umbrella in the last 50 years. Certainly one reason for the growing popularity of this field is that sports are entertaining. This has important consequences for the broader OR field. First, OR approaches can enhance the entertainment value of sports by providing analytical insight into issues that are otherwise not available. This in turn enhances the visibility of OR. Also, the inherent entertainment value of sports makes them ideal for introducing students to OR techniques and thought processes. The use of OR in the classroom is widespread and growing. Cochran [3] provides a good introduction to this topic, focusing on articles that were included in a special issue of INFORMS Transactions on Education devoted to the topic of Sports in the OR Classroom. We believe our paper addresses an entertaining topic in a rigorous analytical fashion that provides insights that will enhance the entertainment value. We also think this topic and our approach are very well suited for use as an example in the OR classroom.

Our approach utilizes gambling line information. We believe gambling line information is ideal because it needs to take into account as many factors as possible so that gamblers will not be able to identify and exploit factors that are not considered (known as inefficiencies) in order to systematically make profits. As we began exploring ways to use this information, we realized that there are qualitative, perhaps even philosophical, issues to consider as well as quantitative factors. It became clear that there could be no single measure on which everyone would agree. We therefore developed several measures that provide different and complementary insights. We will briefly discuss all the measures we developed but we will focus our attention on two measures that most directly address the issues brought up in the media discussion of the streaks of the Heat and the Blackhawks and in our own considerations of those and other streaks such as the Hawks' recent streak. We then present results from computing these two measures for all actual streaks in the following professional sports and seasons for which we have comprehensive gambling line data:

- Baseball (1999 through 2014)
- Basketball (1990-91 through 2014-15)
- Football (1985-86 through 2014-15)
- Hockey (2005-6 through 2014-15)

Note that hockey started using shootouts in 2005-6 so that each game has a winner from that season forward. Winning streaks and points streaks, as well as the nature of the gambling line information, are very different prior to that season. For ease of comparability with the other sports, we have limited our treatment of hockey to data from 2005 forward. Also, in computing our results, we restrict our attention to regular season streaks and do not consider streaks that cross seasons. The approach for computing the measures can, for the most part, be extended to consider other sports and other types of streaks as long as gambling line data are available.

We will present the most impressive streaks according to our measures during the time frames shown above for each sport. One of the nice features of our approach, however, is that it allows streaks to be compared across different sports. To that end, we will present the most impressive streaks across all four sports. Our measures also allow insights into why streaks occur more frequently in some sports than others so we will discuss those issues as well. Finally, we will compare the streaks of the Heat, the Blackhawks, and the Hawks and offer our take on which one was most impressive, all things considered.

2. Why Use Gambling Lines?

Streaks have always fascinated sports fans and, not surprisingly, they have received considerable attention in the literature. We believe that previous approaches for evaluating streaks (team or individual), however, have never dealt specifically and effectively with the types of issues raised in the media debate comparing the Heat's and Blackhawks' streaks. To see why we think this, consider what is arguably the most famous streak in all of sports-Joe DiMaggio's 56 game hitting streak in baseball. This streak has been analyzed in many outlets using various approaches (see, for example, Arbesman and Strogatz [4] and Rockoff and Yates [5]). The primary goal has always been to estimate the probability that **a streak of that length** would occur.

No approach, however, considers the actual games that the Yankees played during the streak and all the factors in those games that might affect the impressiveness of that streak. For example, who were the opposing pitchers, were the games at home or away, were the stadiums ones in which DiMaggio had success in the past, which Yankee hitter batted in front of and behind DiMaggio in that particular game, etc.?

Similarly, approaches for evaluating team streaks have focused on the likelihood that streaks of certain lengths would occur. Note that, with this focus, all streaks of the same length in a particular sport would be considered equivalent, essentially ignoring all the other factors that might differentiate one streak from another. A key component of much of this type of research has been to try to assess whether there is "streakiness" at all or whether the streaks observed are consistent with random variation (associated with sometimes quite sophisticated models). We think this research has been very interesting and valuable but does not deal directly with the issues raised in the media debate about the streaks of the Heat and the Blackhawks. Therefore, we established two criteria for our analysis of team streaks. First, we wanted to use actual games played in all of our measures. Second, we wanted to take into account as many factors as possible that were present during those actual games that would affect the likelihood that each team would win. These factors would include all the factors that fans might bring up in discussing and comparing actual streaks (quality of competition, home vs. away games, travel, back-to-back games, injuries, illnesses, starting pitchers or goalies, weather, etc.). This led to our decision to use gambling line data to evaluate the impressiveness of win streaks. Gambling lines theoretically will comprehensively consider all game-specific factors (more on this later) without specifically quantifying the effect of each factor. Importantly, these factors can change greatly from game to game, and properly recognizing that fact is crucial for comparing streaks in terms of their impressiveness. For example, at the start of the 2013 season, a five game winless streak by the Green Bay Packers in weeks 9-13 would have been considered quite unlikely. They were considered at the start of the season to be one of the stronger teams in the NFL. Prior to the injury to Packers quarterback Aaron Rodgers, the Green Back Packers were favored in seven of eight games. They won five of the seven before the injury. After his injury, the Packers were favored in only two of five games against weak opponents and did not win a single game during this period. This winless streak of five games against weak opponents would have been rather surprising at the start of the season but was not that surprising given the injury situation for Aaron Rodgers. Furthermore, winning streaks by other teams that included beating the Packers with Aaron Rodgers would be far more impressive than those that included beating the Packers without Aaron Rodgers. The gambling lines for either set of games properly reflect this fact.

An alternative approach that would satisfy our two criteria would be to develop a statistical model that explicitly includes as many factors as possible and apply that model to the factors as they existed for each actual game that was part of the streak. Historically, this alternative has not been feasible but, with the proliferation of data that are being gathered and made accessible, this might very well be feasible in the future. If such attempts are successful, our paper should prove quite helpful in providing a systematic approach for incorporating the individual game probabilities from the statistical models (in place of the ones we get from the gambling lines) into several complementary measures that together yield a range of insights.

At the heart of our measures is the use of gambling lines to provide us with the probability (prior to each game) that each team will win the game. The use of gambling lines in this manner has a relatively long and extensive history (see, for examples Peel and Thomas [6] [7]). All studies (such as ours) that use gambling lines as predictors of game outcomes are essentially assuming that the markets are at least reasonably efficient. This is closely tied to our claim that the gambling lines theoretically will comprehensively consider the game-specific factors that we believe are crucial to consider in evaluating team winning streaks. The idea is that, if the gambling lines do not properly include one or more these factors, informed bettors could exploit their knowledge of these factors (the market inefficiency) to consistently make money. There is also a long stream of sports research that examines whether sports betting markets are efficient. Although the literature on this subject is nuanced and complex, the broad conclusion is that the gambling lines generally provide unbiased estimators of game outcomes and market inefficiencies are corrected over time (see, for example, Sauer [8], and Gandar, Zuber, and Dare [9]).

3. Measures for Assessing Winning Streaks

The first measure we will describe is the most straightforward. We will refer to this simply as the streak probability (SP). The first step to compute SP is to convert the gambling line for each actual game in the streak to the probability of each team winning the game. The favorite win probability is $\dfrac{\dfrac{f}{100+f}}{\dfrac{f}{100+f}+\dfrac{100}{100+u}}$, where f is the absolute value of the favorite money line and u is the underdog money line. Of course, the underdog win probability is 1—the favorite win probability. In the cases of the NFL and NBA, the gambling line historical information that was most available came in the form of spreads. This required a preliminary step to convert the spread into the probability that each team would win. For evaluating hockey point streaks, it was necessary to first convert the probability each team would win to the probability that each team would win in regulation (interestingly, favored teams get a larger percentage of their wins in regulation than do underdogs), and the probability each team gets a point is one minus the probability the other wins in regulation. Details on how we accomplished these preliminary steps (and the actual numeric models) are available on request.

SP, the probability that the team would win all the games in the streak, is simply the product of the individual game win probabilities:

$$\text{SP} = \prod_{i=1}^{n} p_i,$$ where p_i = probability that the team with the winning streak would win game i of the actual streak (computed from the money line for that game as shown above).

Note that, since we use the gambling line for each game at the time of the game, we are not computing the probability that the team will win all the games in the streak prior to the streak (or prior to the season) nor are we assuming independence. Instead, we are multiplying the conditional probabilities that they win each game (conditioned on all the factors known and taken into consideration in the gambling line prior to each game).

SP considers the team that actually had the winning streak in the computations. As we will discuss shortly, it might be desirable to have a measure that is independent of the team involved but includes all the other factors in the games. Therefore, we developed a second measure, which we will refer to as the average team streak probability (ATSP). ATSP requires the win probability for each game in the streak to be adjusted to what it would be if the average team in the league were playing each game rather than the team that actually had the streak. How this is done is shown in the Appendix. ATSP is then the product of the probabilities that the average team would win each of the games in the streak:

$$\text{ATSP} = \prod_{i=1}^{n} atp_i,$$ where atp_i = probability that the average team in the league would win game i of the actual streak (computed using the money line for that game as shown above and the adjustment procedure described in the appendix).

In terms of evaluating streaks, both SP and ATSP take into consideration the quality of the opponents played during the streak, whether the games were played home or away, and any other game-specific factors that gamblers would consider such as the ones we identified in our earlier discussion. SP considers the quality of the team that actually had the streak whereas ATSP does not. In that sense, a good team that is highly regarded in the gambling lines would end up with a higher probability of the streak, hence the streak would be considered less impressive. One could argue that this should be true and that a long streak from a not-so-great team is more impressive but one could also argue that the measure of the impressiveness of the streak should be independent of the team that had the streak. For these reasons, we believe both measures have value and present results from both.

To illustrate these concepts and introduce an approach for decomposing SP and ATSP into components that we believe provide many useful insights, we present results for a couple of extreme winning streaks in the NFL. From 11/19/2006 through 12/24/2006, the Tennessee Titans won 6 games in a row. In 2007, the New England

Patriots won all of their regular season games (16 games in a row). The SP for the Titans' streak is 0.0009579 compared to 0.04142827 for the Patriots' streak. This would indicate the Titans' streak is far more impressive. ATSP tells a different story. The ATSP is 0.00000285 for the Patriots' streak compared to 0.00944903 for the Titans'. So which streak is really more impressive? We can break it down further for more insights. Note that the geometric mean of the win probabilities for the Titans during their streak (easily computed for any streak as the streak probability raised to the power of the reciprocal of the streak length) is 0.3140. In other words, if they had a 0.3140 chance of winning each of their 6 games, their streak would have a probability of 0.00095790, the same as their SP. We will denote this geometric mean as GMSP. On the other hand, GMSP for the Patriots' 16 games was 0.8196. This seems to indicate that the Patriots were simply a far more impressive team than the Titans. It could be, however, that the Patriots played a much easier schedule. We can break that component out as well. The geometric mean of the win probabilities for the average team in the league playing the Titans' 6 games was 0.4598. We denote this geometric mean as GMATSP. For the Patriots' 16 games, the GMATSP was 0.4502. Both teams faced difficult schedules during their streaks and, in fact, the Patriots' schedule was slightly more difficult.

To enable further insights to be obtained, we now introduce two new concepts–the schedule factor and the dominance factor. If we define the schedule factor for any streak as SF = GMATSP/.5 and the dominance factor for any streak as DF = GMSP/GMATSP, we can represent SP and ATSP as:

$$SP = \left(DF \times SF \times 0.5\right)^{L} \quad \text{and} \quad ATSP = \left(SF \times 0.5\right)^{L}, \text{ where } L = \text{the streak length.}$$

The larger the dominance factor the more impressive the team was deemed to be and the larger the schedule factor the easier the games in the streak were deemed to be (with 1.0 being average for both factors so that, if both factors were average, both probabilities would be the same as evaluating a series of coin flips). The Patriots had a dominance factor of 1.820 and a schedule factor of 0.900 during their streak. The Titans had a dominance factor of 0.683 and a schedule factor of 0.920 during their streak.

We think most people would say that the Patriots streak was more impressive as the reason their SP was not that impressive was because they were heavily favored in most of their games. In that sense, we think most people will find ATSP to be a better measure than SP. For the most part, we agree. On the other hand, the Titans were not deemed to be a very good team and they won 6 games in a row against a difficult schedule. They were underdogs in all 6 games and heavy underdogs in 2 of the 6. This is impressive in its own right and SP reflects this. We believe SP and ATSP are both useful and provide complementary insights.

Note that both the SP and ATSP measures for a streak reflect only the probabilities for that specific streak and **not a generic streak of that length**. A consequence of this is that these two measures do not consider the frequency of opportunities for streaks (because each streak has only one opportunity). Sports with more games, however, clearly have more opportunities for streaks. As we mentioned earlier, much of the literature on streaks has been interested in the likelihood of a generic streak of a specific length and have properly considered the relative availability of opportunities, sometimes with quite sophisticated models. We have not seen gambling lines incorporated in such analyses so we note briefly here that it is possible to do so (and thereby consider actual games and actual factors involved in games) by combining the approach we have described here with computer simulations of actual seasons. We have done so and can provide details of both the approach and the results upon request. We note, however, that this approach considers actual games and factors across an entire season and how those affect generic streak probabilities for the season; it does not directly focus on the games and factors specifically present during the actual streak. We do not believe them to be as directly useful for our objectives here as the SP and ATSP measures.

4. Empirical Results

We first present results for SP and ATSP for the time frames analyzed that we previously identified. **Table 1** and **Table 2** show the five most impressive streaks for each sport as well as overall (obtained by computing SP and ATSP for every single win streak in each sport for their respective time frames and ranking them). Sports fans will, of course, enjoy looking over the specific streaks and thinking about the actual teams involved to see if the measures match up to their memories and opinions. We will focus here on some overall observations and we will close by comparing the streaks of the Heat, the Blackhawks, and the Hawks since that motivated the investigation in the first place.

Table 1. Most impressive streaks: Winning streak probability (SP).

	Streak Start			Streak End			Games	Probability (SP)
All Sports								
Portland (NBA)	12	3	2007	12	30	2007	13	0.00000574
San Diego (MLB)	6	18	1999	7	2	1999	14	0.00003763
Houston (NBA)	11	1	1996	12	15	1996	19	0.00006462
Tampa Bay (MLB)	6	9	2004	6	22	2004	12	0.00006865
Chicago (NHL Points)	1	19	2013	3	6	2013	24	0.00007414
MLB								
San Diego	6	18	1999	7	2	1999	14	0.00003763
Tampa Bay	6	9	2004	6	22	2004	12	0.00006865
Atlanta	7	26	2013	8	9	2013	14	0.00016572
Colorado	6	4	2009	6	14	2009	11	0.00018205
Baltimore	9	7	1999	9	22	1999	13	0.00020362
NBA								
Portland	12	3	2007	12	30	2007	13	0.00000574
Houston	11	1	1996	12	15	1996	19	0.00006462
New York	3	18	2013	4	9	2013	13	0.00020393
Atlanta	12	27	2014	1	31	2015	19	0.00026859
Houston	3	3	1991	4	6	1991	17	0.00028154
NFL								
Pittsburgh	9	26	2004	1	2	2005	14	0.00018249
New England Patriots	10	5	2003	12	27	2003	12	0.00080506
Tennessee Titans	11	19	2006	12	24	2006	6	0.00095790
San Diego	9	20	1987	11	15	1987	8	0.00173198
Indianapolis Colts	9	13	2009	12	17	2009	14	0.00218152
NHL (Winning)								
Ottawa	1	14	2010	2	4	2010	11	0.00015600
Columbus	3	18	2015	4	4	2015	9	0.00034126
Pittsburgh	3	2	2013	3	30	2013	15	0.00037185
New Jersey	3	28	2006	4	18	2006	11	0.00058317
Phoenix	12	28	2006	1	9	2007	7	0.00078838
NHL (Points)								
Chicago	1	19	2013	3	6	2013	24	0.00007414
San Jose	2	21	2008	4	1	2008	20	0.00045401
Columbus	3	18	2015	4	11	2015	13	0.00059309
Columbus	2	26	2013	3	22	2013	12	0.00085031
Washington	3	23	2006	4	5	2006	7	0.00093308

Table 2. Most impressive streaks: Average team winning streak probability (ATSP).

	Streak Start			Streak End			Games	Probability (ATSP)
All Sports								
Miami (NBA)	2	3	2013	3	25	2013	27	0.00000024
Houston (NBA)	11	1	1996	12	15	1996	19	0.00000044
New England (NFL)	9	9	2007	12	29	2007	16	0.00000285
Houston (NBA)	1	29	2008	3	16	2008	22	0.00000433
Seattle (NBA)	1	21	1996	3	18	1996	19	0.00000467
MLB								
Oakland	8	13	2002	9	4	2002	20	0.00000991
Atlanta	4	16	2000	5	2	2000	15	0.00005043
Seattle	5	23	2001	6	8	2001	15	0.00006312
Baltimore	9	7	1999	9	22	1999	13	0.00009513
San Diego	6	18	1999	7	2	1999	14	0.00011538
NBA								
Miami	2	3	2013	3	25	2013	27	0.00000024
Houston	11	1	1996	12	15	1996	19	0.00000044
Houston	1	29	2008	3	16	2008	22	0.00000433
Seattle	1	21	1996	3	18	1996	19	0.00000467
L.A. Lakers	2	4	2000	3	13	2000	19	0.00000570
NFL								
New England	9	9	2007	12	29	2007	16	0.00000285
Pittsburgh	9	26	2004	1	2	2005	14	0.00005137
Indianapolis	9	13	2009	12	17	2009	14	0.00007558
Green Bay	9	8	2011	12	11	2011	13	0.00013417
Washington	9	1	1991	11	17	1991	11	0.00017511
NHL (Winning)								
Pittsburgh	3	2	2013	3	30	2013	15	0.00002684
Washington	1	13	2010	2	7	2010	14	0.00015849
Ottawa	1	14	2010	2	4	2010	11	0.00022054
San Jose	2	21	2008	3	14	2008	11	0.00032518
Buffalo	10	4	2006	10	26	2006	10	0.00041916
NHL (Points)								
Chicago	1	19	2013	3	6	2013	24	0.00002269
San Jose	2	21	2008	4	1	2008	20	0.00020611
Detroit	3	9	2006	4	17	2006	20	0.00063915
Pittsburgh	1	13	2007	2	18	2007	16	0.00065514
Vancouver	12	8	2010	1	11	2011	17	0.00085407

We first note that the NFL does not have any of the top five overall streaks as measured by SP but the Patriots' 2007 unbeaten season does make the top 5 ATSP results. Since we do not consider streaks across seasons, the NFL is at a disadvantage not only because the maximum possible streak is much shorter but also because a larger percentage of streaks will end (because the season ends) without a loss. Perhaps there is a reasonable adjustment for streaks that are active at season's end that could be made. There is another (and we think more interesting) issue involved. This issue can also be seen in the fact that the NBA has 4 of the top 5 ATSP results (the Patriots' being the 5th) whereas the SP streaks are more evenly spread across the NBA, NHL, and MLB. To understand why this occurred, we computed the dominance factors and the schedule factors described in the previous section for all streaks of at least 10 games in each sport. Summarized results are presented in **Table 3**. We see that the average dominance factors are much higher in basketball and football than in hockey and baseball. The average schedule factors are higher than one in all sports, indicating that streaks are more likely during portions of the schedule that are easier, which makes sense. The average schedule factor is much higher for basketball than the other sports. We can see that basketball has more frequent long streaks than football primarily because there are more games per season and more than baseball and hockey primarily because there are more frequently teams that are more dominant and, to a lesser extent, portions of the schedule that are weaker. One note for sports fans is that the New England Patriots had the highest dominance factor (1.820) in all the data for their 16 game undefeated season in 2007. The next highest dominance factor in any sport was 1.696 by the Chicago Bulls for a 15 game win streak in 1997. Impressively, there were 8 streaks with a dominance factor of more than 1.5 in the NBA during the time period analyzed and all eight were achieved by the Bulls during the Michael Jordan era.

5. Conclusions and Our Vote: Heat, Blackhawks, or Hawks?

We have described two different measures for assessing the impressiveness of team streaks in baseball, basketball, football and hockey. The measures use gambling line data, which we feel are ideally suited to take into consideration all the factors that would affect the probability each team would win a particular game. We have computed the measures for all four sports and provided rankings within and across sports. The ranking of the streaks depends upon which measure is chosen and we believe the measures provide complementary insights. Statistical models that incorporate many game-specific factors and that are applied to specific games involved in streaks could be used in place of the gambling line information in the proposed measures. This would be an opportunity for future research.

Let's now take a closer look at the streaks of the Heat, the Blackhawks, and the Hawks and declare a winner (we won't be wishy-washy). These streaks are clearly among the most impressive streaks across all sports during the data period. The Blackhawks' streak has the best SP (0.000074 vs. 0.000269 for the Hawks and 0.000683 for the Heat), whereas the Heat's streak has the best ATSP (0.00000024 vs. 0.0000257 for the Blackhawks and 0.0000856 for the Hawks). The Heat's schedule during their streak was the easiest (schedule factor of 1.113 vs. 1.082 for the Hawks and 1.044 for the Blackhawks) but this is already taken into account in both measures. The Heat's dominance factor was much higher (1.358 vs. 1.199 for the Hawks and 1.045 for the Blackhawks) and this is the reason their SP is the lowest of the three but their ATSP is the highest (by far). In general, we think a team's dominance is part of what makes what they do impressive so we cast our vote for the Heat's streak as being the most impressive of these three recent streaks and, in fact, the most impressive of any streak in the time period for which we have data. For those that disagree, we have presented some quantitative

Table 3. Average dominance and schedule factors for streaks of 10 or more.

Sport	Average Schedule Factor	Average Dominance Factor	Number of Streaks
NBA	1.066	1.314	136
NFL	1.043	1.334	22
MLB	1.030	1.078	58
NHL Winning	1.014	1.111	13
NHL Points	1.048	1.032	77

results that you are welcome to use to support your arguments.

References

[1] Mottley, C. (1954) Letter to the Editor—The Application of Operations-Research Methods to Athletic Games. *Journal of the Operations Research Society of America*, **2**, 335-338. http://dx.doi.org/10.1287/opre.2.3.335

[2] Wright, M. (2009) 50 Years of OR in Sport. *Journal of the Operational Research Society*, **60**, S161-S168. http://dx.doi.org/10.1057/jors.2008.170

[3] Cochran, J. (2005) OR Scores in Sports. *OR/MS Today*, **32**, 32-40.

[4] Arbesman, S. and Strogatz, S. (2008) A Journey to Baseball's Alternate Universe. *The New York Times*, March 30.

[5] Rockoff, D. and Yates, P. (2009) Chasing DiMaggio: Streaks in Simulated Seasons Using Non-Constant At-Bats. *Journal of Quantitative Analysis in Sports*, **5**, Article 4.

[6] Peel, D. and Thomas, D. (1992) The Demand for Football: Some Evidence on Outcome Uncertainty. *Empirical Economics*, **17**, 323-331. http://dx.doi.org/10.1007/BF01206291

[7] Peel, D. and Thomas, D. (1988) Outcome Uncertainty and the Demand for Football: An Analysis of Match Attendance in the English Football League. *Scottish Journal of Political Economy*, **35**, 242-249. http://dx.doi.org/10.1111/j.1467-9485.1988.tb01049.x

[8] Sauer, R. (1998) The Economics of Wagering Markets. *Journal of Economic Literature*, **36**, 2021-2064.

[9] Gandar, J., Zuber, R. and Dare, W. (2000) The Search for Informed Traders in the Totals Betting Markets for National Baksetball Association Games. *Journal of Sports Economics*, **3**, 133-148.

[10] Bowman, R., Ashman, T. and Lambrinos, J. (2013) Prospective Measures of Competitive Balance: Application to Money Lines in Major League Baseball. *Applied Economics*, **45**, 4071-4081. http://dx.doi.org/10.1080/00036846.2012.750421

[11] Bowman, R., Lambrinos, J. and Ashman, T. (2013) Competitive Balance in the Eyes of the Sports Fan: Prospective Measures Using Point Spreads in the NFL and NBA. *Journal of Sports Economics*, **14**, 498-520. http://dx.doi.org/10.1177/1527002511430230

Appendix

Bowman, Ashman and Lambrinos [10] and Bowman Lambrinos, and Ashman [11] presented a conceptual model of the gambling spread (for any NFL or NBA game in any season) as:

$$S_{jk} = N_{ijk} - H$$

where

S_{ijk} = the spread when team i is playing at home against team j in game k,

N_{ijk} = the "neutral" spread—the spread adjusted by the average home advantage, and

H = the average home advantage (averaged across all games in the season).

They further broke down N_{ijk} as follows:

$$N_{ijk} = -\left(R_i - R_j + \varepsilon_{ijk}\right)$$

where

R_i = the rating of team i (how much team i would be favored against the average league team in a neutral site game based on the entire season—note that this means the average of all the ratings is 0), and

ε_{ijk} = the factors specific to team i playing team j in game k.

They suggested estimating the team ratings using the following regression model with the ε_{ijk} terms being the error terms in the model.

$$S_{h(k)a(k)k} = \beta_0 + \sum_{i=1}^{m-1} \beta_i X_i + \varepsilon_{h(k)a(k)k}$$

where

$$X_i = \begin{cases} -1 & \text{if } h(k) = i \left(\text{team } i \text{ is the home team in game } k\right) \\ 1 & \text{if } a(k) = i \left(\text{team } i \text{ is the away team in game } k\right) \\ 0 & \text{otherwise} \end{cases}$$

and m = the number of teams in the league.

Note that one of the teams (team m) must be left out of the model and that the coefficient estimates are relative to that team. To make the ratings estimates invariant to the team left out (and relative to the average team as required by the definition of the ratings terms), the coefficients are converted to the ratings as follows:

$$\widehat{R_i} = \widehat{\beta_i} - \frac{1}{m}\sum_{j=1}^{m-1}\widehat{\beta_j} , i = 1 \text{ to } m-1, \text{ and } \widehat{R_m} = -\frac{1}{m}\sum_{j=1}^{m-1}\widehat{\beta_j}.$$

Bowman, Ashman, and Lambrinos [10] and Bowman, Lambrinos and Ashman [11] showed how to translate this approach to the case where the gambling line information comes in the form of a money line. First, the money line is used to compute the probability that each team will win the game. Then 0.5 is subtracted from each team's win probability and the result is treated as a spread in the regression model just described.

The authors developed the ratings so that they could be used in competitive balance measures but we can use them to adjust the win probabilities in the manner necessary to compute ATSP. In the case where the ratings come from the win probabilities, the probability that the average team in the league would win any particular game in the streak would be the probability the actual team involved in the streak would win the game minus the rating for that actual team (since the average team would have a rating of zero). This new probability would then be adjusted the same as for SP if evaluating a point streak in hockey. If the gambling line information is in the form of a spread, the procedure just described would be used to adjust the spread to reflect the average team playing the game rather than the actual team. The win probabilities would then be computed the same as for SP but now based on the adjusted spread rather than the actual spread.

3

A New Heuristic for the Convex Quadratic Programming Problem

Elias Munapo[1], Santosh Kumar[2*]

[1]Graduate School of Business and Leadership, University of KwaZulu-Natal, Westville Campus, Durban, South Africa
[2]Department of Mathematics and Statistics, University of Melbourne, Parkville, Australia
Email: emunapo@gmail.com, [*]skumar@ms.unimelb.edu.au

Abstract

This paper presents a new heuristic to linearise the convex quadratic programming problem. The usual Karush-Kuhn-Tucker conditions are used but in this case a linear objective function is also formulated from the set of linear equations and complementarity slackness conditions. An unboundedness challenge arises in the proposed formulation and this challenge is alleviated by construction of an additional constraint. The formulated linear programming problem can be solved efficiently by the available simplex or interior point algorithms. There is no restricted base entry in this new formulation. Some computational experiments were carried out and results are provided.

Keywords

Convex Quadratic Programming, Linear Programming, Karush-Kuhn-Tucker Conditions, Simplex Method, Interior Point Method

1. Introduction

There are so many real life applications for the convex quadratic programming (QP) problem. The applications include portfolio analysis, structural analysis, discrete-time stabilisation, optimal control, economic dispatch and finite impulse design; see [1]-[3]. Some of the methods for solving the convex quadratic problem are active set, interior point, branch and bound, gradient projection, and Lagrangian methods, see [4]-[9] for more information on these methods.

In this paper we present a new heuristic to linearise the convex quadratic programming problem. The usual Karush-Kuhn-Tucker conditions are still used but in this case a linear objective function is also formulated from

---

[*]Corresponding author.

the set of linear objective function equations and the complementary slackness conditions. There is an unboundedness challenge that is associated with the proposed linear formulation. To alleviate this challenge, an additional constraint is constructed and added to the linear formulation. The new linear formulation can be solved efficiently by the available simplex and interior point algorithms. There is no restricted base entry in the proposed approach. The time consuming complementarity pivoting is no longer necessary. Some computational experiments have been carried out and the objective of the computational experiments was to determine CPU times of the:

1) Proposed heuristic;
2) Regularised Active Set Method Mae and Saunders [10];
3) Primal-Dual Interior Point Algorithm.

It may be noted that the proposed method is suitable only if the quadratic programming problem satisfies conditions (1) to (5) mentioned in Section 2.1.

2 Mathematical Background

2.1 The Quadratic Programming Problem

Let a quadratic programming (QP) problem be represented by (1).

Minimize $f(X) = CX^{\mathrm{T}} + \dfrac{1}{2}XQX^{\mathrm{T}}$

Subject to:

$$AX^{\mathrm{T}} \leq B^{\mathrm{T}}, X^{\mathrm{T}} \geq 0 \tag{1}$$

where

$$C = (c_1, c_2, \cdots, c_n), X = (x_1, x_2, \cdots, x_n), Q = \begin{pmatrix} q_{11} & \cdots & q_{1n} \\ \vdots & \ddots & \vdots \\ q_{n1} & \cdots & q_{nn} \end{pmatrix},$$

$$A = \begin{pmatrix} a_{11} & \cdots & a_{1n} \\ \vdots & \ddots & \vdots \\ a_{m1} & \cdots & a_{mn} \end{pmatrix} \text{ and } B = (b_1, b_2, \cdots, b_m).$$

It is assumed that:
1) Matrix Q is a $n \times n$ symmetric and positive definite,
2) Function $f(X)$ is strictly convex,
3) The conditions $YX^{\mathrm{T}} = 0$ and $\lambda S^{\mathrm{T}} = 0$ hold. Here Y and S are dual and primal slack variables, respectively.
4) Since constraints are linear then the solution space is convex, and
5) Any maximization quadratic problem can be changed into a minimization and vice versa.

When the function $f(X)$ is strictly convex for all points in the convex region then the quadratic problem has a unique local minimum which is also the global minimum [11].

2.2. Karush-Kuhn-Tucker Conditions

The convex quadratic programming problem has special features that we can capitalize on when solving. All constraints are linear and the only nonlinear expression is the objective function. Let the Lagrangian function for the QP problem be L and in this case

$$L = CX^{\mathrm{T}} + \frac{1}{2}XQX^{\mathrm{T}} + \lambda\left(AX^{\mathrm{T}} - B^{\mathrm{T}}\right) \tag{2}$$

where $\lambda = (\lambda_1, \lambda_2, \cdots, \lambda_m)$ and $\lambda^{\mathrm{T}} \geq 0$. In this case we exclude the non-negativity conditions $X^{\mathrm{T}} \geq 0$. If $Y = (y_1, y_2, \cdots, y_n), Y^{\mathrm{T}} \geq 0$ and $S = (s_1, s_2, \cdots, s_m), S \geq 0$ then the Karush-Kuhn-Tucker conditions as given in [11] for a local minimum are:

$$QX^{\mathrm{T}} + A^{\mathrm{T}}\lambda^{\mathrm{T}} - Y^{\mathrm{T}} = -C^{\mathrm{T}} \tag{3}$$

$$AX^{\mathrm{T}} + S^{\mathrm{T}} = B^{\mathrm{T}}, \tag{4}$$

$$Y^{\mathrm{T}}, S^{\mathrm{T}} \geq 0.$$

Complementary slackness conditions are given in (5) and are only satisfied at the optimal point. These conditions are:

$$YX^{\mathrm{T}} = 0 \text{ and } \lambda S^{\mathrm{T}} = 0 \tag{5}$$

Note Y and S are n and m dimensional vectors representing the slack variables. At this stage, we are unable to apply the simplex algorithm due to restricted base entry and this makes the simplex method approximately 8 times slower than its full speed compared to its unrestricted basis version.

2.3. Some Matrix Operations

Suppose $D = (d_1, d_2, \cdots, d_m)$ and $E = (e_1, e_2, \cdots, e_m)$ are single row matrices of the same dimension m and $H = (h_{ij}), i = 1, 2, \cdots, m$ and $j = (1, 2, \cdots, m)$ is an $m \times m$ dimensional matrix, the following must hold.

$$DE^{\mathrm{T}} = ED^{\mathrm{T}} \tag{6}$$

$$DHE^{\mathrm{T}} = EHD^{\mathrm{T}} \tag{7}$$

Equations (6) and (7) can be easily verified. These simple results are used to eliminate the complementary slackness conditions.

3. Elimination of Complementary Slackness Conditions

3.1. Elimination of $YX^{\mathrm{T}} = 0$

Pre-multiply (3) by X, we have:

$$X\left(QX^{\mathrm{T}} + A^{\mathrm{T}}\lambda^{\mathrm{T}} - Y^{\mathrm{T}}\right) = X\left(-C^{\mathrm{T}}\right) \tag{8}$$

$$XQX^{\mathrm{T}} + XA^{\mathrm{T}}\lambda^{\mathrm{T}} - XY^{\mathrm{T}} = -XC^{\mathrm{T}} \tag{9}$$

From (6) $YX^{\mathrm{T}} = XY^{\mathrm{T}}$ and from (5) $XY^{\mathrm{T}} = 0$, then

$$XQX^{\mathrm{T}} + XA^{\mathrm{T}}\lambda^{\mathrm{T}} = -XC^{\mathrm{T}} \tag{10}$$

By rearranging, we have

$$XQX^{\mathrm{T}} + XA^{\mathrm{T}}\lambda^{\mathrm{T}} + XC^{\mathrm{T}} = 0 \tag{11}$$

3.2. Elimination of $\lambda S^{\mathrm{T}} = 0$

Pre-multiply (4) by λ, we have

$$\lambda\left(AX^{\mathrm{T}}\right) + \lambda S^{\mathrm{T}} = \lambda B^{\mathrm{T}} \tag{12}$$

$$\lambda\left(AX^{\mathrm{T}}\right) + \lambda S^{\mathrm{T}} = \lambda B^{\mathrm{T}} \tag{13}$$

Since from (4) $\lambda S^{\mathrm{T}} = 0$, then

$$\lambda AX^{\mathrm{T}} - \lambda B^{\mathrm{T}} = 0 \tag{14}$$

3.3. Elimination of λAX^{T} or $XA^{\mathrm{T}}\lambda^{\mathrm{T}}$

From (7), we have: $\lambda AX^{\mathrm{T}} = XA^{\mathrm{T}}\lambda^{\mathrm{T}}$, hence we can replace $XA^{\mathrm{T}}\lambda^{\mathrm{T}}$ by λAX^{T} in relations (11) to get t(15):

$$XQX^{\mathrm{T}} + \lambda AX^{\mathrm{T}} + XC^{\mathrm{T}} = 0 \tag{15}$$

Subtracting (14) from (15), we obtain (16):

$$XQX^{\mathrm{T}} + XC^{\mathrm{T}} + \lambda B^{\mathrm{T}} = 0 \tag{16}$$

3.4. Linear Objective Function for the Quadratic Programming Problem

Note that the expression in relation (13) is nonlinear but it can be rearranged so that the original quadratic programming objective function becomes a linear quantity. This can be achieved as follows:
Divide relation (16) by two, one obtains:

$$\frac{1}{2}XQX^{\mathrm{T}} + \frac{1}{2}XC^{\mathrm{T}} + \frac{1}{2}\lambda B = 0 \tag{17}$$

Rearranging (17), we obtain (18):

$$\frac{1}{2}XQX^{\mathrm{T}} + XC^{\mathrm{T}} - \frac{1}{2}XC^{\mathrm{T}} + \frac{1}{2}\lambda B = 0 \tag{18}$$

From (1) $f(X) = CX^{\mathrm{T}} + \frac{1}{2}XQX^{\mathrm{T}}$, then, (18) becomes (19) or equivalently (20):

$$f(X) - \frac{1}{2}CX^{\mathrm{T}} + \frac{1}{2}\lambda B = 0 \tag{19}$$

$$f(X) = \frac{1}{2}CX^{\mathrm{T}} - \frac{1}{2}\lambda B \tag{20}$$

Thus the nonlinear objective function of the QP problem is now linearised but it creates a new challenge. We will discuss this in the next section.

3.5. LP Equivalent to the Given QP

From (1), (3) and (20), we have the following LP problem:

Minimize $\frac{1}{2}CX^{\mathrm{T}} - \frac{1}{2}\lambda B$

Subject to:

$$QX^{\mathrm{T}} + A^{\mathrm{T}}\lambda^{\mathrm{T}} - Y^{\mathrm{T}} = -C^{\mathrm{T}}, AX^{\mathrm{T}} + S^{\mathrm{T}} = B^{\mathrm{T}}, X^{\mathrm{T}}, \lambda^{\mathrm{T}}, Y^{\mathrm{T}}, S^{\mathrm{T}} \geq 0 \tag{21}$$

The minimisation problem (21) will have an unbounded solution due to negative coefficient of λ in the objective function and negative coefficients of the slack variable Y^{T} in the constraints. These are the only source of unboundedness in the LP (21). Here, we let: $\omega = \frac{1}{2}\lambda B$ and $\phi = \mu Y^{\mathrm{T}}$ where $\mu = (1,1,\cdots,1)$ a row vector of dimension n. The objective function is now modified as :

Minimize $\frac{1}{2}CX^{\mathrm{T}} - \frac{1}{2}\lambda B + l_1\omega + l_2\phi$, where l_1 and l_2 are very large constants relative to all other objective

coefficients. Both of these constants do not have to assume the same large values. A large number of experiments were done on a large number of quadratic programming problems and and it was observed that $l_1 \ll l_2$ seems to work well. These experiments have been recorded later in this paper. In these experiments, it was noted that values of l_1 and l_2 on higher side can be as much as $l_1 = 1000\left(|c_1| + |c_2| + \cdots + |c_n|\right)$ and $l_2 = 50000\left(|c_1| + |c_2| + \cdots + |c_n|\right)$.

3.6. Existence of a Linear Objective Function and Verification of Optimality

The optimal solution of a convex quadratic programming model is unique and it satisfies the complementary conditions $YX^{\mathrm{T}} = 0$ and $\lambda S^{\mathrm{T}} = 0$. The unique optimal solution to the convex quadratic programming is a corner point P_Q. Since the KKT conditions can be expressed as a linear objective function that can make P_Q

exist.

4. Numerical Illustrations

4.1. Example 1

Minimize $-8x_1 - 16x_2 + x_1^2 + 4x_2^2$
   Subject to:

$$x_1 + x_2 \leq 5, x_1 \leq 3, x_1, x_2 \geq 0 \tag{22}$$

This example was taken from Jensen and Bard (2012) without any modifications.
Linear formulation of the above QP
In this case we took $l_1 = 1000$ and $l_2 = 50000$ which are very large compared to coefficients 4; 8; 2.5; and 1.5. The LP problem is given by:
Maximize $4x_1 + 8x_2 + 2.5\lambda_1 + 1.5\lambda_2 - 1000\omega - 50000\phi$
   Subject to

$$2x_1 + \lambda_1 + \lambda_2 - y_1 = 8,$$

$$8x_2 + \lambda_1 - y_2 = 16, x_1 + x_2 + s_1 = 5,$$

$$x_1 + s_2 = 3,$$

$$2.5\lambda_1 + 1.5\lambda_2 = \omega,$$

$$y_1 + y_2 = \phi$$

$$x_1, x_2, \lambda_1, \lambda_2, y_1, y_2, s_1, s_2, \omega, \phi \geq 0 \tag{23}$$

The solution of (23) by the simplex method is given by:

$$x_1 = 3, x_2 = 2, \lambda_2 = 2, \omega = 3, \lambda_1 = y_1 = y_2 = s_1 = s_2 = \phi = 0. \tag{24}$$

From the original QP objective function, we have the objective value given in (25).

$$f(3,2) = -31 \tag{25}$$

Verification of optimality
The solution is optimal because complementary slackness conditions are satisfied as given in (26).

$$\lambda_1 s_1 = \lambda_2 s_2 = y_1 x_1 = y_2 x_2 = 0 \tag{26}$$

4.2. Two More Examples

Two more examples are solved to illustrate how the large constants are selected. Example 2 is taken from [12] and example 3 is from [13].
   Example 2 from [12]
   Minimize $(x_1 - 1)^2 + (x_2 - 2.5)^2$
   Subject to:

$$-x_1 + 2x_2 \leq 2, x_1 + 2x_2 \leq 6, x_1 - 2x_2 \leq 2, x_1, x_2 \geq 0 \tag{27}$$

The linear formulation of (27) becomes as given in (28).
Maximize $2x_1 + 5x_2 + 2\lambda_1 + 6\lambda_2 + 2\lambda_3 - 1000\omega - 300000\phi$
Such that:

$$2x_1 - \lambda_1 + \lambda_2 + \lambda_3 - y_1 = 2,$$

$$2x_2 + 2\lambda_1 + 2\lambda_2 - 2\lambda_3 - y_2 = 5,$$

$$-x_1 + 2x_2 + s_1 = 2,$$

$$x_1 + 2x_2 + s_2 = 6,$$

$$x_1 - 2x_2 + s_3 = 2,$$

$$2\lambda_1 + 5\lambda_2 = \omega,$$

$$y_1 + y_2 = \phi,$$

$$x_1, x_2, \lambda_1, \lambda_2, \lambda_3, y_1, y_2, s_1, s_2, s_3, \omega, \phi \geq 0 \tag{28}$$

The solution of (28) is as given in (29) and once again it is optimal as all complementary slackness conditions are satisfied.

$$x_1 = 1.4, x_2 = 1.7, \lambda_1 = 0.8, s_2 = 1.2, s_3 = 4, \omega = 1.8, \lambda_2 = \lambda_3 = y_1 = y_2 = s_1 = \phi = 0 \tag{29}$$

Example 3 from [13]

Minimize $-x_1 - x_2 + \dfrac{1}{2}x_1^2 + x_2^2 - x_1 x_2$

Subject to: $-x_1 + x_2 \leq 3, -2x_1 - 3x_2 \leq -6; x_1, x_2 \geq 0.$
The linear formulation of the above example is given by (30).

Maximize $\dfrac{1}{2}x_1 + \dfrac{1}{2}x_2 + 1.5\lambda_1 - 3\lambda_2 - 500\omega - 80000\phi$

Subject to:

$$x_1 - x_2 + \lambda_1 - 2\lambda_2 - y_1 = 1; \quad -x_1 + 2x_2 + \lambda_1 - 3\lambda_2 - y_2 = 1; \quad x_1 + x_2 + s_1 = 3;$$

$$2x_1 + 3x_2 + s_2 = 6; 1.5\lambda_1 - 3\lambda_2 = \omega; y_1 + y_2 = \phi; \text{ where } x_1, x_2, \lambda_1, \lambda_2, y_1, y_2, s_1, s_2, \omega, \phi \geq 0. \tag{30}$$

The solution is given by: $x_1 = 1.8, x_2 = 1.2, \lambda_1 = 0.4, s_2 = 1.2, s_3 = 4, \omega = 0.6, \lambda_2 = y_1 = y_2 = s_1 = \phi = 0$. This solution is once again optimal as all complementary slackness conditions are satisfied.

5. Computational Experiments

A set of convex quadratic programming test problems are given in [14]. All these test problems were used in testing the proposed approach. The objective of the computational experiments was:

1) To determine that the LP optimal solution is also optimal to the given QP.

2) Compare CPU times of the proposed heuristics with Regularized Active Set Method and Primal-Dual Interior Point Method

The results are tabulated in **Table 1**. MATLAB R2013 (version 8.2) running on an Intel Pentium Dual desktop

Table 1. Computational experiments on the set of QP test problems.

Exp. No.	Prob. Name	No. of constraints (m)	No. of Variables (n)	CPU secs Proposed Heuristic	CPU secs Active Set	CPU secs Interior Point
1	AUG2D	10,000	20,200	29.34	0.55	15.12
2	AUG2DC	10,000	20,200	34.39	0.57	14.25
3	AUG2DCQP	10,000	20,200	21.89	240.73	14.63
4	AUG2DQP	10,000	20,200	37.19	228.72	14.76
5	AUG3D	1000	3873	0.29	0.07	1.65
6	AUG3DC	1000	3873	0.45	0.06	1.69
7	AUG3DCQP	1000	3873	0.82	3.84	1.39
8	AUG3DQP	1000	3873	0.53	5.02	1.56
9	BOYD1	18	93,261	89.67	214.24	107.10
10	BOYD2	0	93,263	*	4168.93	2245.64

Continued

11	CONT-050	2401	2597	0.31	0.84	3.37
12	CONT-100	9801	10197	1.88	26.37	19.12
13	CONT-101	10,098	10197	112.7	35855.97	20.66
14	CONT-200	39,601	40,397	76.16	277.68	136.22
15	CONT-201	40,198	40,397	51.12	285.50	143.56
16	CONT-300	90,298	90,597	219.47	2449.75	721.76
17	CVXQP1-L	5000	10,000	413.18	4516.19	2488.97
18	CVXQP1-M	500	1000	0.65	5.94	1.85
19	CVXQP1-S	50	100	0.01	0.04	0.21
20	CVXQP2-L	2500	10,000	218.82	670.50	443.34
21	CVXQP2-M	250	1000	0.63	4.24	1.52
22	CVXQP2-S	25	100	0.02	0.04	0.26
23	CVXQP3-L	7500	10,000	76.22	14069.08	736.74
24	CVXQP3-M	750	1000	0.65	17.10	2.63
25	CVXQP3-S	75	100	0.02	0.40	0.23
26	DPKLO1	77	133	0.02	0.01	0.17
27	DTOC3	9998	1499	74.1	0.32	107.10
28	DUAL1	1	85	~0.00	0.03	0.46
29	DUAL2	1	96	~0.00	0.01	0.43
30	DUAL3	1	111	~0.00	0.03	0.58
31	DUAL4	1	75	~0.00	0.01	0.37
32	DUALC1	215	9	~0.00	0.02	0.60
33	DUALC2	229	7	~0.00	0.01	0.44
34	DUALC5	278	8	~0.00	0.01	0.24
35	DUALC8	503	8	~0.00	0.02	0.70
36	EXDATA	3001	3000	154.08	~0.00	200.08
37	GENH28	8	10	~0.00	7.76	0.05
38	GOULDQP2	349	699	0.31	0.87	0.78
39	GOULDQP3	349	699	0.08	0.01	0.65
40	HS118	17	15	~0.00	~0.00	0.14
41	HS21	1	2	~0.00	~0.00	0.14
42	HS268	5	5	~0.00	0.00	0.16
43	HS35	1	3	~0.00	~0.00	0.05
44	HS35MOD	1	3	~0.00	~0.00	0.08
45	HS51	3	5	~0.00	0.00	0.05
46	HS52	3	5	~0.00	~0.00	0.04

Continued

47	HS53	3	5	~0.00	~0.00	0.09
48	HS76	3	4	~0.00	5.44	0.06
49	HUES-MOD	2	10,000	2.44	5.32	3.78
50	HUESTIC	2	10,000	1.15	9.64	3.42
51	KSIP	1001	20	0.64	6.58	2.55
52	LASER	1000	1002	0.56	213.02	1.99
53	LISWET1	10,000	10,002	26.87	215.75	14.89
54	LISWET10	10,000	10,002	26.12	223.19	15.12
55	LISWET11	10,000	10,002	24.97	234.96	15.03
56	LISWET12	10,000	10,002	28.67	223.23	15.01
57	LISWET2	10,000	10,002	46.96	212.28	135.27
58	LISWET3	10,000	10,002	33.16	214.23	145.28
59	LISWET4	10,000	10,002	39.17	212.05	150.86
60	LISWET5	10,000	10,002	28.18	211.28	153.74
61	LISWET6	10,000	10,002	40.06	203.76	136.04
62	LISWET7	10,000	10,002	23.12	217.98	14.63
63	LISWET8	10,000	10,002	24.05	230.47	14.54
64	LISWET9	10,000	10,002	35.08	~0.00	15.05
65	LOTSCHD	7	12	~0.00	7.14	0.09
66	MOSARQP1	700	2500	1.08	3.10	0.57
67	MOSARQP2	600	900	2.17	190.45	0.47
68	POWELL 20	10,000	10,000	23.18	0.13	10.88
69	PRIMAL 1	85	325	0.02	0.25	0.46
70	PRIMAL 2	96	649	0.03	0.64	0.64
71	PRIMAL 3	111	745	0.02	0.34	1.23
72	PRIMAL 4	75	1489	0.02	0.34	1.18
73	PRIMALC1	9	230	~0.01	0.34	0.42
74	PRIMAL C2	7	231	~0.01	0.36	0.26
75	PRIMALC5	8	287	~0.01	0.87	0.20
76	PRIMALC8	8	520	~0.02	0.08	0.31
77	Q25FV47	820	1571	32.16	0.01	11.85
78	QADLITTL	56	97	0.01	2.18	0.17
79	QAFIRO	27	32	0.01	0.31	0.14
80	QBANDM	305	472	0.03	0.36	0.54
81	QBEACONF	173	262	0.04	0.82	0.35
82	QBORE3D	233	315	0.06	1.27	0.51

Continued

83	QBRANDY	220	249	0.05	1.86	0.35
84	QCAPRI	271	353	0.06	4.22	1.18
85	QE226	223	282	0.04	5.39	0.50
86	QETAMACR	400	688	0.06	1.26	1.86
87	QFFFFF80	524	854	0.07	6.24	1.54
88	QFORPLAN	161	421	0.09	3.77	1.13
89	QGFRDXPN	616	1092	1.06	8.42	2.04
90	QGROW15	300	645	0.08	0.84	1.32
91	QGROW22	440	946	0.05	0.66	2.09
92	QGROW7	140	301	0.04	0.10	0.81
93	QISRAEL	174	142	0.02	3.68	0.71
94	QPCBLEND	74	83	0.01	0.78	0.22
95	QPCBOE11	351	384	0.02	1.61	1.24
96	QPCBOE12	166	143	0.05	55.42	0.69
97	QPCSTAIR	356	467	0.08	~0.00	0.86
98	QPILOTNO	975	2172	8.15	0.02	4.76
99	QPTEST	2	2	~0.00	0.30	0.08
100	QRECIPE	91	180	0.08	2.58	0.41
101	QSC205	205	203	0.09	0.13	0.30
102	QSCAGR25	471	500	0.05	1.61	0.63
103	QSCAGR7	129	140	0.08	5.82	0.35
104	QSEFXM1	330	457	0.08	12.18	0.85
105	QSEFXM2	660	914	0.13	0.99	1.55
106	QSEFXM3	990	1371	1.12	5.25	2.38
107	QSCRPIO	388	358	0.05	0.95	0.35
108	QSCRS8	490	1169	0.08	4.64	1.14
109	QSCSD1	77	760	0.87	29.78	6.87
110	QSCSd6	147	1350	0.09	2,71	0.68
111	QSCSD8	397	2750	0.04	22.41	1.13
112	QSETAP1	300	480	0.08	39.33	0.50
113	QSETAP2	1090	1880	0.23	1.58	1.17
114	QSETAP3	1480	2480	0.07	0.40	1.51
115	QSEBA	515	1028	0.06	0.11	1.80
116	QSHARE1B	117	225	0.04	23.37	0.44
117	QSHARE2B	96	79	0.02	6.36	0.27
118	QSHELL	536	1775	0.03	3.55	3.03

Continued

119	QSHIP04L	402	2118	0.08	48.37	1.05
120	QSHIP04S	402	1458	0.11	13.18	0.72
121	QSHIP08L	778	4283	1.03	23.35	6.10
122	QSHIP08S	778	2387	0.79	12.19	1.75
123	QSHIP12L	1151	5247	1.26	1.90	11.76
124	QSHIP12S	1151	2763	0.16	2.68	2.24
125	SIERRA	1227	2036	0.12	~0.00	3.79
126	QSTAIR	356	467	0.07	36.94	0.87
127	QSTANDAT	359	1075	0.05	39.34	0.98
128	S268	5	5	~0.00	95.97	0.16
129	STADAT1	3999	2001	0.08	12.28	6.61
130	STADAT2	3999	2001	0.13	1.87	8.12
130	STADAT3	7999	4001	0.09	~0.00	14.16
131	STCQP1	2052	4097	0.15	759.14	1.87
132	STCQP2	2052	4097	0.06	0.61	3.89
133	TAME	1	2	~0.00	9.38	0.03
134	UBH1	12,000	18,009	34.54	~0.00	62.83
135	VALUES	1	202	~0.00	0.55	0.51
136	YAO	2000	2002	0.08	0.57	3.66
137	ZECEVIC2	2	2	~0.00	240.73	0.83

(Dual core G2020 2.9 GHz CPU, 2GB DDR3 1333 RAM) was used in these experiments. There were no advanced processing techniques embedded within the three methods. The set up time was excluded from the CPU times in all three methods. The zero (~0.00) means CPU time is less than 0.01 second. In all the test problems, it was found that the LP optimal solution was optimal to the QP problem. However, in the CPU time challenges were observed with the BODYD2 for the proposed heuristic and as a result we could not accurately obtain the necessary CPU time for these two cases. There was no challenge with the other two methods on the same BODYD2 problem. This experiment was conducted twice, but the same observation. We have no reason to support this behaviour but we believe it may be due to some local computational environment.

6. Conclusion

The convex QP problem can be solved like a linear programming problem efficiently either by the simplex method or the interior point algorithm. The restricted base entry is not necessary by the proposed approach. Complementary slackness can retard the simplex method, which is roughly eight times slower than the full speed simplex method. Taking complementary slackness conditions away itself is a big reduction in the number of constraints in the proposed linear formulation of the quadratic programming problem. More experiments are likely to give more insight and advantages of the proposed approach. The proposed method is in fact the usual simplex method applied to solving an ordinary LP that was obtained from the given convex QP. Also note that a large number of Maros-Maszaros test problems are giving rise to small to medium size LPs and therefore the proposed method dominates solving a large number of QPs, as is reflected in **Table 1**. From these results, it may be noted that, for example in the case of medium sized problems at serial 118 to 124 and large sized problems at serial number 125 to 132, the proposed heuristic outperformed the other two with respect to the cpu time.

Acknowledgements

The authors are thankful to the referees for their helpful and constructive comments.

References

[1] Gupta, O.K. (1995) Applications of Quadratic Programming. *Journal of Information and Optimization Sciences*, **16**, 177-194. http://dx.doi.org/10.1080/02522667.1995.10699213

[2] Horst, R., Pardalos, P.M. and Thoai, N.V. (2000) Introduction to Global Optimization: Non-Convex Optimization and Its Applications. Kluwer Academic Publishers, Dordrecht. http://dx.doi.org/10.1007/978-1-4615-0015-5

[3] McCarl, B.A., Moskowitz, H. and Furtan, H. (1977) Quadratic Programming Applications. *Omega*, **5**, 43-55. http://dx.doi.org/10.1016/0305-0483(77)90020-2

[4] Burer, S.D. and Vandenbussche, D. (2008) A Finite Branch and Bound Algorithm for Non-Convex Quadratic Programs with Semidefinite Relaxations. *Mathematical Programming Series A*, **113**, 259-282. http://dx.doi.org/10.1007/s10107-006-0080-6

[5] Burer, S.D. and Vandenbussche, D. (2009) Globally Solving Box-Constrained Non Convex Quadratic Programs with Semidefinite-Based Finite Branch-and-Bound. *Computational Optimisation and Applications*, **43**, 181-195.

[6] Freund, R.M. (2002) Solution Methods for Quadratic Optimization. Lecture Notes, Massachusetts Institute of Technology, Cambridge, MA.

[7] Gondzio, J. (2012) Interior Point Methods 25 Years Later. *European Journal of Operational Research*, **218**, 587-601. http://dx.doi.org/10.1016/j.ejor.2011.09.017

[8] More, J.J. and Toraldo, G. (1989) Algorithms for Bound Constrained Quadratic Programming Problems. *Numerische Mathematik*, **55**, 377-400. http://dx.doi.org/10.1007/BF01396045

[9] Liu, S.T. and Wang, R.T. (2007) A Numerical Solution Method to Interval Quadratic Programming. *Applied Mathematics and Computations*, **189**, 1274-1281. http://dx.doi.org/10.1016/j.amc.2006.12.007

[10] Maes, C. and Saunders, M. (2012) A Regularized Active-Set Method for Sparse Convex Quadratic Programming. *21st International Symposium on Mathematical Programming*, Berlin, 19-24 August 2012.

[11] Jensen, P.A. and Bard, J.F. (2012) Operations Research Models and Methods. John Wiley & Sons Inc., Hoboken.

[12] Lee, C.R. (2011) Unit 8: Quadratic Programming Active set Method and Sequential Quadratic Programming. http://www.cs.nthu.edu.tw/~cherung/teaching/2011cs5321/handout8.pdf

[13] Winston, W.L. (2004) Operations Research Applications and Algorithms. 4th Edition, Duxbury Press, Pacific Grove, CA.

[14] Maros, I. and Meszaros, C. (1999) A Repository of Convex Quadratic Programming Problems. *Optimization Methods and Software*, **11-12**, 671-681.

4

Searching for a Target Traveling between a Hiding Area and an Operating Area over Multiple Routes

Hongyun Wang[1*], Hong Zhou[2*]

[1]Department of Applied Mathematics and Statistics, Baskin School of Engineering, University of California, Santa Cruz, USA
[2]Department of Applied Mathematics, Naval Postgraduate School, Monterey, USA
Email: *hongwang@soe.ucsc.edu, *hzhou@nps.edu

Abstract

We consider the problem of searching for a target that moves between a hiding area and an operating area over multiple fixed routes. The search is carried out with one or more cookie-cutter sensors, which can detect the target instantly once the target comes within the detection radius of the sensor. In the hiding area, the target is shielded from being detected. The residence times of the target, respectively, in the hiding area and in the operating area, are exponentially distributed. These dwell times are mathematically described by Markov transition rates. The decision of which route the target will take on each travel to and back from the operating area is governed by a probability distribution. We study the mathematical formulation of this search problem and analytically solve for the mean time to detection. Based on the mean time to capture, we evaluate the performance of placing the searcher(s) to monitor various travel route(s) or to scan the operating area. The optimal search design is the one that minimizes the mean time to detection. We find that in many situations the optimal search design is not the one suggested by the straightforward intuition. Our analytical results can provide operational guidances to homeland security, military, and law enforcement applications.

Keywords

Optimal Search Design, Moving Target with Constrained Pathways, Single or Multiple Searchers, Escape Probability, Mean Time to Detection

---

*Corresponding authors.

1. Introduction

Since World War II, search theory has provided many valuable tools to decision makers in both civilian and military operations including search and rescue (SAR), intelligence, surveillance, and reconnaissance (ISR), and enhanced situational awareness [1]-[4]. The prevalence of mobile robotic search agents such as unmanned aerial vehicles (UAVs) or unmanned underwater vehicles (UUVs) has stimulated various search problems in recent years [5]-[10].

In this article, we consider a search problem as depicted in **Figure 1** where a target follows constrained pathways, moving between a hiding area and an operating area. The target can stay in the hiding area (e.g. port or foliage) where the sensors cannot penetrate and consequently the target is not detectable; it can travel along one of the routes connecting the hiding area and the operating area; and it can spend time in the operating area to carry out certain activities/tasks before returning to the hiding area via one of the routes. In **Figure 1**, the hiding area is denoted by A; the collection of all routes is represented by B; and the operating area is marked by C. The target is detectable by a searcher both along the set of routes B and in the operating area C. In the hiding area A, the target is not detectable. We refer this scenario as the A-B-C search problem and it is practically relevant to homeland security, military, counter-drug UAV patrol, and law enforcement operations. To the authors' knowledge, the A-B-C search problem has not been examined systematically in the open literature.

We consider the situation where one or more searchers are deployed. Each searcher possesses an ideal sensor. The definite-range law, or cookie-cutter sensor is an ideal sensor which detects the target instantly once the target comes within distance R to the center of the sensor but cannot detect the target outside that range. The radius R is called the detection radius of the searcher. For the present discussion, we assume that the detection radius of the searcher is large enough to cover the full width of any route in the problems considered here. Under this assumption, if the target chooses to travel along a route that is monitored by a searcher, then the target will definitely be detected by the searcher along that route. Our goal here is to evaluate the performance of placing the searcher(s) at various routes/areas and obtain a search plan that minimizes the mean time to detection of the target.

The rest of this paper proceeds as follows. In Section 2, we consider a moving target which follows the constrained pathways defined above in the presence of one searcher and for which the probability of taking a given route is the same for both travel directions. Section 3 extends the results of Section 2 to the case where the target has different probability of taking a given route for the two travel directions. In Section 4 we further generalize the discussions in Section 3 to include two searchers. Section 5 presents a discussion of three or more searchers. Finally, Section 6 concludes the paper and points out avenues for future study.

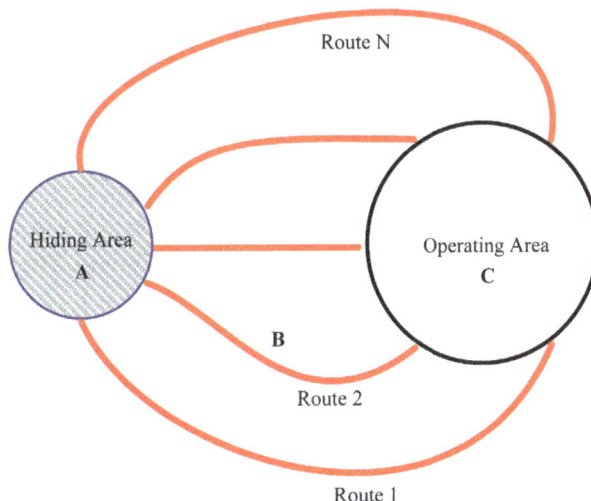

Figure 1. The A-B-C search problem. The target may stay in the hiding area A where it is not detectable; it may travel along one of the routes in collection B; it may spend time in the operating area C before returning to the hiding area.

2. Mathematical Formulation and Solution for a Target Following Constrained Pathways in the Presence of One Searcher

We consider the situation where there is one searcher and the target behaves as follows.

- The dwell time of the target in the hiding area is exponentially distributed with rate r_f, the forward rate of going from the hiding area to the operating area.
- On its travels between the hiding area and the operating area, the target takes route k with probability p_k. In particular, the probability of choosing route k is the same for both directions. This assumption will be re-laxed in the later discussion.
- The dwell time of the target in the operating area is exponentially distributed with rate r_b, the backward rate of going from the operating area back to the hiding area.
- The travel time between the operating area and the hiding area is negligible in comparison with the dwell times in the hiding area and the operating area. Mathematically, we treat the travel time along a route as ze-ro.

After this setup, we have several options of placing the cookie-cutter searcher.

2.1. Case 2A: The Searcher Is Placed on Route k

First, we investigate the situation where the searcher is placed on route k. In this case, the target is detected if and only it travels via route k. As mentioned above, we assume that the detection radius of the searcher is large enough to cover the full width of the route and the detection is instantaneous once the target enters into the detection area of the searcher.

Before the arrival of the searcher, the target jumps between two Markov states with forward rate r_f and backward rate r_b, as shown in **Figure 2**. The steady state probabilities $p_h^{(s)}$ (hiding area) and $p_o^{(s)}$ (operating area) are given by

$$p_h^{(s)} = \frac{r_b}{r_f + r_b}$$

$$p_o^{(s)} = \frac{r_f}{r_f + r_b}$$

(1)

Once the searcher arrives at route k to commence monitoring there, the diagram of Markov transitions contains 3 states. As shown in **Figure 3**, the third state is the target being detected. When the target leaves the hiding area, it has probability p_k of taking route k (and thus being detected). The transition rate from the hiding area to being detected is $r_f p_k$; the rate of going from the hiding area to the operating area (without being detected on the way) is $r_f (1 - p_k)$. Similarly, the transition rate from the operating area to being detected is $r_b p_k$; the rate of traveling from the operating area back to the hiding area (without being detected on the way) is $r_b (1 - p_k)$.

Let $p_h(t)$ and $p_o(t)$ denote, respectively, the probability of being in the hiding area and that of being in the operating area at time t. The probability of the target having been detected by time t is $1 - p_h(t) - p_o(t)$ (*i.e.*, the probability of the third state). Notice that there is no transition from the third state back to the other two states. The time evolution of probabilities $p_h(t)$ and $p_o(t)$ is governed by

$$\frac{dp_h}{dt} = -r_f p_h + r_b (1 - p_k) p_o$$

$$\frac{dp_o}{dt} = r_f (1 - p_k) p_h - r_b p_o$$

(2)

with initial conditions ($t = 0$ being the time when the searcher arrives at route k)

$$p_h(0) = \frac{r_b}{r_f + r_b}, \quad p_o(0) = \frac{r_f}{r_f + r_b}.$$

(3)

The initial conditions (3) correspond to the equilibrium distribution of the target between the hiding area and the

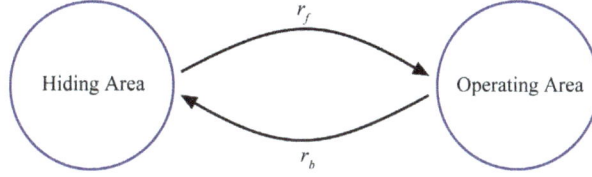

Figure 2. Markov transitions of the target between the hiding area and the operating area before the arrival of the searcher.

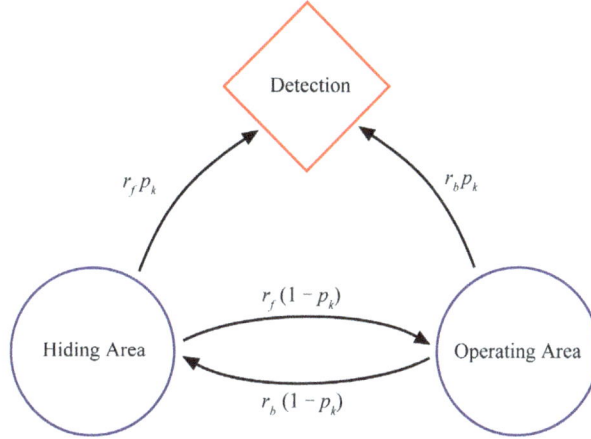

Figure 3. Markov transitions after the searcher is placed on route k.

operating area before the start of search. Equation (2) and **Figure 3** describe the evolution of probability distribution after the start of search.

Both $p_h(t)$ and $p_o(t)$ have the general form of

$$p(t) = c_1 \exp(-\lambda_1 t) + c_2 \exp(-\lambda_2 t) \tag{4}$$

where $(-\lambda_1)$ and $(-\lambda_2)$ are the eigenvalues of the coefficient matrix of the linear ODE system (2). That is,

$$\lambda_1 = \frac{(r_f + r_b) + \sqrt{(r_f + r_b)^2 - 4r_f r_b (2p_k - p_k^2)}}{2}$$

$$\lambda_2 = \frac{(r_f + r_b) - \sqrt{(r_f + r_b)^2 - 4r_f r_b (2p_k - p_k^2)}}{2}. \tag{5}$$

The probability that the target is not detected by time t (*i.e.* the non-detection probability, also called the escape probability) has the same form which reads

$$p_{nd}(t) = p_h(t) + p_o(t) = c_1 \exp(-\lambda_1 t) + c_2 \exp(-\lambda_2 t) \tag{6}$$

where the coefficients c_1 and c_2 are calculated below and are contained in Equation (8).

The non-detection probability is constrained by two other conditions, namely,

$$p_{nd}(0) = 1$$

$$-p'_{nd}(0) = -p'_h(0) - p'_o(0) = p_h(0)r_f p_k + p_o(0)r_b p_k = \frac{2r_f r_b}{r_f + r_b} p_k \tag{7}$$

where the second condition is derived from differential equations (2) and initial conditions (3). The two constraints in (7) allow us to calculate the coefficients c_1 and c_2 in (6) explicitly. After some algebra, we find the non-detection probability has the expression

$$p_{nd}(t) = \frac{1-\alpha}{2}\exp(-\lambda_1 t) + \frac{1+\alpha}{2}\exp(-\lambda_2 t) \qquad (8)$$

where the coefficient α is given by

$$\alpha = \frac{\left(r_f + r_b\right)^2 - 4r_f r_b p_k}{\left(r_f + r_b\right)\sqrt{\left(r_f + r_b\right)^2 - 4r_f r_b\left(2p_k - p_k^2\right)}}.$$

Note that $\alpha > 0$. Observe also that according to (8), the non-detection probability is a linear combination of two exponential modes: one decreases with fast rate λ_1, the other with slow rate λ_2. Since $\alpha > 0$, the slow decaying mode carries more weight, usually a lot more than the weight of the fast decaying mode.

When p_k is moderately small, we obtain the following asymptotic expansions:

$$\lambda_1 = \left(r_f + r_b\right)\left(1 - \frac{2r_f r_b}{\left(r_f + r_b\right)^2}p_k + O\left(p_k^2\right)\right)$$

$$\lambda_2 = \left(r_f + r_b\right)\left(\frac{2r_f r_b}{\left(r_f + r_b\right)^2}p_k + O\left(p_k^2\right)\right) \qquad (9)$$

$$\alpha = 1 - O\left(p_k^2\right)$$

$$p_{nd}(t) = O\left(p_k^2\right)\exp(-\lambda_1 t) + \left(1 - O\left(p_k^2\right)\right)\exp(-\lambda_2 t).$$

Therefore, when p_k is moderately small (for example, $p_k \leq 0.1$), the decay of the non-detection probability is mainly described by the slow decay rate

$$\lambda_2 \approx \frac{2r_f r_b}{r_f + r_b}p_k$$

which is approximately proportional to probability p_k. Consequently, the non-detection probability simplifies approximately to

$$p_{nd}(t) \approx \exp\left(\frac{-2r_f r_b}{r_f + r_b}p_k t\right).$$

In an effort to use one quantity to characterize the decay of the non-detection probability, we compute the mean time to detection. Let $T_{\text{detection}}$ denote the time to detection, which is a random variable. We calculate the mean time to detection from $p_{nd}(t)$ given in (8):

$$E\left[T_{\text{detection}}\right] = \int_0^\infty p_{nd}(t)\,dt = \frac{1-\alpha}{2}\cdot\frac{1}{\lambda_1} + \frac{1+\alpha}{2}\cdot\frac{1}{\lambda_2} = \frac{r_f + r_b}{2r_f r_b p_k}\cdot\frac{1 - \dfrac{2r_f r_b}{\left(r_f + r_b\right)^2}p_k}{1 - \dfrac{1}{2}p_k}. \qquad (10)$$

The derivative of $E\left[T_{\text{detection}}\right]$ with respect to p_k is

$$\frac{dE\left[T_{\text{detection}}\right]}{dp_k} = -\left(\frac{r_f + r_b}{2r_f r_b p_k^2}\right)\cdot\frac{\left(1 - p_k + \dfrac{r_f r_b}{\left(r_f + r_b\right)^2}p_k^2\right)}{\left(1 - \dfrac{1}{2}p_k\right)^2} < 0. \qquad (11)$$

Thus, the mean time to detection is a strictly decreasing function of p_k. Among all possible routes, placing the searcher on the route with the largest probability p_k will yield the fastest decay of the non-detection probability and the shortest mean time to detection. In other words, if the searcher is placed on one of the pathways, then the best choice is to place the searcher on the route that the target is most likely to journey. This is consistent with intuition. As we will see later, in the case of the target having different probabilities of taking a given route for the two travel directions, and/or in the case of more than one searcher, the straightforward intuition often fails to yield the optimal placement.

When p_k is small, it follows from (10) that the mean time to detection can be approximated by

$$E\left[T_{\text{detection}}\right] \approx \frac{r_f + r_b}{2r_f r_b p_k} \tag{12}$$

which is inversely proportional to p_k.

2.2. Case 2B: The Searcher Is Placed in the Operating Area to Scan Search the Area

Now we consider the situation where the searcher looks over only the operating area. Since the searcher is committed to searching the operating area, the target can only be detected when it is in the operating area.

Here we do not specify how the searcher conducts search over the operating area, which may include random search, parallel sweeps (mowing the lawn), spiral-in or spiral-out paths. We consider the conditional probability of the non-detection given that the target is in the operating area. We model the conditional non-detection probability as a single exponential decay.

$$p_{nd}\left(t|O\right) \equiv p_{nd}\left(t|\text{target in operating area}\right) = \exp\left(-r_d t\right) \tag{13}$$

where r_d is the detection rate in the operating area.

In addition to transitions between the hiding area and the operating area, there is now a new transition of the target going from the operating area to being detected. Upon the arrival of the searcher in the operating area, the Markov transitions of the target are depicted in the diagram shown in **Figure 4**.

For mathematical convenience, we represent the detection rate r_d as

$$r_d = \eta \cdot r_b$$

where $\eta : 1$ are the odds of being detected in a trip to the operating area.

The time evolution of probabilities $p_h(t)$ and $p_o(t)$ is now given by

$$\frac{dp_h}{dt} = -r_f p_h + r_b p_o$$
$$\frac{dp_o}{dt} = r_f p_h - r_b\left(1+\eta\right)p_o \tag{14}$$

subject to initial conditions

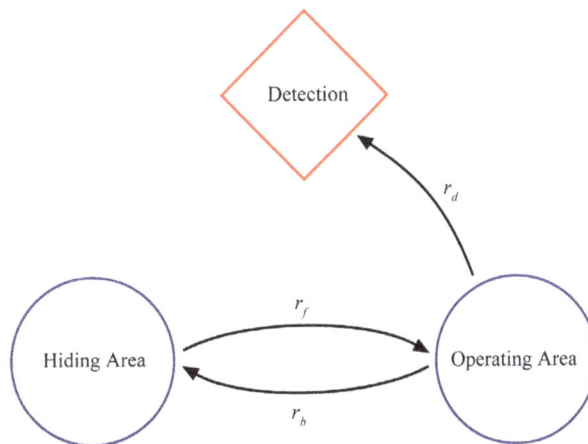

Figure 4. Markov transitions after the searcher is placed in the operating area.

$$p_h(0) = \frac{r_b}{r_f + r_b}, \quad p_o(0) = \frac{r_f}{r_f + r_b}.$$

Proceeding exactly as for Case 2A, we can succinctly express the non-detection probability as

$$p_{nd}(t) = c_1 \exp(-\lambda_1 t) + c_2 \exp(-\lambda_2 t) \tag{15}$$

where λ_1 and λ_2 are the eigenvalues of the coefficient matrix of the linear ODE system (14):

$$\lambda_1 = \frac{(r_f + r_b(1+\eta)) + \sqrt{(r_f + r_b(1+\eta))^2 - 4r_f r_b \eta}}{2}$$

$$\lambda_2 = \frac{(r_f + r_b(1+\eta)) - \sqrt{(r_f + r_b(1+\eta))^2 - 4r_f r_b \eta}}{2}. \tag{16}$$

The non-detection probability also satisfies two other conditions:

$$p_{nd}(0) = 1$$

$$-p'_{nd}(0) = p_o(0) r_b \eta = \frac{r_f r_b}{r_f + r_b} \eta. \tag{17}$$

We solve for coefficients c_1 and c_2 in (15) from these two conditions. We obtain

$$p_{nd}(t) = \frac{1-\alpha}{2} \exp(-\lambda_1 t) + \frac{1+\alpha}{2} \exp(-\lambda_2 t) \tag{18}$$

where the coefficient α is defined by

$$\alpha = \frac{(r_f + r_b)^2 + r_b(r_b - r_f)\eta}{(r_f + r_b)\sqrt{(r_f + r_b(1+\eta))^2 - 4r_f r_b \eta}} > 0.$$

As discussed before, the non-detection probability contains two exponential modes: one decays with fast rate λ_1, the other with slow rate λ_2. Since $\alpha > 0$, the slow decaying mode bears more weight. When η is moderately small (*i.e.*, the probability of being detected in one trip to the operating area is mall), we arrive at the following asymptotic expansions:

$$\lambda_1 = (r_f + r_b)\left(1 + \frac{r_b^2}{(r_f + r_b)^2}\eta + O(\eta^2)\right)$$

$$\lambda_2 = (r_f + r_b)\left(\frac{r_f r_b}{(r_f + r_b)^2}\eta + O(\eta^2)\right) \tag{19}$$

$$\alpha = 1 - O(\eta^2)$$

$$p_{nd}(t) = O(\eta^2)\exp(-\lambda_1 t) + (1 - O(\eta^2))\exp(-\lambda_2 t).$$

From the results of these asymptotic expansions, we see that when η is moderately small, the decay of the non-detection probability can be characterized by the slow decay rate

$$\lambda_2 \approx \frac{r_f r_d}{r_f + r_b}.$$

Accordingly, the non-detection probability takes the approximate form

$$p_{nd}(t) \approx \exp\left(\frac{-r_f r_d}{r_f + r_b} t\right).$$

Again, we calculate the mean time to detection from $p_{nd}(t)$ in (18):

$$
\begin{aligned}
E[T_{\text{detection}}] = \int_0^\infty p_{nd}(t)\,\mathrm{d}t &= \frac{1-\alpha}{2}\cdot\frac{1}{\lambda_1} + \frac{1+\alpha}{2}\cdot\frac{1}{\lambda_2}\\
&= \frac{r_f + r_b}{r_f r_b \eta} + \frac{r_b}{r_f(r_f + r_b)} = \frac{r_f + r_b}{r_f r_d} + \frac{r_b}{r_f(r_f + r_b)} \equiv g(r_d).
\end{aligned}
\tag{20}
$$

Notice that the mean time to detection contains a constant part that does not decrease with increasing detection rate r_d. This reflects the fact that no matter how fast the target can be detected in the operating area, the detection can occur only when the target moves to the operating area. The constant term $\dfrac{r_b}{r_f(r_f + r_b)}$ is the mean elapsed time until the first arrival of the target in the operating area. This constant term gives the lower bound on the mean time to detection, which can be achieved when the detection rate in the operating area is much larger than other transition rates: $r_d \gg \max(r_f, r_b)$.

Combining the results of Case 2A and Case 2B, we can now determine the optimal placement of the searcher (on a particular route vs. in the operating area) in order to attain the minimum mean time to detection of the target. We introduce the probability and identify of the route that is most likely to be travelled by the target:

$$
p_{\max} = \max_j\left(p_j\right)
$$

$$
k_{\max} = \arg\max_j\left(p_j\right).
$$

We compare the mean times to detection in the two cases

$$
\frac{r_f + r_b}{2 r_f r_b p_{\max}} \cdot \frac{1 - \dfrac{2 r_f r_b}{\left(r_f + r_b\right)^2} p_{\max}}{1 - \dfrac{1}{2} p_{\max}} \quad \text{vs.} \quad \frac{r_f + r_b}{r_f r_d} + \frac{r_b}{r_f\left(r_f + r_b\right)}.
$$

If the former is smaller, we place the searcher on route $k_{\max}$, the route with the highest probability of being traveled by the target; if the later is smaller, we place the searcher in the operating area to scan search the area. Note that when the probability of taking a given route is the same for the two travel directions, the minimum mean time to detection among all routes is the one corresponding to the route most likely to be traveled by the target. This is consistent with intuition. This intuitive strategy, however, breaks down when the probability of taking a given route has different values for the two travel directions. That situation is discussed in the next section.

3. A Target Having Different Probabilities of Taking a Given Route for the Two Travel Directions

In this section, we consider the target which behaves the same as in Section 2 except that the probability of taking route k may have different values for the forward travel (from the hiding area to the operating area) and the backward travel (from the operating area to the hiding area). We use p_k and q_k to denote, respectively, the probabilities of taking route k in the two opposite travel directions. More specifically,

- On its travel from the hiding area to the operating area, the target takes route k with probability p_k.
- On its travel from the operating area back to the hiding area, the target chooses route k with probability q_k.

We study the performance of placing the searcher on a route, which is analogous to Case 2A. Note that regardless of the target's behavior in selecting which route to take, the results of Case 2B apply directly here, which is affected only by how frequently the target visits the operating area and how fast the target is detected while in the operating area.

Case 3A: The Searcher Is Placed on Route k

To analyze the case where the searcher is placed on route k, we observe that the Markov transitions are illustrated in **Figure 5**. When the target leaves the hiding area, it has probability p_k of taking route k (and thus

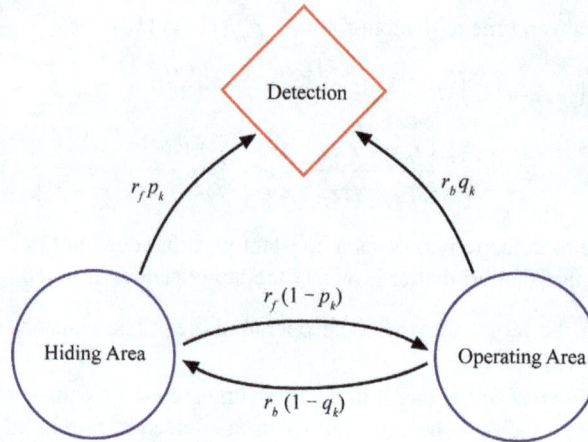

Figure 5. Markov transitions after the searcher is placed on route k in Case 3A where the probability of taking route k is p_k for the forward travel and q_k for the backward travel.

being detected). The transition rate from the hiding area to being detected is $r_f p_k$; the rate of going from the hiding area to the operating area (without being detected on the way) is $r_f(1-p_k)$. Likewise, on the target's returning trip to the hiding area, the transition rate from the operating area to being detected is $r_b q_k$; the rate of traveling from the operating area back to the hiding area (without being detected on the way) is $r_b(1-q_k)$.

The time evolution of probabilities $p_h(t)$ and $p_o(t)$ follows a set of ODEs:

$$\frac{dp_h}{dt} = -r_f p_h + r_b(1-q_k)p_o$$

$$\frac{dp_o}{dt} = r_f(1-p_k)p_h - r_b p_o \tag{21}$$

with initial conditions

$$p_h(0) = \frac{r_b}{r_f+r_b}, \quad p_o(0) = \frac{r_f}{r_f+r_b}.$$

The non-detection probability is of the form

$$p_{nd}(t) = c_1 \exp(-\lambda_1 t) + c_2 \exp(-\lambda_2 t) \tag{22}$$

where λ_1 and λ_2 are the eigenvalues of the coefficient matrix of the linear ODE system (21)

$$\lambda_1 = \frac{(r_f+r_b) + \sqrt{(r_f+r_b)^2 - 4r_f r_b(p_k+q_k-p_k q_k)}}{2}$$

$$\lambda_2 = \frac{(r_f+r_b) - \sqrt{(r_f+r_b)^2 - 4r_f r_b(p_k+q_k-p_k q_k)}}{2}.$$

The non-detection probability needs to meet two more conditions:

$$p_{nd}(0) = 1$$

$$-p'_{nd}(0) = p_h(0)r_f p_k + p_o(0)r_b q_k = \frac{r_f r_b}{r_f+r_b}(p_k+q_k).$$

With the help of these two constraints, we can determine the coefficients c_1 and c_2. We write out the non-detection probability as

$$p_{nd}(t) = \frac{1-\alpha}{2}\exp(-\lambda_1 t) + \frac{1+\alpha}{2}\exp(-\lambda_2 t) \tag{23}$$

with the coefficient α satisfying

$$\alpha = \frac{\left(r_f + r_b\right)^2 - 2r_f r_b\left(p_k + q_k\right)}{\left(r_f + r_b\right)\sqrt{\left(r_f + r_b\right)^2 - 4r_f r_b\left(p_k + q_k - p_k q_k\right)}} > 0.$$

As in the earlier discussion, the non-detection probability is a linear combination of two exponential modes: one decreases with fast rate λ_1 while the other with slow rate λ_2. The fact that α is positive suggests that the slow decaying component weighs more than the fast decaying one.

When both p_k and q_k are moderately small, we carry out the asymptotic analysis to obtain

$$\lambda_1 = \left(r_f + r_b\right)\left(1 - \frac{r_f r_b}{\left(r_f + r_b\right)^2}\left(p_k + q_k\right) + O\left(p_k^2 + q_k^2\right)\right)$$

$$\lambda_2 = \left(r_f + r_b\right)\left(\frac{r_f r_b}{\left(r_f + r_b\right)^2}\left(p_k + q_k\right) + O\left(p_k^2 + q_k^2\right)\right) \tag{24}$$

$$\alpha = 1 - O\left(p_k^2 + q_k^2\right)$$

$$p_{nd}(t) = O\left(p_k^2 + q_k^2\right)\exp\left(-\lambda_1 t\right) + \left(1 - O\left(p_k^2 + q_k^2\right)\right)\exp\left(-\lambda_2 t\right).$$

It therefore follows that when p_k and q_k are moderately small (for example, $p_k \le 0.1$ and $q_k \le 0.1$), the decay of the non-detection probability is dominated by the slow decay rate

$$\lambda_2 \approx \frac{r_f r_b}{r_f + r_b}\left(p_k + q_k\right)$$

which is approximately proportional to probability $\left(p_k + q_k\right)$. Furthermore, the non-detection probability is well approximated by

$$p_{nd}(t) \approx \exp\left(\frac{-r_f r_b}{r_f + r_b}\left(p_k + q_k\right)t\right).$$

Next we find the mean time to detection. Using the formula (23), we compute the mean time to detection:

$$E\left[T_{\text{detection}}\right] = \frac{1 - \alpha}{2} \cdot \frac{1}{\lambda_1} + \frac{1 + \alpha}{2} \cdot \frac{1}{\lambda_2} = \frac{r_f + r_b}{r_f r_b} \cdot \frac{1 - \frac{r_f r_b}{\left(r_f + r_b\right)^2}\left(p_k + q_k\right)}{p_k + q_k - p_k q_k} \equiv h\left(p_k, q_k\right). \tag{25}$$

The most important thing to notice about this result is that the mean time to detection is not just a function of $\left(p_k + q_k\right)$, the sum of probabilities of taking route k in the two travel directions. Consequently, the optimal route is not necessarily the one that has the largest value of $\left(p_k + q_k\right)$. To illustrate this point, we consider a simple and interesting example in the following.

An Example: Let us consider a target traveling between the hiding area and the operating area. The transition rates and the probabilities of taking individual routes in each travel direction are listed below.

$$r_f = 0.1, \ r_b = 1$$

$$\left(p_1, p_2, p_3, p_4\right) = \left(0.47, 0.35, 0.13, 0.05\right)$$

$$\left(q_1, q_2, q_3, q_4\right) = \left(0.15, 0.3, 0.0, 0.55\right)$$

The small value for r_f corresponds to the fact that the target stays in the hiding area for most time. If we compute the values of $p_k + q_k$, we have

$$p_1 + q_1 = 0.62, \ p_2 + q_2 = 0.65, \ p_3 + q_3 = 0.13, \ p_4 + q_4 = 0.6.$$

Intuitively, the optimal route for placing the searcher seems to be $k = 2$, which yields the largest value of

$p_k + q_k$; the second candidate for the optimal route seems to be $k = 1$, which leads to the second largest value of $p_k + q_k$. However, to select the optimal route for placing the searcher, we appeal to (25) and calculate the mean time to detection for each route:

$$E\left[T_{\text{detection}}\big|_{\text{route 1}}\right] = h(p_1, q_1) = 18.99$$

$$E\left[T_{\text{detection}}\big|_{\text{route 2}}\right] = h(p_2, q_2) = 19.10$$

$$E\left[T_{\text{detection}}\big|_{\text{route 3}}\right] = h(p_3, q_3) = 83.71$$

$$E\left[T_{\text{detection}}\big|_{\text{route 4}}\right] = h(p_4, q_4) = 18.26$$

It is clear that route 4 has the smallest mean time to detection, and thus, is the optimal route for placing the searcher. This example highlights that when the probability of taking a given route has different values for the two travel directions, the optimal placement of the searcher is not the one suggested by straightforward intuition.

4. A Moving Target and Two Searchers

We study the effect of two searchers where each searcher acts and detects independently. We assume the same setup for the target as in Section 3. The difference here is that there are two cookie-cutter searchers deployed in search for the target. We consider several options of placing the two cookie-cutter searchers.

4.1. Case 4A: Both Searchers Are Placed on Routes: One on Route k and the Other on Route j

Since we assume that one searcher is capable of covering the full width of a route, there is absolutely no benefit for placing more than one searchers on any one route. Thus, the two searchers are to be placed on two different routes. In this case, the mathematical formulation is exactly the same as in Case 3A in Section 3 except that probability p_k in Section 3 is now replaced with the effective probability $p_k + p_j$; and in place of q_k in Case 3A we have $q_k + q_j$. We introduce two short notations that will be convenient throughout this section:

$$p \equiv p_k + p_j, \quad q \equiv q_k + q_j. \tag{26}$$

Then, the non-detection probability is given by

$$P_{nd}(t) = \frac{1-\alpha}{2}\exp(-\lambda_1 t) + \frac{1+\alpha}{2}\exp(-\lambda_2 t) \tag{27}$$

where eigenvalues λ_1 and λ_2, and coefficient α are

$$\lambda_1 = \frac{(r_f + r_b) + \sqrt{(r_f + r_b)^2 - 4r_f r_b((p+q-pq)}}{2}$$

$$\lambda_2 = \frac{(r_f + r_b) - \sqrt{(r_f + r_b)^2 - 4r_f r_b(p+q-pq)}}{2}$$

$$\alpha = \frac{(r_f + r_b)^2 - 2r_f r_b(p+q)}{(r_f + r_b)\sqrt{(r_f + r_b)^2 - 4r_f r_b(p+q-pq)}} > 0$$

Recalling (25) obtained in Case 3A of Section 3, one finds the mean time to detection in the form

$$E[T_{\text{detection}}] = h(p, q) \tag{28}$$

where function $h(p, q)$ is defined as

$$h(p,q) \equiv \frac{r_f + r_b}{r_f r_b} \cdot \frac{1 - \frac{r_f r_b}{\left(r_f + r_b\right)^2}(p+q)}{p + q - pq} \tag{29}$$

The optimal routes for placing the two searchers are selected by minimizing the mean time to detection:

$$(k,j) = \arg\min_{(k,j)} h\left(p_k + p_j, q_k + q_j\right), \quad k \neq j.$$

It is attempting to simplify the selection process by using the optimal route for a single searcher plus the second optimal route for a single searcher. The example below demonstrates that this intuitive approach does not necessarily give us the optimal routes for placing the two searchers.

An example: We consider a target traveling between the hiding area and the operating area over 5 routes. The transition rates and the probabilities of taking route k in the two travel directions are given below for the target

$$r_f = 0.1, \quad r_b = 1$$

$$(p_1, p_2, p_3, p_4, p_5) = (0.45, 0.0, 0.25, 0.25, 0.05)$$

$$(q_1, q_2, q_3, q_4, q_5) = (0.0, 0.46, 0.25, 0.25, 0.04).$$

When only a single searcher is present, the mean times to detection for individual routes are calculated from (25)

$$E\left[T_{\text{detection}}\big|_{\text{route 1}}\right] = h(p_1, q_1) = 23.54$$

$$E\left[T_{\text{detection}}\big|_{\text{route 2}}\right] = h(p_2, q_2) = 23.0$$

$$E\left[T_{\text{detection}}\big|_{\text{route 3}}\right] = h(p_3, q_3) = 24.10$$

$$E\left[T_{\text{detection}}\big|_{\text{route 4}}\right] = h(p_4, q_4) = 24.10$$

$$E\left[T_{\text{detection}}\big|_{\text{route 5}}\right] = h(p_5, q_5) = 124.1$$

The optimal route and the second optimal route for placing a single searcher are, respectively, route 2 and route 1. It seems reasonable to conclude that in the case of two searchers, the optimal routes for placing the two searchers are routes 2 and 1. However, this is not true. When the two searchers are placed respectively on routes 1 and 2, the mean time to detection is then obtained from (28) as

$$E\left[T_{\text{detection}}\big|_{\text{route }(1,2)}\right] = h(p_1 + p_2, q_1 + q_2) = 14.47.$$

In comparison, when the two searchers are placed respectively on routes 1 and 3, the mean time to detection becomes

$$E\left[T_{\text{detection}}\big|_{\text{route }(1,3)}\right] = h(p_1 + p_3, q_1 + q_3) = 13.81$$

which is still not yet the optimal. It turns out that the smallest mean time to detection among all possible pairs of routes is achieved by placing the two searchers respectively on routes 3 and 4:

$$E\left[T_{\text{detection}}\big|_{\text{route }(3,4)}\right] = h(p_3 + p_4, q_3 + q_4) = 13.46.$$

In this example, the optimal pair of routes for placing the two searchers are routes 3 and 4, not the collection of optimal and second optimal routes for a single searcher.

4.2. Case 4B: Both Searchers Are Placed in the Operating Area to Scan Search the Area

When both searchers are put to search only the operating area, the mathematical formulation is exactly the same as in Case 2B of Section 2 except that the detection rate r_d in Section 2 is now doubled to $2r_d$ because the operating area is patrolled by two searchers.

Based on the results in Case 2B of Section 2, the mean time to detection can be written as

$$E\left[T_{\text{detection}}\right] = g\left(2r_d\right) \tag{30}$$

where function $g(\beta)$ is defined in (20). That is,

$$g(\beta) \equiv \frac{r_f + r_b}{r_f \beta} + \frac{r_b}{r_f\left(r_f + r_b\right)}.$$

4.3. Case 4C: One Searcher Is Placed on Route k and the Other in the Operating Area

The last mathematical formulation is for the case where one searcher is put on a pathway while the other scans the operating area. After the arrival of one searcher at route k and the other in the operating area, the target evolves stochastically according to the Markov process shown in **Figure 6**.

For mathematical convenience, we express the detection rate r_d in the operating area as $r_d = \eta \cdot r_b$. The governing equations for the time evolution of probabilities $p_h(t)$ and $p_o(t)$ consist of

$$\frac{dp_h}{dt} = -r_f p_h + r_b\left(1 - q_k\right) p_o$$

$$\frac{dp_o}{dt} = r_f\left(1 - p_k\right) p_h - r_b\left(1 + \eta\right) p_o. \tag{31}$$

These probability functions satisfy the initial conditions

$$p_h(0) = \frac{r_b}{r_f + r_b}, \quad p_o(0) = \frac{r_f}{r_f + r_b}.$$

Once again, one finds that the non-detection probability has the form

$$p_{nd}(t) = c_1 \exp\left(-\lambda_1 t\right) + c_2 \exp\left(-\lambda_2 t\right) \tag{32}$$

where λ_1 and λ_2 correspond to the two eigenvalues of the coefficient matrix of the linear ODE system (31):

$$\lambda_1 = \frac{\left(r_f + r_b\left(1 + \eta\right)\right) + \sqrt{\left(r_f + r_b\left(1 + \eta\right)\right)^2 - 4r_f r_b\left(\eta + p_k + q_k - p_k q_k\right)}}{2}$$

$$\lambda_2 = \frac{\left(r_f + r_b\left(1 + \eta\right)\right) - \sqrt{\left(r_f + r_b\left(1 + \eta\right)\right)^2 - 4r_f r_b\left(\eta + p_k + q_k - p_k q_k\right)}}{2}$$

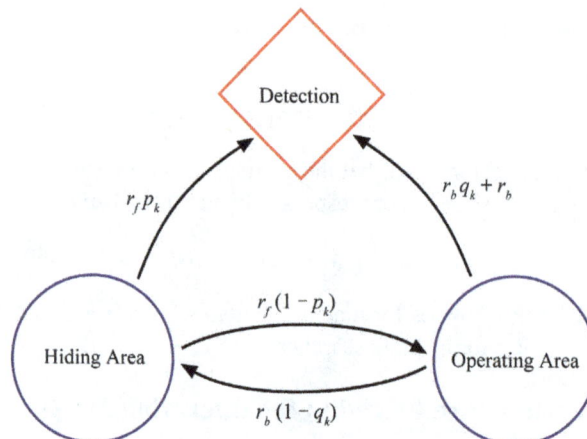

Figure 6. Markov transitions after one searcher is placed on route k and a second searcher is placed in the operating area.

The coefficients c_1 and c_2 can be derived by imposing two constraints on the non-detection probability:

$$p_{nd}(0) = 1$$

$$-p'_{nd}(0) = p_h(0) r_f p_k + p_o(0) r_b (\eta + q_k) = \frac{r_f r_b}{r_f + r_b}(\eta + p_k + q_k).$$

From these two constraints, one derives that the non-detection probability is given by

$$p_{nd}(t) = \frac{1-\alpha}{2} \exp(-\lambda_1 t) + \frac{1+\alpha}{2} \exp(-\lambda_2 t) \tag{33}$$

where coefficient α is

$$\alpha = \frac{(r_f + r_b)^2 + r_b(r_b - r_f)\eta - 2r_f r_b(p_k + q_k)}{(r_f + r_b)\sqrt{(r_f + r_b(1+\eta))^2 - 4r_f r_b(\eta + p_k + q_k - p_k q_k)}}.$$

The mean time to detection is calculated from the non-detection probability $p_{nd}(t)$ given in (33). The resulting mean time to detection is equal to:

$$
\begin{aligned}
E[T_{\text{detection}}] &= \frac{1-\alpha}{2} \cdot \frac{1}{\lambda_1} + \frac{1+\alpha}{2} \cdot \frac{1}{\lambda_2} \\
&= \left(\frac{r_f + r_b}{r_f r_b}\right) \cdot \frac{1 + \dfrac{r_b^2}{(r_f + r_b)^2}\eta - \dfrac{r_f r_b}{(r_f + r_b)^2}(p_k + q_k)}{\eta + p_k + q_k - p_k q_k} \\
&= \left(\frac{r_f + r_b}{r_f}\right) \cdot \frac{1 + \dfrac{r_b}{(r_f + r_b)^2}r_d - \dfrac{r_f r_b}{(r_f + r_b)^2}(p_k + q_k)}{r_d + r_b(p_k + q_k - p_k q_k)} \\
&\equiv f(r_d, p_k, q_k).
\end{aligned}
\tag{34}
$$

This result lies at the heart of everything else that is going on and it includes all of these as special cases. In particular, function $h(p, q)$ and function $g(r_d)$ can be expressed in terms of function $f(r_d, p_k, q_k)$ as:

$$h(p, q) = f(0, p, q)$$

$$g(r_d) = f(r_d, 0, 0).$$

We are now ready to summarize what we have found so far.

4.4. Optimal Placement of Two Searchers

When the two searchers are placed on two routes k and j (one searcher on each route), the shortest mean time to detection is given by

$$\min_{(k,j)} f(0, p_k + p_j, q_k + q_j), \quad k \ne j.$$

When one searcher is placed on route k and the second searcher is placed in the operating area, the smallest mean time to detection is similarly defined by

$$\min_k f(r_d, p_k, q_k).$$

When both searchers are placed in the operating area, the mean time to detection is

$$f(2r_d, 0, 0).$$

By examining these cases, we obtain the overall minimum mean time to detection over these three options:

$$\min\left(\min_{(k,j)} f\left(0, p_k + p_j, q_k + q_j\right), \min_k f\left(r_d, p_k, q_k\right), f\left(2r_d, 0, 0\right)\right)$$

where function $f\left(\beta, p, q\right)$ is defined by

$$f\left(\beta, p, q\right) \equiv \left(\frac{r_f + r_b}{r_f}\right) \cdot \frac{1 + \dfrac{r_b}{\left(r_f + r_b\right)^2}\beta - \dfrac{r_f r_b}{\left(r_f + r_b\right)^2}\left(p + q\right)}{\beta + r_b\left(p + q - pq\right)}. \tag{35}$$

5. Three or More Searchers

We now turn to the case where three cookie-cutter searchers are deployed in the search for the target. Probabilistic independence is assumed for each searcher. Suppose the target behaves exactly the same as described in Sections 3 and 4. We apply function $f\left(\beta, p, q\right)$ to calculate the optimal placement of the three searchers.

When all three searchers are placed on three routes (one searcher on each route), the shortest mean time to detection is found to be

$$\min_{(k,j,i)} f\left(0, p_k + p_j + p_i, q_k + q_j + q_i\right).$$

Similarly, when two searchers are placed on two routes (one searcher on each route) and the third searcher is placed in the operating area, the shortest mean time to detection is

$$\min_{(k,j)} f\left(r_d, p_k + p_j, q_k + q_j\right).$$

When one searcher is placed on route k and the two other searchers are placed in the operating area, the smallest mean time to detection is equal to

$$\min_k f\left(2r_d, p_k, q_k\right).$$

Finally, when all three searchers are placed in the operating area, it follows that the mean time to detection is

$$f\left(3r_d, 0, 0\right).$$

We conclude that the overall minimum mean time to detection over these options can be computed by comparing the results in these cases

$$\min\left(\min_{(k,j,i)} f\left(0, p_k + p_j + p_i, q_k + q_j + q_i\right), \min_{(k,j)} f\left(r_d, p_k + p_j, q_k + q_j\right), \min_k f\left(2r_d, p_k, q_k\right), f\left(3r_d, 0, 0\right)\right).$$

It is also evident that this method can be extended to finding the optimal placement for any number of searchers.

6. Conclusions

This paper presented a mathematical approach to investigate the search problem of detecting a mobile target that travels between a hiding area and an operating area over a collection of fixed pathways in the presence of single or multiple searchers with cookie-cutter sensors. Both the dwell time of the target in the hiding area and the dwell time of the target in the operating area were assumed to be exponentially distributed and were modeled mathematically by Markov transition rates. The travel time of the target on a pathway was assumed to be short in comparison with the dwell times in the hiding area and the operating area. The target took a route to and from the operating area based on a probability distribution. Under these assumptions, a mathematical formulation was developed and solved analytically.

The main results can be summarized as follows.

1) When only one searcher is present, one can compute the mean time to detection, respectively, when the searcher is placed on route k or when the search is placed in the operating area. By comparing the mean times to detection, we can decide whether to put the searcher on a certain route or in the operating area.

2) In a similar fashion, when multiple searchers are deployed, we can calculate the mean times to detection for various scenarios and thereby obtain the overall minimum time to detection and the optimal placement of

searchers.

It was discovered that in many cases the optimal search design was not necessarily the one indicated by straightforward intuition.

There are a few possible future research directions. For example, extending current study to include time-dependent transition rates and detection rates would provide a more sophisticated modeling. Multiple targets could also be considered in the future studies.

Acknowledgements and Disclaimer

Hong Zhou would like to thank Naval Postgraduate School Center for Multi-INT Studies for supporting this work. Special thanks go to Professor Jim Scrofani and Deborah Shifflett. The views expressed in this document are those of the authors and do not reflect the official policy or position of the Department of Defense or the U.S. Government.

References

[1]   Koopman, B.O. (1999) Search and Screening: General Principles with Historical Applications. The Military Operations Research Society, Inc., Alexandria.

[2]   Stone, L.D. (1989) Theory of Optimal Search. 2nd Edition, Operations Research Society of America, Arlington.

[3]   Wagner, D.H., Mylander, W.C. and Sanders, T.J. (1999) Naval Operations Analysis. 3rd Edition, Naval Institute Press, Annapolis.

[4]   Washburn, A. (2014) Search and Detection. 5th Edition, Create Space Independent Publishing Platforms.

[5]   Chung, H., Polak, E., Royset, J.O. and Sastry, S. (2011) On the Optimal Detection of an Underwater Intruder in a Channel Using Unmanned Underwater Vehicles. *Naval Research Logistics*, **58**, 804-820. http://dx.doi.org/10.1002/nav.20487

[6]   Chung, T.H., Hollinger, G.A. and Isler, V. (2011) Search and Pursuit-Evasion in Mobile Robotics. *Autonomous Robots*, **31**, 299-316. http://dx.doi.org/10.1007/s10514-011-9241-4

[7]   Chung, T.H. and Silvestrini, R.T. (2014) Modeling and Analysis of Exhaustive Probabilistic Search. *Naval Research Logistics*, **61**, 164-178. http://dx.doi.org/10.1002/nav.21574

[8]   Kress, M. and Royset, J.O. (2008) Aerial Search Optimization Model (ASOM) for UAVs in Special Operations. *Military Operations Research*, **13**, 23-33. http://dx.doi.org/10.5711/morj.13.1.23

[9]   Wang, H. and Zhou, H. (2015) Computational Studies on Detecting a Diffusing Target in a Square Region by a Stationary or Moving Searcher. *American Journal of Operations Research*, **5**, 47-68. http://dx.doi.org/10.4236/ajor.2015.52005

[10]   Wilson, K.E., Szechtman, R. and Atkinson, M.P. (2011) A Sequential Perspective on Searching for Static Targets. *European Journal of Operational Research*, **215**, 218-226. http://dx.doi.org/10.1016/j.ejor.2011.05.045

5

An Inventory Model for Perishable Items with Time Varying Stock Dependent Demand and Trade Credit under Inflation

Sushil Kumar*, U. S. Rajput

Department of Mathematics & Astronomy, University of Lucknow, Lucknow, India
Email: *sushilmath4444@gmail.com

Abstract

In the classical inventory models, it is assumed that the retailer pays to the supplier as soon as he received the items and in such cases the supplier offers a cash discount or credit period (permissible delay) to the retailer. In this paper we presented an inventory model for perishable items with time varying stock dependent demand under inflation. It is assumed that the supplier offers a credit period to the retailer and the length of credit period is dependent on the order quantity. The purpose of our study is to minimize the present value of retailer's total cost. Numerical examples are also given to demonstrate the presented mode.

Keywords

Inventory, Deterioration, Perishable, Credit Period, Time Varying Stock Dependent Demand

1. Introduction

In the classical inventory models payment for the items paid by the supplier depends on the payment paid by the retailer and in such cases the supplier offers a fixed credit period to the retailer during which no interest will be charged by the supplier so there is no need to pay the purchasing cost by the retailer and after this credit period up to the end of a period interest charged and paid by the retailer. In such situations the retailer starts to accumulate revenue on his sale and earn interest on his revenue. If the revenue earned by the retailer up to the end of credit period is enough to pay the purchasing cost or there is a budget then the balance is settled and the supplier does not charge any interest, otherwise the supplier charges interest for unpaid balance after the credit period. The interest and the remaining payment are made at the end of replenishment cycle.

---

*Corresponding author.

In traditional EOQ models the payment time does not affect the profit and replenishment policy. If we consider the inflation then order quantity and payment time can influence both the supplier's and retailer's decisions. A large pile of perishable foods such as fruits, vegetables, milk, bread, chocklet etc. attract the consumers to buy more. Buzacott [1] considered an EOQ model with different type of pricing policies under inflation. Silver and Peterson [2] developed an inventory model and show that the consumption rate is proportional to the displayed stock level. Baker and Urban [3] proposed an inventory model for deteriorating items with the demand rate is a polynomial function of instantaneous stock level. Mandal and Phaujdar [4] presented an inventory model for deteriorating items with stock level dependent consumption rate. Vrat and Padmanabhan [5] considered an inventory model with stock dependent demand under constant inflation rate. Padmanabhan and Vrat [6] developed an EOQ model for perishable products under stock dependent selling rate. Bose *et al.* [7] considered an inventory model for deteriorating items with time dependent demand and shortages under inflation and time discounting. Mandal and Maiti [8] developed an inventory model for damageable items with stock dependent demand and variable replenishment rate. Chung and Lin [9] determine an optimal replenishment policy for an inventory model of deteriorating items by considering inflation and credit period. Chang [10] proposed an EOQ model for deteriorating items under inflation and time discounting assuming that the supplier offers a trade credit policy if the retailer order size is larger than a certain level. Dye and Ouyang [11] developed an EOQ model for perishable items with stock dependent selling rate by allowing shortages. Hou [12] presented an inventory model for deteriorating items with stock dependent consumption rate and shortages under inflation and credit period. Jaggi *et al.* [13] determine an optimal ordering policy for deteriorating items under inflation induced demand. Sana and Chaudhuri [14] developed a deterministic EOQ model with stock dependent demand and delay in payments. Valliaththal and Uthayakumar [15] presented an inventory model for perishable items under stock and time dependent selling rate with shortages. Roy *et al.* [16] considered an inventory model for deteriorating items with stock dependent demand under fuzzy inflation and time discounting over a random planning horizon. Sana [17] proposed a lot size inventory model with stock dependent demand and time varying deterioration and partial backlogging. Chang *et al.* [18] determine an optimal replenishment policy for an inventory model of non-instantaneous deteriorating items with stock dependent demand. Sarkar *et al.* [19] presented an EMQ (economic manufacturing quantity) model of an imperfect production process with time dependent demand and time value of money under inflation. Yan [20] considered an EOQ model for perishable items with freshness dependent demand and partial backlogging. Nagrare and Dutta [21] developed a continuous review inventory model for perishable products with inventory dependent demand. Sana [22] proposed a control policy for a production system inflation assuming a stock dependent demand and sales team promotional effort.Shuai *et al.* [23] considered an inventory model for perishable products with stock dependent demand and trade credit under inflation.

Table 1 and Table 2 show the variation of the parameters r and M when $M \geq T$ and Table 3 & Table 4 show the variation of parameters *r* and *M* when $M \prec T$. Figure 1 & Figure 2 are correspond to the developed model. Figure 3 & Figure 4 show the variation of retailer's total cost with respect to the parameters r and M when $M \geq T$ and Figure 5 & Figure 6 show the variation of retailer's total cost with respect to the parameters r and M when $M \prec T$.

In the present paper we presented an inventory model for perishable items with time varying stock dependent demand and trade credit under inflation. Although there are so many research papers related to the perishable

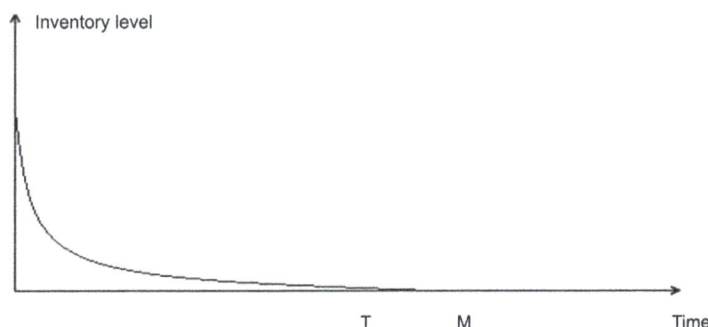

Figure 1. Corresponding to developed model.

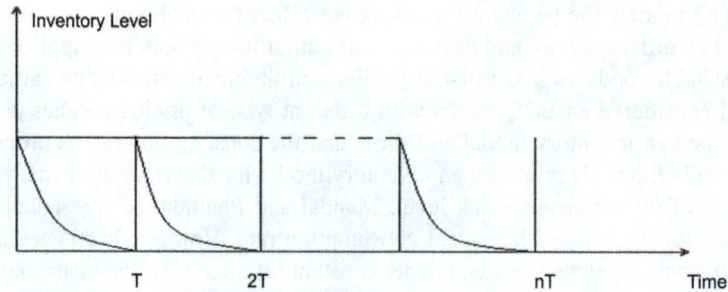

Figure 2. With n cycles in the developed model.

Table 1. Variation of retailer's total cost with respect to the change of parameter r.

r	T	$TCH_1(T)$
1.2	0.809153	-3.18984×10^6
1.4	1.29391	-7.38628×10^6
1.6	1.70349	-1.360695×10^7
1.8	2.07928	-2.238335×10^7
2	2.43731	-3.4213250×10^7
2.2	2.78705	-4.963000×10^7
2.5	3.30941	-8.063075×10^7

Table 2. Variation of retailer's total cost with respect to the change of parameter M.

M	T	$TCH_1(T)$
2	0.809153	-3.18984×10^6
4	3.71242	-6.82474×10^7
6	6.66987	-2.96128×10^8
8	9.64061	-7.88637×10^8
10	12.6158	-1.64743×10^9
12	15.5929	-2.97412×10^9

Table 3. Variation of retailer's total cost with respect to the change of parameter r.

r	T	$TCH_2(T)$
0.05	0.494393	-4.59988×10^7
0.1	0.500148	-4.30252×10^7
0.15	0.500591	-4.02547×10^7
0.2	0.495351	-3.78662×10^7
0.25	0.483842	-3.58238×10^7
0.3	0.465167	-3.41820×10^7
0.35	0.437937	-3.30523×10^7
0.4	0.399872	-3.26667×10^7

Table 4. Variation of retailer's total cost with respect to the change of parameter M.

M	T	$TCH_2(T)$
2	0.494393	-4.59988×10^7
4	1.02028	-3.86733×10^8
6	1.54603	-1.32099×10^9
8	2.07174	-3.13999×10^9
10	2.59744	-6.03636×10^9
12	3.12314	-1.05591×10^{10}

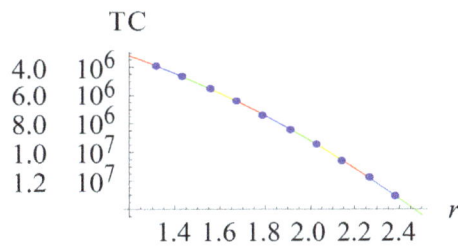

Figure 3. Variation in TC with respect to r.

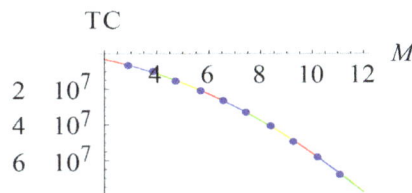

Figure 4. Variation in TC with respect to M.

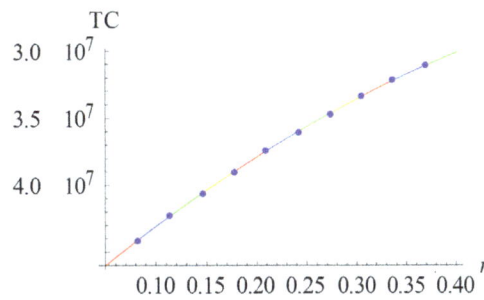

Figure 5. Variation in TC with respect to r.

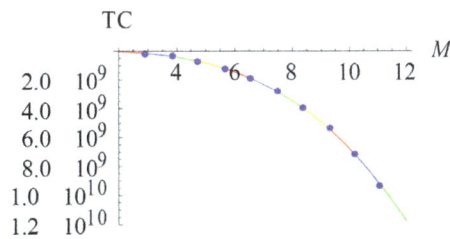

Figure 6. Variation in TC with respect to M.

products with stock dependent demand under inflation. This paper deals with the same type problem and it provides an approximate solution procedure of this problem for minimizing the present value of retailer's total cost.

2. Assumptions and Notations

We consider the following assumptions and notations corresponding to the developed model
1) The demand rate $R(t)$ is $R(t) = a + bt + kI(t)$, $a \geq 0, 0 \leq b \leq 1, k \succ 0$
2) θ is the constant deterioration rate.
3) o_C is the ordering cost per order.
4) h_C is the holding cost per unit.
5) s_C is the shortage cost.
6) M is the credit period.
7) T is the replenish cycle length.
8) r is the inflation rate.
9) I_C is the interest charged per \$ per unit time when $T \succ M$.
10) C is the purchasing cost per unit.
11) P is the selling price per unit with $P \succ C$.
12) Q is the initial inventory level.
13) L is the planning horizon.
14) The supplier sells one single item to the retailer.
15) The items are replenished when the stock level becomes zero.
16) The supplier provides a credit period, which is dependent on the order quantity.
17) The lead time is zero.
18) Shortages are not allowed.
19) The inventory planning horizon is finite and the numbers of cycles are finite in the planning horizon.
20) $I(t)$ is the inventory level at any time t.

3. Mathematical Formulation

Suppose an inventory system consists the maximum inventory level at any time $t = 0$ and due to both demand and deterioration the inventory level decreases in the interval $[0,T]$. The replenishment cycle starts with the initial maximum inventory level Q and ends with zero stock level. The retailer's instantaneous inventory level at any time t in the interval $[0,T]$ is governed by the following differential equation

$$\frac{dI}{dt} + \theta I = -\left[a + bt + kI(t)\right], \quad 0 \leq t \leq T \tag{1}$$

With the boundary condition

$$I(T) = 0$$

The equation (1) can also be written as

$$\frac{dI}{dt} + \alpha I = -\left[a + bt\right], \quad 0 \leq t \leq T \tag{2}$$

where $\alpha = (\theta + k)$
With the boundary condition

$$I(T) = 0$$

For a 2nd order approximation of $e^{-\alpha t}$ and $e^{\alpha t}$, the solution of Equation (2) is

$$I = \left[\frac{(2a\alpha + b\alpha - 2b)}{2\alpha}T - \frac{(2a\alpha + b\alpha - 2b)}{2\alpha}t + \frac{(a\alpha + b\alpha - b)}{2}T^2 + \frac{(a\alpha - b)}{2}t^2 - \frac{(2a\alpha + b\alpha - 2b)}{2}Tt \right.$$
$$\left. + \frac{(2a\alpha^2 + b\alpha^2 - 2b\alpha)}{2}t^3 - \frac{(a\alpha^2 + b\alpha - b\alpha)}{2}T^2t - \frac{(a\alpha^2 + b\alpha^2 - b\alpha)}{2}Tt^2 \right], \tag{3}$$

Using the boundary condition, $I(0) = Q$ the initial order quantity is

$$Q = \left[\frac{(2a\alpha + b\alpha - 2b)}{2\alpha} T + \frac{(a\alpha + b\alpha - b)}{2} T^2 \right], \tag{4}$$

Now we discuss the following two cases
(1) $M \geq T$ and (2) $M \prec T$

3.1. Case I

When $M \geq T$ then in this case the retailer can sell all the items before the end of credit period M because the credit period M is greater than the replenishment cycle length so no interest will be charged by the retailer. Since the purchasing cost is paid at the end of credit period M.

During the 1st cycle the present value of ordering cost is

$$O_C = A, \tag{5}$$

During the 1st cycle the present value of purchasing cost is

$$P_C = CQe^{-rM}$$

$$P_C = \frac{Ce^{-rM}}{2} \left[\frac{(2a\alpha + b\alpha - 2b)}{2} T + (a\alpha + b\alpha - b)T^2 \right], \tag{6}$$

During the 1st cycle the present value of holding cost is

$$H_C = h_C \int_0^T I(t)e^{-rt}\,dt$$

$$H_C = h_C \left[\frac{(2a\alpha + b\alpha - 2b)}{4\alpha} T^2 + \frac{(2a\alpha + 3b\alpha - 2b)}{12} T^3 - \frac{r(2a\alpha + b\alpha - 2b)}{12\alpha} T^3 \right], \tag{7}$$

Therefore during the 1st cycle the present value of retailer's total cost is

$$TC_1(T) = \left[O_C + H_C + P_C \right]$$

$$TC_1(T) = \left[A + \frac{h_C}{12\alpha} \left\{ (6a\alpha + 3b\alpha - 6b)T^2 + (2a\alpha^2 + 3b\alpha^2 - 2b\alpha)T^3 \right. \right.$$
$$\left. - r(2a\alpha + b\alpha - 2b)T^3 \right\} + \frac{Ce^{-rM}}{2\alpha} \left\{ (2a\alpha + b\alpha - 2b)T \right. \tag{8}$$
$$\left. \left. + (a\alpha^2 + b\alpha^2 - b\alpha)T^2 \right\} \right],$$

Since there are m cycles in the planning horizon L then the present value of retailer's total cost over the planning horizon L is

$$TCH_1(T) = \sum_{n=0}^{m-1} TC_1(T)e^{-rnT}$$

$$TCH_1(T) = \frac{(1 - e^{-rL})}{(1 - e^{-rT})} \left[A + \frac{h_C}{12\alpha} \left\{ (6a\alpha + 3b\alpha - 6b)T^2 + (2a\alpha^2 + 3b\alpha^2 - 2b\alpha)T^3 \right. \right.$$
$$\left. - r(2a\alpha + b\alpha - 2b)T^3 \right\} + \frac{Ce^{-rM}}{2\alpha} \left\{ (2a\alpha + b\alpha - 2b)T \right.$$
$$\left. \left. + (a\alpha^2 + b\alpha^2 - b\alpha)T^2 \right\} \right]$$

$$TCH_1(T) = L\left[\frac{A}{T} + Ar + \frac{Ar^2T}{4} + \frac{h_C}{12\alpha}\{(6a\alpha + 3b\alpha - 6b)T + \alpha(2a\alpha + 3b\alpha - 2b)T^2\right.$$

$$-r(2a\alpha + b\alpha - 2b)T^2\} + \frac{Ce^{-rM}}{2\alpha}\{(2a\alpha + b\alpha - 2b) + \alpha(a\alpha + b\alpha - b)T\}$$

$$+\frac{rh_C}{12\alpha}\{(6a\alpha + 3b\alpha - 6b)T^2 + \alpha(2a\alpha + 3b\alpha - 2b)T^3 - r(2a\alpha + b\alpha - 2b)T^3\}$$

$$\tag{9}$$

$$+\frac{rCe^{-rM}}{2\alpha}\{(2a\alpha + b\alpha - 2b)T^2 + \alpha(a\alpha + b\alpha - b)T^2\}$$

$$+\frac{Cr^2e^{-rM}}{8\alpha}\{(2a\alpha + b\alpha - 2b)T^2 + \alpha(a\alpha + b\alpha - b)T^3\}$$

$$\left.+\frac{r^2h_C}{48\alpha}(6a\alpha + 3b\alpha - 6b)T^3\right],$$

The necessary condition for $TCH_1(T)$ to be minimum is that $\dfrac{dTCH_1(T)}{dT} = 0$ and the sufficient condition

is $\dfrac{d^2TCH_1(T)}{dT^2} \succ 0$ at the optimum value of T.

$$\frac{dTCH_1(T)}{dT} = L\left[\frac{Ar^2}{4} - \frac{A}{T^2} + \frac{h_C}{12\alpha}\{(6a\alpha + 3b\alpha - 6b) + 2\alpha(2a\alpha + 3b\alpha - 2b)T\right.$$

$$-2r(2a\alpha + b\alpha - 2b)T\} + \frac{Ce^{-rM}}{2}(a\alpha + b\alpha - b) + \frac{rh_C}{12\alpha}\{2(6a\alpha + 3b\alpha - 6b)T$$

$$+3\alpha(2a\alpha + 3b\alpha - 2b)T^2 - 3r(2a\alpha + b\alpha - 2b)T^2\} + \frac{Cre^{-rM}}{\alpha}\{(2a\alpha + b\alpha - 2b)T \tag{10}$$

$$+\alpha(a\alpha + b\alpha - b)T\} + \frac{Cr^2e^{-rM}}{8\alpha}\{2(2a\alpha + b\alpha - 2b) + 3\alpha(a\alpha + b\alpha - b)T^2\}$$

$$\left.+\frac{r^2h_C}{16\alpha}(6a\alpha + 3b\alpha - 6b)T^2\right],$$

$$\frac{d^2TCH_1(T)}{dT^2} = L\left[\frac{2A}{T^3} + \frac{h_C}{6\alpha}\{\alpha(2a\alpha + 3b\alpha - 2b) - r(2a\alpha + b\alpha - 2b)\}\right.$$

$$+\frac{rh_C}{6\alpha}\{(6a\alpha + 3b\alpha - 6b) + 3\alpha(2a\alpha + 3b\alpha - 2b)T - 3r(2a\alpha + b\alpha - 2b)T\}$$

$$+\frac{Cre^{-rM}}{\alpha}\{(2a\alpha + b\alpha - 2b) + \alpha(a\alpha + b\alpha - b)\} \tag{11}$$

$$\left.+\frac{3Cr^2e^{-rM}}{4}(a\alpha + b\alpha - b)T + \frac{r^2h_C}{8\alpha}(6a\alpha + 3b\alpha - 6b)\right],$$

3.2. Case II

When $M \prec T$ then there are three possibilities

1) Let $Pe^{rM}\int_0^M R(t)e^{-rt}dt \succ CQ$ then at M the revenue earned by the retailer is more than the purchasing cost

so in this case no interest will be charged by the supplier although the credit period M is smaller than the replenishment cycle length T so the present value of retailer's total will be same as that in case I.

$$
\begin{aligned}
TCH_1(T) = L\Bigg[& \frac{A}{T} + Ar + \frac{Ar^2T}{4} + \frac{h_C}{12\alpha}\big\{(6a\alpha + 3b\alpha - 6b)T + \alpha(2a\alpha + 3b\alpha - 2b)T^2 \\
& -r(2a\alpha + b\alpha - 2b)T^2\big\} + \frac{Ce^{-rM}}{2\alpha}\big\{(2a\alpha + b\alpha - 2b) + \alpha(a\alpha + b\alpha - b)T\big\} \\
& + \frac{rh_C}{12\alpha}\big\{(6a\alpha + 3b\alpha - 6b)T^2 + \alpha(2a\alpha + 3b\alpha - 2b)T^3 - r(2a\alpha + b\alpha - 2b)T^3\big\} \\
& + \frac{rCe^{-rM}}{2\alpha}\big\{(2a\alpha + b\alpha - 2b)T^2 + \alpha(a\alpha + b\alpha - b)T^2\big\} \\
& + \frac{Cr^2e^{-rM}}{8\alpha}\big\{(2a\alpha + b\alpha - 2b)T^2 + \alpha(a\alpha + b\alpha - b)T^3\big\} \\
& + \frac{r^2h_C}{48\alpha}(6a\alpha + 3b\alpha - 6b)T^3\Bigg],
\end{aligned}
\tag{12}
$$

2) Let $Pe^{rM}\int_0^M R(t)e^{-rt}dt \prec CQ$ then at M the revenue earned by the retailer is less than the purchasing cost and the retailer has a budget to pay the remaining short purchasing cost so in this case there is still no interest charged by the supplier although the credit period M is smaller than the replenishment cycle length T so the present value of retailer's total will be same as that in case I.

$$
\begin{aligned}
TCH_1(T) = L\Bigg[& \frac{A}{T} + Ar + \frac{Ar^2T}{4} + \frac{h_C}{12\alpha}\big\{(6a\alpha + 3b\alpha - 6b)T + \alpha(2a\alpha + 3b\alpha - 2b)T^2 \\
& -r(2a\alpha + b\alpha - 2b)T^2\big\} + \frac{Ce^{-rM}}{2\alpha}\big\{(2a\alpha + b\alpha - 2b) + \alpha(a\alpha + b\alpha - b)T\big\} \\
& + \frac{rh_C}{12\alpha}\big\{(6a\alpha + 3b\alpha - 6b)T^2 + \alpha(2a\alpha + 3b\alpha - 2b)T^3 - r(2a\alpha + b\alpha - 2b)T^3\big\} \\
& + \frac{rCe^{-rM}}{2\alpha}\big\{(2a\alpha + b\alpha - 2b)T^2 + \alpha(a\alpha + b\alpha - b)T^2\big\} \\
& + \frac{Cr^2e^{-rM}}{8\alpha}\big\{(2a\alpha + b\alpha - 2b)T^2 + \alpha(a\alpha + b\alpha - b)T^3\big\} \\
& + \frac{r^2h_C}{48\alpha}(6a\alpha + 3b\alpha - 6b)T^3\Bigg],
\end{aligned}
\tag{13}
$$

3) Let $Pe^{rM}\int_0^M R(t)e^{-rt}dt \prec CQ$ then at M the revenue earned by the retailer is less than the purchasing cost and the retailer has no budget to pay the remaining short purchasing cost so in this case for unpaid balance the interest will be charged by the supplier from M to T. The interest and the remaining payments are made at the end of replenishment cycle. So in this case the retailer's total cost containing the ordering cost, holding cost, purchasing cost paid at M, the interest and the remaining payments are made at the end of replenishment cycle.

The present values of retailer's ordering and holding are same cost as in case I

During the first cycle the purchasing cost paid at M is equal to the amount of revenue earned by the retailer up to M so

$$
P_C = P\int_0^M R(t)e^{-rt}dt
$$

$$P_C = P\left[aM + \frac{bM^2}{2} - \frac{arM^2}{2} - \frac{brM^3}{3} + k\left\{\frac{(2a\alpha + b\alpha - 2b)}{2\alpha}TM\right.\right.$$

$$-\frac{(2a\alpha + b\alpha - 2b)}{4\alpha}M^2 + \frac{(a\alpha + b\alpha - b)}{2}T^2M + \frac{(a\alpha - b)}{6}M^3$$

$$\left.-\frac{(2a\alpha + b\alpha - 2b)}{4}TM^2\right\} - rk\left\{\frac{(2a\alpha + b\alpha - 2b)}{4\alpha}TM^3\right.$$

$$\left.\left.-\frac{(2a\alpha + b\alpha - 2b)}{6\alpha}M^3 + \frac{(a\alpha + b\alpha - b)}{4}T^2M^2\right\} + \frac{ar^2M^3}{6}\right],$$

(14)

During the first cycle the present values of remaining payments and interest paid at the end of replenishment cycle are

$$RP_I = \left\{CQ - Pe^{rM}\int_0^M R(t)e^{-rt}dt\right\}\left[1 + I_C(T - M)\right]e^{-rT}$$

$$RP_I = \left[\frac{C}{2\alpha}(1 - I_C M)(2a\alpha + b\alpha - 2b)T + \frac{C}{2}(1 - I_C M)(a\alpha + b\alpha - b)T^2\right.$$

$$-P(1 - I_C M)\left\{aM + \frac{arM^2}{2} + \frac{bM^2}{2} + \frac{brM^3}{6} + \frac{a^2r^2M^3}{6} - \frac{k(2a\alpha + b\alpha - 2b)M^2}{4\alpha}\right.$$

$$\left.+\frac{k(a\alpha - b)M^3}{6} - \frac{kr(2a\alpha + b\alpha - 2b)M^3}{12\alpha}\right\} - \frac{Pk(1 - I_C M)(2a\alpha + b\alpha - 2b)}{2\alpha}TM$$

$$-\frac{Pkr(1 - I_C M)}{2\alpha}(2a\alpha + b\alpha - 2b)TM^2 + \frac{Pk(r + 1)(1 - I_C M)(2a\alpha + b\alpha - 2b)}{4}TM^2$$

$$+\frac{Pkr(1 - I_C M)}{4}(2a\alpha + b\alpha - 2b)TM^3 - \frac{Pkr(1 - I_C M)(a\alpha + b\alpha - b)}{4}T^2M^2$$

$$+\frac{C}{2\alpha}(I_C + rMI_C - r)(2a\alpha + b\alpha - 2b)T^2 + \frac{C(I_C + rMI_C - r)(a\alpha + b\alpha - b)}{2}T^3$$

$$-P(I_C + rMI_C - r)\left\{aM + \frac{arM^2}{2} + \frac{bM^2}{2} + \frac{brM^3}{6} + \frac{a^2r^2M^3}{6} - \frac{k(2a\alpha + b\alpha - 2b)}{4\alpha}M^2\right.$$

$$\left.+\frac{k(a\alpha - b)M^3}{6} - \frac{kr(2a\alpha + b\alpha - 2b)M^3}{12\alpha}\right\}T - \frac{Pk(I_C + rMI_C - r)}{2\alpha}(2a\alpha + b\alpha - 2b)T^2M$$

$$-\frac{Pkr(I_C + rMI_C - r)(2a\alpha + b\alpha - 2b)}{2\alpha}T^2M^2$$

$$+\frac{Pk(r + 1)(I_C + rMI_C - r)}{4}(2a\alpha + b\alpha - 2b)T^2M^2$$

$$+\frac{C(r^2 - 2rI_C - r^2MI_C)(2a\alpha + b\alpha - 2b)}{4\alpha}T^3$$

(15)

$$-\frac{P}{2}(r^2 - 2rI_C - r^2MI_C)\left\{aM + \frac{bM^2}{2} - \frac{k(2a\alpha + b\alpha - 2b)}{4\alpha}M^2\right\}T^2$$

$$\left.-\frac{Pk(r^2 - 2rI_C)(2a\alpha + b\alpha - 2b)}{4\alpha}T^3M\right],$$

During the 1st cycle the present value of retailer's total cost is

$$TC_2(T) = \left[A + \frac{h_C}{12\alpha} \left\{ (6a\alpha + 3b\alpha - 6b)T^2 + (2a\alpha^2 + 3b\alpha^2 - 2b\alpha)T^3 \right. \right.$$

$$-r(2a\alpha + b\alpha - 2b)T^3 \Big\} - P \left\{ arM^2 + \frac{brM^3}{2} \right.$$

$$+ \frac{a(a-1)r^2M^3}{6} - \frac{rk(2a\alpha + b\alpha - 2b)M^3}{4\alpha}$$

$$- \frac{k}{2\alpha}(2a\alpha + b\alpha - 2b)M^2I_CT - \frac{k(a\alpha + b\alpha - b)}{2}T^2M$$

$$- \frac{k(r - MrI_C - I_CM)}{4}(2a\alpha + b\alpha - 2b)TM^2$$

$$+ \frac{kr(3 - 2MI_C)(2a\alpha + b\alpha - 2b)}{4\alpha}TM^2$$

$$+ \frac{kr(2 - MI_C)(a\alpha + b\alpha - b)}{4}T^2M^2 \Big\}$$

$$+ \frac{C(1 - MI_C)(2a\alpha + b\alpha - 2b)}{2\alpha}T$$

$$+ \frac{C(1 - MI_C)(a\alpha + b\alpha - b)}{2}T^2$$

$$+ P \left\{ MI_C - (I_C + rMI_C - r) - \frac{(r^2 - 2rI_C - Mr^2I_C)}{2}T^2 \right\}$$

$$\times \left(aM + \frac{bM^2}{2} + \frac{arM^2}{2} + \frac{brM^3}{6} + \frac{a^2r^2M^3}{6} \right.$$

$$- \frac{k(2a\alpha + b\alpha - 2b)M^2}{4\alpha} + \frac{k(a\alpha - b)M^3}{6}$$

$$\left. - \frac{k(2a\alpha + b\alpha - 2b)M^3}{12\alpha} \right) + \frac{Pkr(1 - MI_C)(2a\alpha + b\alpha - 2b)}{4}TM^3$$

$$+ \frac{C(I_C + rMI_C - r)(2a\alpha + b\alpha - 2b)}{2\alpha}T^2$$

$$+ \left\{ \frac{C(I_C + rMI_C - r)(a\alpha + b\alpha - b)}{2} + \frac{C(r^2 - 2rI_C - Mr^2I_C)(2a\alpha + b\alpha - 2b)}{4\alpha} \right\}T^3$$

$$- \frac{Pk(I_C + rMI_C - r)(2a\alpha + b\alpha - 2b)}{2\alpha}T^2M - \frac{Pk(I_C - r)(2a\alpha + b\alpha - 2b)}{2\alpha}T^2M^2$$

$$- \frac{Pk(r^2 - 2rI_C)(2a\alpha + b\alpha - 2b)}{4\alpha}T^3M \right], \tag{16}$$

Since there are m cycles in the planning horizon L then the present value of retailer's total cost over the planning horizon L is

$$TCH_2(T) = \sum_{n=0}^{m-1} TC_2(T)e^{-rnT}$$

$$TCH_2(T) = \frac{(1 - e^{-rL})}{(1 - e^{-rT})}TCH_2(T)$$

$$TCH_2(T) = L\left[\frac{Ar}{2} + P(r - I_C)\left\{aM + \frac{bM^2}{2} + \frac{arM^2}{2} + \frac{brM^3}{6} + \frac{a^2r^2M^3}{6} - \frac{k(2a\alpha + b\alpha - 2b)M^2}{4\alpha}\right.\right.$$

$$+ \frac{k(a\alpha - b)M^3}{6} - \frac{kr(2a\alpha + b\alpha - 2b)M^3}{12\alpha}\bigg\} + \frac{PrMI_C}{2}\left\{aM + \frac{bM^2}{2} + \frac{arM^2}{2}\right.$$

$$\left.- \frac{kM^2(2a\alpha + b\alpha - 2b)}{4\alpha}\right\} + \left\{\frac{PkM^2I_C}{2\alpha} + \frac{PkM^2(r - MrI_C - MI_C)}{4} - \frac{PkrM^3(3 - 2MI_C)}{4\alpha}\right.$$

$$+ \left(\frac{PkrM^3}{4} + \frac{C}{2\alpha} - \frac{CMI_C}{2\alpha}\right)\bigg\}(2a\alpha + b\alpha - 2b) + \left[\frac{Ar^2}{4} - \frac{Par^3M^2}{4} - \frac{Pbr^3M^3}{8} - \frac{Pa(a-1)r^4M^3}{6}\right.$$

$$+ \left\{\frac{h_C}{4\alpha} + \frac{C(I_C + MrI_C - r)}{2\alpha} - \frac{PkM(I_C + MrI_C - r)}{2\alpha} - \frac{PkM^2(I_C + MrI_C - r)}{2\alpha}\right.$$

$$+ \frac{PkrM^3(I_C - r)}{4} + \frac{PkrM^2I_C}{4\alpha} + \frac{PkrM^2(r - MrI_C - MI_C)}{8} - \frac{Pkr^2M^2(3 - 2MI_C)}{8\alpha}$$

$$+ \frac{Cr(1 - MI_C)}{4\alpha} + \frac{Pkr^2M^3}{8} + \frac{Pkr^3M^3}{16\alpha}\bigg\}(2a\alpha + b\alpha - 2b) + \left\{\frac{PkM}{2} + \frac{C(1 - MI_C)}{2}\right.$$

$$- \frac{PkrM^2(1 - MI_C)}{4} + \frac{Pkr^2M^3}{4}\bigg\}(a\alpha + b\alpha - b) + \frac{P}{4}\left\{Mr^2I_C - 2(2r^2 - rI_C)\right\}\left(aM + \frac{bM^2}{2}\right.$$

$$+ \frac{arM^2}{2} + \frac{brM^3}{6} + \frac{a^2r^2M^3}{6} - \frac{kM^2(2a\alpha + b\alpha - 2b)}{4\alpha} + \frac{k(a\alpha - b)M^3}{6} - \frac{krM^3(2a\alpha + b\alpha - 2b)}{12\alpha}\bigg)$$

$$+ 2Mr^2I_C\left(aM + \frac{bM^2}{2} + \frac{arM^2}{2} - \frac{kM^2(2a\alpha + b\alpha - 2b)}{4\alpha}\right)\bigg\}\bigg]T + \left[\frac{h_C(2a\alpha + 3b\alpha - 2b)}{12}\right.$$

$$+ \left\{\frac{C(r^2 - 2rI_C - Mr^2I_C)}{4\alpha} - \frac{PkrM^2(r^2 - 2rI_C)}{4\alpha} + \frac{PkrM^2(r^2 - 2rI_C)}{4} + \frac{PkM^2(r^2 - 2rI_C)}{4}\right.$$

$$- \frac{PkM(r^2 - 2rI_C - Mr^2I_C)}{4} + \left(\frac{Cr}{4} - \frac{PkrM}{4\alpha}\right)(I_C + MrI_C - r) + \frac{Cr^2(1 - MI_C)}{8\alpha} + \frac{rh_C}{24\alpha}$$

$$+ \frac{Pkr^2M^2I_C}{8\alpha} + \frac{Pkr^3M^2}{16}\bigg\}(2a\alpha + b\alpha - 2b) + \left\{\frac{C(I_C + rMI_C - r)}{2} - \frac{PkrM^2(I_C - r)}{4}\right.$$

$$+ \frac{Cr(1 - MI_C)}{4} - \frac{Pkr^2M^2}{4} + \frac{PkrM}{4}\bigg\}(a\alpha + b\alpha - b) + \frac{Pr^2I_C}{4}\left(aM + \frac{bM^2}{2} + \frac{arM^2}{2}\right.$$

$$\left.- \frac{kM^2(2a\alpha + b\alpha - 2b)}{4\alpha}\right)\bigg]T^2 + \left[\left\{\frac{rh_C}{24} + \frac{r^2h_C}{48\alpha} + \frac{Cr(r^2 - 2rI_C - Mr^2I_C)}{8\alpha} - \frac{PkrM(r^2 - 2rI_C)}{8\alpha}\right.\right.$$

$$+ \frac{C(I_C + rMI_C - r)}{2\alpha} - \frac{Pkr^2M(I_C - r)}{8\alpha}\bigg\}(2a\alpha + b\alpha - 2b) + \left\{\frac{Cr(I_C + rMI_C - r)}{4}\right.$$

$$+ \left(\frac{Cr^2}{8} + \frac{Pkr^4}{16}\right)(1 - MI_C) + \frac{Pkr^2M}{8}\bigg\}(a\alpha + b\alpha - b) - \frac{Par^2M(r^2 - 2rI_C)}{8}\bigg]T^3$$

$$+ \frac{1}{T}\left[A - \frac{P(r+2)}{2}\left\{arM^2 + \frac{brM^3}{2} + \frac{a(a-1)^2M^3}{6} - \frac{krM^3(2a\alpha + b\alpha - 2b)}{4\alpha}\right\}\right.$$

$$\left.\left.\left.+ PMI_C\left(aM + \frac{bM^2}{2} + \frac{arM^2}{2} - \frac{kM^2(2a\alpha + b\alpha - 2b)}{4\alpha}\right)\right]\right]\right],$$

$$(17)$$

The necessary condition for $TCH_2(T)$ to be minimum is that $\dfrac{dTCH_2(T)}{dT} = 0$ and the sufficient condition is $\dfrac{d^2TCH_2(T)}{dT^2} \succ 0$ at the optimum value of T.

$$
\begin{aligned}
\frac{dTCH_2(T)}{dT} = L\Bigg[& \frac{Ar^2}{4} - \frac{Par^3M^2}{4} - \frac{Pbr^3M^3}{8} - \frac{Pa(a-1)r^4M^3}{6} + \Bigg\{ \frac{h_C}{4\alpha} + \frac{C}{2\alpha}\left(I_C + MrI_C - r\right) \\
& - \frac{PkM\left(I_C + MrI_C - r\right)}{2\alpha} - \frac{PkM^2\left(I_C + MrI_C - r\right)}{2\alpha} + \frac{PkrM^3\left(I_C - r\right)}{4} + \frac{PkrM^2I_C}{4\alpha} \\
& + \frac{PkrM^2\left(r - MrI_C - MI_C\right)}{8} - \frac{Pkr^2M^2\left(3 - 2MI_C\right)}{8\alpha} + \frac{Cr\left(1 - MI_C\right)}{4\alpha} + \frac{Pkr^2M^3}{8} \\
& + \frac{Pkr^3M^3}{16\alpha} \Bigg\}\left(2a\alpha + b\alpha - 2b\right) + \Bigg\{ \frac{PkM}{2} + \frac{C\left(1 - MI_C\right)}{2} - \frac{PkrM^2\left(1 - MI_C\right)}{4} \\
& + \frac{Pkr^2M^3}{4} \Bigg\}\left(a\alpha + b\alpha - b\right) + \frac{P}{4}\Bigg\{ Mr^2I_C - \left(2r^2 - rI_C\right)\left(aM + \frac{bM^2}{2} + \frac{arM^2}{2} + \frac{brM^3}{6} \right. \\
& \left. + \frac{a^2r^2M^3}{6} - \frac{kM^2\left(2a\alpha + b\alpha - 2b\right)}{4\alpha} + \frac{k\left(a\alpha - b\right)M^3}{6} - \frac{krM^3\left(2a\alpha + b\alpha - 2b\right)}{12\alpha} \right) \\
& + 2Mr^2I_C\left(aM + \frac{bM^2}{2} + \frac{arM^2}{2} - \frac{kM^2\left(2a\alpha + b\alpha - 2b\right)}{4\alpha} \right) \Bigg\}\Bigg] + 2\Bigg[\frac{h_C\left(2a\alpha + 3b\alpha - 2b\right)}{12} \\
& + \Bigg\{ \frac{C\left(r^2 - 2rI_C - Mr^2I_C\right)}{4\alpha} - \frac{PkrM^2\left(r^2 - 2rI_C\right)}{4\alpha} + \frac{PkrM^2\left(r^2 - 2rI_C\right)}{4} + \frac{PkM^2\left(r^2 - 2rI_C\right)}{4} \\
& - \frac{PkM\left(r^2 - 2rI_C - Mr^2I_C\right)}{4} + \left(\frac{Cr}{4} - \frac{PkRM}{4\alpha} \right)\left(I_C + MrI_C - r\right) + \frac{Cr^2\left(1 - MI_C\right)}{8\alpha} + \frac{rh_C}{24\alpha} \\
& + \frac{Pkr^2M^2I_C}{8\alpha} + \frac{Pkr^3M^2}{16} \Bigg\}\left(2a\alpha + b\alpha - 2b\right) + \Bigg\{ \frac{C\left(I_C + rMI_C - r\right)}{2} - \frac{PkrM^2\left(I_C - r\right)}{4} \\
& + \frac{Cr\left(1 - MI_C\right)}{4} - \frac{Pkr^2M^2}{4} + \frac{PkrM}{4} \Bigg\}\left(a\alpha + b\alpha - b\right) + \frac{Pr^2I_C}{4}\left(aM + \frac{bM^2}{2} + \frac{arM^2}{2} \right. \\
& \left. - \frac{kM^2\left(2a\alpha + b\alpha - 2b\right)}{4\alpha} \right)\Bigg]T + \Bigg[\Bigg\{ \frac{rh_C}{24} + \frac{r^2h_C}{48\alpha} + \frac{Cr\left(r^2 - 2rI_C - Mr^2I_C\right)}{8\alpha} - \frac{PkrM\left(r^2 - 2rI_C\right)}{8\alpha} \\
& + \frac{C\left(I_C + rMI_C - r\right)}{2\alpha} - \frac{Pkr^2M\left(I_C - r\right)}{8\alpha} \Bigg\}\left(2a\alpha + b\alpha - 2b\right) + \Bigg\{ \frac{Cr\left(I_C + rMI_C - r\right)}{4} \\
& + \left(\frac{Cr^2}{8} + \frac{Pkr^4}{16} \right)\left(1 - MI_C\right) + \frac{Pkr^2M}{8} \Bigg\}\left(a\alpha + b\alpha - b\right) - \frac{Par^2M\left(r^2 - 2rI_C\right)}{8} \Bigg]T^2 \\
& - \frac{1}{T^2}\Bigg[A - \frac{P\left(r+2\right)}{2}\Bigg\{ arM^2 + \frac{brM^3}{2} + \frac{a\left(a-1\right)^2 M^3}{6} - \frac{krM^3\left(2a\alpha + b\alpha - 2b\right)}{4\alpha} \Bigg\} \\
& + PMI_C\left(aM + \frac{bM^2}{2} + \frac{arM^2}{2} - \frac{kM^2\left(2a\alpha + b\alpha - 2b\right)}{4\alpha} \right) \Bigg]\Bigg],
\end{aligned}
\tag{18}
$$

$$\frac{d^2 TCH_2(T)}{dT^2} = 2L\left[\frac{h_C(2a\alpha + 3b\alpha - 2b)}{12} + \left\{\frac{C(r^2 - 2rI_C - Mr^2 I_C)}{4\alpha} - \frac{PkrM^2(r^2 - 2rI_C)}{4\alpha}\right.\right.$$

$$+\frac{PkrM^2(r^2 - 2rI_C)}{4} + \frac{PkM^2(r^2 - 2rI_C)}{4} - \frac{PkM(r^2 - 2rI_C - Mr^2 I_C)}{4}$$

$$+\left(\frac{Cr}{4} - \frac{PkRM}{4\alpha}\right)(I_C + rMI_C - r) + \frac{Cr^2(1 - MI_C)}{8\alpha} + \frac{rh_C}{24\alpha} + \frac{Pkr^2 M^2 I_C}{8\alpha}$$

$$\left.+\frac{Pkr^3 M^2}{16}\right\}(2a\alpha + b\alpha - 2b) + \left\{\frac{C(I_C + rMI_C - r)}{2} - \frac{PkrM^2(I_C - r)}{4}\right.$$

$$+\frac{Cr(1 - MI_C)}{4} - \frac{Pkr^2 M^2}{4} + \frac{PkrM}{4}\right\}(a\alpha + b\alpha - b) + \frac{Pr^2 I_C}{4}\left(aM + \frac{bM^2}{2}\right)$$

$$\left.+\frac{arM^2}{2} - \frac{kM^2(2a\alpha + b\alpha - 2b)}{4\alpha}\right)\right] + 6\left[\left\{\frac{rh_C}{24} + \frac{r^2 h_C}{48\alpha} + \frac{Cr(r^2 - 2rI_C - Mr^2 I_C)}{8\alpha}\right.\right.$$

$$\left.-\frac{PkrM(r^2 - 2rI_C)}{8\alpha} + \frac{C(I_C + rMI_C - r)}{2\alpha} - \frac{Pkr^2 M(I_C - r)}{8\alpha}\right\}(2a\alpha + b\alpha - 2b)$$

$$+\left\{\frac{Cr(I_C + rMI_C - r)}{4} + \left(\frac{Cr^2}{8} + \frac{Pkr^4}{16}\right)(1 - MI_C) + \frac{Pkr^2 M}{8}\right\}(a\alpha + b\alpha - b)$$

$$\left.-\frac{Par^2 M(r^2 - 2rI_C)}{8}\right]T + \frac{2}{T^3}\left[A - \frac{P(r+2)}{2}\left\{arM^2 + \frac{brM^3}{2} + \frac{a(a-1)^2 M^3}{6}\right.\right.$$

$$\left.\left.\left.-\frac{krM^3(2a\alpha + b\alpha - 2b)}{4\alpha}\right\} + PMI_C\left(aM + \frac{bM^2}{2} + \frac{arM^2}{2} - \frac{kM^2(2a\alpha + b\alpha - 2b)}{4\alpha}\right)\right]\right], \qquad (19)$$

4. Numerical Parameters

Let us consider the following parameters in the appropriate units

$a = 100$, $b = 3$, $k = 10$, $C = 6$, $\alpha = 12$, $A = 4$, $h_C = 5$, $s_C = 8$, $P = 20$, $r = 1.20$, $M = 2$, $I_C = 0.8$, $L = 50$

4.1. Numerical Example I

When $M \geq T$ then solving the equation $\dfrac{dTCH_1}{dT} = 0$, we find the optimum value of T satisfying the condition

$\dfrac{d^2 TCH_1}{dT^2} \succ 0$.

Since $\dfrac{d^2 TCH_1}{dT^2} = 2.11046 \times 10^7$

As we increase the parameter r then the value of total cost decreases.
As we increase the parameter M then the value of total cost decreases.

4.3. Numerical Parameters

Let us consider the following parameters in the appropriate units

$a = 100$, $b = 3$, $k = 10$, $C = 6$, $\alpha = 12$, $A = 4$, $h_C = 5$, $s_C = 8$, $P = 20$, $r = 0.05$, $M = 2$, $I_C = 0.8$, $L = 50$

4.4. Numerical Example II

When $M \prec T$ then solving the equation $\dfrac{\mathrm{d}TCH_2}{\mathrm{d}T} = 0$, we find the optimum value of T satisfying the condi-

tion $\dfrac{\mathrm{d}^2 TCH_2}{\mathrm{d}T^2} \succ 0$. Since $\dfrac{\mathrm{d}^2 TCH_2}{\mathrm{d}T^2} = -146586 \times 10^9$ so the total cost is maximum.

As we increase the parameter r then the value of total cost increases.

As we increase the parameter M then the value of total cost decreases.

5. Conclusion

In this paper, we proposed an inventory model for perishable items with time varying stock dependent demand under inflation and time discounting. In the numerical analysis we study the effect of the change of the parameters r and M on the optimal solution. From **Table 1** and **Table 2** we observe that as we increase the parameters r and M then the replenishment cycle length increases and the corresponding total cost decreases since the total cost decreases and the revenue increases on his sell to pay the purchasing cost and in the case when $M \geq T$ no interest will be charged by the supplier from the retailer. When the inflation rate increases then the retailer wants to short the length of replenishment cycle. From **Table 3** we see that as we increase the parameter r then the value of replenishment cycle length and total cost decreases. From **Table 4** we see that as we increase the parameter M then the value of replenishment cycle length increases and the value of total cost decreases. Thus we see that when the credit period is short then the retailer wants to order less and decrease the chargeable interest. When the credit period is large enough then the retailer wants to order more and he earns enough revenue on his sell to pay the purchasing cost therefore the credit period attracts the retailer to buy more or less.

References

[1] Buzacott, J.A. (1975) Economic Order Quantity Model under Inflation. *Operational Research Quarterly*, **26**, 553-558. http://dx.doi.org/10.2307/3008214

[2] Silver, E.A. and Peterson, R. (1985) Decision Systems for Inventory Management and Production Planning. 2nd Edition, John Wiley & Sons, New York.

[3] Baker, R.C. and Urban, T.L. (1988) A Deterministic Inventory System with an Inventory Level Dependent Demand Rate. *Journal of Operational Research Society*, **39**, 823-831. http://dx.doi.org/10.1057/jors.1988.142

[4] Mandal, B.N. and Phaujdar, S. (1989) An Inventory Model for Deteriorating Items and Stock Dependent Consumption Rate. *Journal of Operational Research Society*, **40**, 483-488. http://dx.doi.org/10.1057/jors.1989.75

[5] Vrat, P. and Padmanabhan, G. (1990) An Inventory Model with Stock Dependent Consumption Rate under Inflation. *Engineering Costs and Production Economics*, **19**, 379-383. http://dx.doi.org/10.1016/0167-188X(90)90068-S

[6] Padmanabhan, G. and Vrat, P. (1995) EOQ Models for Perishable Items with Stock Dependent Selling Rate. *European Journal of Operational Research*, **86**, 281-292. http://dx.doi.org/10.1016/0377-2217(94)00103-J

[7] Bose, S., Goswami, A. and Choudhuri, K.S. (1995) An EOQ Model for Deteriorating Items with Linear Time Dependent Demand Rate and Shortages under Inflation and Time Discounting. *Journal of Operational Research Society*, **46**, 771-782. http://dx.doi.org/10.1057/jors.1995.107

[8] Mandal, M. and Maiti, M. (1999) An Inventory Model of Damageable Products Having Some Units in Hand with Stock Dependent Demand, Variable Replenishment Rate. *Applied Mathematical Modeling*, **23**, 799-807. http://dx.doi.org/10.1016/S0307-904X(99)00018-9

[9] Chung, K.-J. and Lin, C.-N. (2001) An Optimal Replenishment Policy for an Inventory Model for Deteriorating Items with Time Discounting. *Computers and Operations Research*, **28**, 67-83. http://dx.doi.org/10.1016/S0305-0548(99)00087-8

[10] Chang, C.T. (2004) An EOQ Model for Deteriorating Items under Inflation When Supplier Credits Linked to Order Quantity. *International Journal of Production Economics*, **88**, 307-316.

[11] Dye, C.Y. and Ouyang, L.Y. (2005) An EOQ Model for Perishable Items with Stock Dependent Selling Rate by Allowing Shortages,. *European Journal of Operational Research*, **163**, 776-783. http://dx.doi.org/10.1016/j.ejor.2003.09.027

[12] Hou, K.L. (2006) An Inventory Model for Deteriorating Items with Stock Dependent Consumption Rate and Shortages under Inflation and Time Discounting. *European Journal of Operational Research*, **168**, 463-474.

http://dx.doi.org/10.1016/j.ejor.2004.05.011

[13] Jaggi, C.K., Aggarwal, K.K. and Goel, S.K. (2006) An Optimal Ordering Policy for an Inventory Model of Deteriorating Items under Inflation Induced Demand. *International Journal of Production Economics*, **103**, 707-714. http://dx.doi.org/10.1016/j.ijpe.2006.01.004

[14] Sana, S.S. and Chaudhuri, K.S. (2008) A Deterministic EOQ Model for Deteriorating Items with Delay in Payments Offered by the Supplier. *European Journal of Operational Research*, **184**, 509-533. http://dx.doi.org/10.1016/j.ejor.2006.11.023

[15] Valliathal, M. and Uthayakumar, R. (2009) An EOQ Model for Perishable Items with Stock and Time Dependent Selling Rate by Allowing Shortages. *APRN Journal of Engineering and Applied Sciences*, **4**, 8-13.

[16] Roy, A., Maiti, M.K., Kar, S. and Maiti, M. (2009) An Inventory Model for Deteriorating Items with Stock Dependent Demand under Fuzzy Inflation and Time Discounting over a Random Planning Horizon. *Applied Mathematical Modeling*, **33**, 744-759. http://dx.doi.org/10.1016/j.apm.2007.12.015

[17] Sana, S.S. (2010) An Inventory Model for Time Varying Deteriorating Items with Optimal Selling Price and Lot Size by Allowing Shortages. *Applied Mathematics and Computation*, **217**, 185-194. http://dx.doi.org/10.1016/j.amc.2010.05.040

[18] Chang, C.T., Teng, J.T. and Goyal, S.K. (2010) Optimal Replenishment Policies for Non Instantaneous Deteriorating Items with Stock Dependent Demand. *International Journal of Production Economics*, **123**, 62-68. http://dx.doi.org/10.1016/j.ijpe.2009.06.042

[19] Sarkar, B., Sana, S.S. and Chaudhuri, K. (2011) An EMQ Model for an Imperfect Production Process for Time Varying Demand under Inflation and Time Value of Money. *Expert Systems with Applications*, **38**, 13543-13548.

[20] Yan, X. (2012) An EOQ Model for Perishable Items with Freshness Dependent Demand and Partial Backlogging. *International Journal of Control and Automation*, **5**, 19-38.

[21] Nagare, M. and Dutta, P. (2012) A Continuous Review Inventory Model for Perishable Products with Inventory Dependent Demand. *Proceeding of the International Multi Conference of Engineers and Computer Scientists Hong-Kong*, 14-16 March 2012, 1513-1517.

[22] Sana, S.S. (2013) A Sales Team's Initiatives and Stock Sensitive Demand under Production Control Policy. *Economic Modeling*, **31**, 783-788. http://dx.doi.org/10.1016/j.econmod.2013.01.026

[23] Yang, S., Lee, C. and Zhang, A.M. (2013) An Inventory Model for Perishable Products with Stock Dependent Demand and Trade Credit under Inflation. *Mathematical Problems in Engineering*, **2013**, 1-10. http://dx.doi.org/10.1155/2013/702939

6

Basic Limit Theorems for Light Traffic Queues & Their Applications

Onkabetse A. Daman[1], Sulaiman Sani[1,2]*

[1]Department of Mathematics, University of Botswana, Gaborone, Botswana
[2]Department of Mathematics and Computer Science, Umaru Musa Yar'Adua University, Katsina, Nigeria
Email: damanoa@mopipi.ub.bw, *man15j@yahoo.com

Abstract

In this paper, we study some basic limit theorems characterizing the stationary behavior of light traffic queuing systems. Beginning with limit theorems for the simple M/M/1 queuing system, we demonstrate the methodology for applying these theorems for the benefit of service systems. The limit theorems studied here are dominant in the literature. Our contribution is primarily on the analysis leading to the application of these theorems in various problem situations for better operations. Relevant Examples are included to aid the application of the results studied in this work.

Keywords

Queue, Light Traffic Queues, M/M/C Model, M/G/C Model and Occupation Rates

1. Introduction

The word **queue** derives its meaning from the Latin word "cauda" which means tail. Literally, to queue is to wait, of course for a reason which is to receive service. On the other hand, queuing theory is the mathematical study of formation and behavior of queues involving problems connected with traffic congestions and storage systems. This definition extends the relevance of modeling queuing systems to a wide variety of contentious situations for instance, how customers of a service shop checkout lines (arrival process), how lines can be minimized (queuing analysis), how many servers a business person should employ (server capacity), how long a customer waits for service (waiting time analysis) and so on. Generally, the objective of queuing theory is relieving problems in business settings primarily, in operational management and operations research.

In this note, we study business queuing systems with arrival rates less than service rates. This type of queuing system is called **light traffic queuing** system and is found in businesses such as manufacturing, transportation,

---

*Corresponding author.

shops and telecommunications today. Because the approach here is practical, the note is restricted to those business queues where arrivals occur as Poisson inputs and services are exponential.

Our motivations stemmed from the need to bring Mathematics closer to business in both methodology and application. A large number of business persons believe that Mathematical proofs covering any subject of application are too abstract for real life business transactions and operations. This assertion is evident in one of the many discussions held between ourselves and some petty business persons on how Mathematics will help improve their day to day business transactions. For instance, in a unique discussion session, one business person argues that Mathematics is extremely hard and devoid of practical applications. Similarly, another one argues that translating the difficult results of Mathematics is not an easy task even to students and academics not to talk of a business person who is engaged with day to day selling of commodities. While this believe is strong among business persons, and is valid, it may not necessarily reflect the truth of the situation. It is well known that Mathematics contributes greatly to the advancement of systems and societies through understanding and application of results to various fields of endeavors such as business. However, the achievement of this goal is possible if our results are brought down to a level that is understood by many. In this note, we are motivated to simplify the level of difficulty in an aspect of Mathematics of varied applications in the business field in order that understanding the goodness of Mathematics as it relates to business development will be appreciated. This is for the purpose of achieving business advancement amongst business persons without Mathematical appreciation for better operations and improvements.

The note starts with a brief historical background of queuing theory and some preliminaries required for easy understanding of the Mathematics used here. In Section three, general methodologies leading to some interesting results are discussed. These methods are not new. Similar ones can be found in the literature. Section four provides applications of the general concepts and in Section five; we realize them through relevant Examples. Here, instances covering several aspects of business transactions are considered. The note is concluded in Section six with summaries.

2. Historical Background

From evolutionary perspective, Medhi [1] dated the origin of queuing modeling as far back as 1909 when the Danish Mathematician Agner Krarup Erlang published his fundamental paper on congestion in telephone traffic. Erlang, in addition to formulating analytic practical problems and solutions, laid a solid foundation to queuing system modeling in terms of enacting basic assumptions and techniques of analysis. Interestingly to date, we are using them even in the wider areas of modern communications and computer systems. For instance, using Erlang basic assumptions and techniques, Ericsson telecom developed a programming language called Erlang used in programming concurrent processes and verifications such as the conditional term rewriting systems (CTRS). His works led to the development of queuing models vital for analyzing lost and delay in operation systems (similar to that of the M/M/C discussed here.). The first model is called the Erlang-B and is used to compute probability that, an incoming arrival is rejected. Similarly, the Erlang-C model gives the probability that an arrival has to wait before service.

Though, Erlang pioneered Mathematical modeling in queuing systems especially its applications to operations research, the pioneer of modeling from the perspective of stochastic process was Kendal. In 1951, Kendal developed and introduced certain notations which to date are adopted to denote queuing systems. The Kendal's A/B/C notation specifies three basic characteristics in a given queuing system namely; the arrival process (A), the service distribution (B) and the number of servers in a system (C). Similarly, Kendal's integral model relating the Laplace-Stieljes transformations of the busy period and that of the arrival process is a remarkable achievement and breakthrough in the field of queuing modeling. To date, lots of priority models covering peak and rush hour (steady state models) service distributions are computed using the model and several numerical models to analyze it have been developed. The Kendal's era in queuing studies marked the beginning of mathematical modeling of queuing processes as stochastic processes.

After these breakthroughs, modeling trends shifted from conditioning and design to rigorous applications in form of averages and other statistics of operational significance. A good Example is the work of D.C. Little. In 1961, Little came up with a model relating the averages of three quantities in every queuing system in what is known in the queuing parlance as little's formula. The formula relates the average number of customers in the system or in the queue to the average sojourn or waiting times. To date, the model is applied in analyzing expectations in light traffic queues (similar to the ones discussed here) in manufacturing and other service systems as well as in decision making to quantify expectations of parameters for better service delivery. The Erlang-Kendal-Little's models (EKL models) are the initiating models in queuing studies.

After this era, a lot of sub areas of interest continue to emerge especially in the 40's. For instance, Franken *et al.* [2] indicated that in the early 50's, Mathematicians were faced with the onerous challenge of developing appropriate Mathematical tools to describe sequence of arrivals in a given system. The first stage of this development was taken by Conny Palm in 1943 and was made Mathematically precise and well expanded by Aleksandr Khinchine. In 1955 precisely, Khinchine studied point processes on the positive real line which he addressed as stream of homogenous events. This development opened further examination of similar areas among others including that of insensitivity of queuing systems. Here, existence and continuity statements (stability conditions for a model) and relationships between time and customer stationery quantities with special inputs were emphasized. This led to the emergence of a new class of random processes connected with point processes which seemed most suitable for describing queuing systems. The new class is termed random processes with embedded Marked Point Process (PMP)[1]. The development of this class of process was attributed to Khinchine and Kendal in 1976 and was the base for modeling in every sense of the word, as far as modeling light traffic queuing systems is concerned.

The last four decades to date symbolizes an era of model development[2] as Medhi [1] implies. Queuing theorists today seemed more interested in model developments, applications and extensions. This era is of modeling and lots of queuing systems were modeled and developed. What we witnessed of late is a shift in paradigm that centers on creating models to capture every bit of system development, advancement and conjecture. Models may be classified under two categories; deterministic models (stable models such as the case here) and stochastic models. The methodology today is to pose a problem and model it in transient or limiting sense (Here, our analysis is purely a limiting case analysis). However, it requires the mathematical analysis of existence of solution so that the constructed model is realistic. The central limit theorem and the maximum principle form the basis for proofing existence of queuing models today.

Limit theorems over the years such as the ones used in this note, have been developed for various queuing models to show limiting behavior. For instance, on light traffic models, the series of deterministic results obtained over the years on these models are enormous. Federgruen & Tijms [3] obtained the stationery distribution of the queue length for the M/G/1. Hoksad [4] and Hoksad [5] worked on a more general queue called the M/G/m in terms of its limiting state solution and its specific case. Smith [6] specified the system performance of a finite capacity queue called the M/G/C/K. Also, Tijms *et al.* [7] approximated the steady state probabilities for the M/G/C queue. For the classical M/G/1 queue with two priority classes under preemptive and non-preemptive-resume disciplines, Abate & Whitt [8] derived these limit theorems. They proved that the low-priority limiting waiting-time is a geometric random sum of independent and identically distributed random variables like the M/G/1 first come first served waiting-time distribution. Similarly, the asymptotic behavior of tail probabilities is such that there is routinely a region where these probabilities have non-exponential asymptotics even if the service time distributions are exponential. In addition, the asymptotic formed tends to be determined by the non-exponential asymptotics for the high-priority busy-period distribution etc.

Finally, light traffic modeling is near complete today and modeling is basically given application outlook for better system performance and evaluation.

3. Preliminaries

3.1. The M/M/C Model

By the M/M/C queuing model[3], we refer to all service stations where customers arrive according to a Poisson process independently with a mean arrival rate of λ, then receive service in a time that is exponentially distributed on any of the C servers available in the system.

The number of servers C specifies the type of the M/M/C queue. For Instance, if C = 1 or infinite then, the corresponding queuing model is called the M/M/1 or the M/M/∞ respectively. Researches in this area have shown that in steady state, the M/M/C model is a continuous-time birth-death process.

---

[1]A special case of the PMP is a piece wise Markov process.

[2]Quantum of light traffic and heavy traffic models has been formulated and is still on-going.

[3]First M means Markovian; second M means Exponential and C means number of servers.

3.2. Occupation Rate

The occupation rate of a server is the fraction of time the server is busy.

It is the ratio of the arrival rate to the service rate of customers in the system. Denoted normally by ρ, the occupation rate provides information on the stability and class of a queuing system. For instance, if $\rho < 1$ then, a queuing system is a stable light traffic system otherwise, it is a heavy traffic queuing system.

3.3. Queuing Schedule

This is the ordering principle of a queuing system. It is the rule describing the manner in which customers access the server for service purposes.

A queuing schedule may be First-Come-First-Served (FCFS), a Last-Come-First-Served (LCFS), processor sharing (ps) or priority discipline (pd) etc.

3.4. Poisson Distribution

A random variable W with parameter λ denoting the mean number of arrivals in a fixed interval of time say $[0,T]$ is said to be Poisson if its probability density function is given by

$$P(W = k) = \frac{\lambda^k}{k!}e^{-\lambda}, \quad k = 0,1,2,\cdots,K$$

The Poisson distribution is commonly used in queuing modeling to capture inter-arrival times of customers because of thememory-less property therein. This makes it unique and suitable for depicting the realities of arrivals or departures in physical systems such as the telephone and similar traffic systems[4].

3.5. Exponential Distribution

A random variable W with parameter μ denoting the mean number of departures in a queuing system is exponential in distribution if its probability density function is given by

$$P(W = k) = \mu e^{-\mu k}, \quad \mu > 0.$$

Similarly, the exponential distribution is used in modeling light tail distributions such as the service process in a queuing system of shops, telephone centers etc.

3.6. Erlang Distribution

A random variable W with a mean w/μ is said to have an Erlang-w $(w = 1,2,3,\cdots)$ distribution if it is a sum of w-independent random variables each exponentially distributed.

Basic Limit Theorems:

Theorem 3.1: *The limiting probability distribution for a general **birth-death process** in a state say k is given by*

$$p_k = \frac{\lambda_0 \lambda_1 \cdots \lambda_{k-1}}{\mu_1 \mu_2 \cdots \mu_k} p_0 \tag{1}$$

where λ_i and μ_j are the individual birth and death rates and p_0 is the probability that the generating process is in the idle state.

In steady state[5], a birth-death process is time independent almost surely. Thus, if it is generated by a unit service point then, its continuous-time distribution is strictly that of the **M/M/1 queue**.

Theorem 3.2: *Denote by p_k the stationary customer distribution of a shop queue with a Poisson arrival process and exponential service times operating under the FCFS queue schedule. If ρ is the occupation rate of the server then, p_k is given by the geometric distribution*

---

[4]Though, the Poisson controversy in the internet and communications modeling depicting Poisson based models as obsolete exist today; see [9], recent researches for instance, [10] have shown that, the observed long-range dependence in the internet traffic does not make the Poisson based models obsolete.

[5]A state when arrival rate equals to service rate in a given system.

$$p_k = (1 - \rho)\rho^k \tag{2}$$

The performance measures such as the mean and variance of the model can be computed. We will demonstrate this subsequently.

Now, if the single server here, is replaced by a general value say C where $C > 1$ then, a new model called the M/M/C queue analogous to a multi-server service system with Poisson arrival rate and exponential service rate is obtained. The parameter for this model is the size of ρ. If ρ is less than unity that is, under light traffic then, steady state could be attained.

Theorem 3.3: *For the M/M/C queuing system under light traffic and FCFS service schedule, the stationary customer distribution p_k is given by*

$$p_k = \begin{cases} \dfrac{\rho^k}{k!} p_0, & \text{for } 0 \leq k \leq C-1 \\[3mm] \dfrac{\rho^k}{C!(k-C)!} p_0, & \text{for } k \geq C \end{cases} \tag{3}$$

In this case, p_0 gives the probability that the system is idle.

The two limiting distributions for the M/M/C model in (3) above give the stationary distribution for the **Erlang-B and Erlang-C models**[6] respectively. The first model gives the blocking probability that a customer is lost by the system and the second model gives the probability that an arriving customer has to wait before being served. There are other relevant cases of the M/M/C queuing model. Of immense importance is the case when the waiting space is bounded. This model is applied in dimensioning service points for instance, a parking lot, a lift etc. In this case, the stationery customer distribution is a slight modification of the above mentioned distributions represented in (3).

Theorem 3.4:

For the M/M/C/K queuing model with finite waiting capacity K, let p_k be the steady state customer distribution and ρ, the occupation rate of the C-servers in the system. Under the FCFS schedule, the two-case stationery customer distribution p_k is given by

$$p_k = \begin{cases} \dfrac{1}{k!}\left(\dfrac{\lambda}{\mu}\right)^k p_0 & 1 \leq k \leq C \\[4mm] \dfrac{1}{C^{k-C}C!}\left(\dfrac{\lambda}{\mu}\right)^k p_0 & C \leq k \leq K \end{cases} \tag{4}$$

and p_0 is the idle state distribution given by

$$p_0 = \begin{cases} \left[\displaystyle\sum_{k=0}^{C-1}\dfrac{A^k}{k!} + \dfrac{A^C}{C!}\dfrac{1-\rho^{K-C+1}}{1-\rho}\right]^{-1} & \rho \neq 1 \\[6mm] \left[\displaystyle\sum_{k=0}^{C-1}\dfrac{A^k}{k!} + \dfrac{A^C}{C!}(K-C+1)\right]^{-1} & \rho = 1 \end{cases} \tag{5}$$

Here, A^7 is the ratio λ/μ where λ is the average arrival rate of customers in the system and μ is the average service rate of a single exponential server in the system. A basic assumption here is that the service rate of all the C- servers is assumed approximately equal.

Theorem 3.5: Suppose that a FCFS-M/M/1/K queuing system is given, where K is the capacity of the system. If ρ is the occupation rate of the server, then the light traffic model for the steady state customer distribution p_k is given by

$$p_k = \dfrac{\rho^k}{\displaystyle\sum_{i=0}^{K}\rho^i} p_0, \quad k = 0, 1, 2, \cdots, K \tag{6}$$

---

[6]These models are to date, relevant in light traffic modeling.

[7]The parameter A provides information on a unit-server rate of service.

where p_0 is given by

$$p_0 = \begin{cases} \dfrac{1}{K+1} & \text{if } \rho = 1 \\[2mm] \dfrac{1-\rho}{1-\rho^{K+1}}, & \text{if } \rho \neq 1 \end{cases} \tag{7}$$

The stability condition for this model is that the server occupation rate $\rho = 1$. If ρ is less than one, the model reverts to the M/M/1 infinite capacity queue.

Similarly, there are real life scenarios where priorities[8] are observed[9]. If priorities are observed then, the model is pre-emptive. In this case, the customer with the highest priority is serviced even when the low priority customer is receiving service.

Theorem 3.6: *In an M/M/1 queuing system with priority customers say type 1 and type 2. Let ρ_1 and ρ_2 denotes the occupation rates for both customer classes respectively. Under the pre-emptive priority scheduling, the stationery customer distribution for type 1 customers is given by the geometric distribution*

$$p_k^1 = \left(1 - \rho_1\right)\rho_1^k \tag{8}$$

For the type 2 customers, it is a little bit tricky and we ignore it completely.

Suppose, we wish to model the distribution of customers in a river with Poisson arrival process and exponential service times. In this case, we are talking of a model whose server size is infinite. The model is depicted as the M/M/∞ and is applied in modeling completely open-server queuing systems.

Theorem 3.7: *For the M/M/∞ queuing model with occupation rate ρ and FCFS service schedule, the stationery customer distribution is given by*

$$p_k = \frac{\rho^k}{k!}\mathrm{e}^{-\rho} \tag{9}$$

Theorem 3.8: *Suppose $W = S_1 + S_2 + \cdots + S_N$ is the stationery waiting time of an arbitrary customer where each S_i denotes the waiting time of a customer ahead of the arbitrary customer and S_i is exponential at a rate μ. If N is geometric with zero idle state then, the stationery waiting time distribution of customers in the M/M/1 queuing model having λ as a mean arrival rate and μ as mean service rate is the light tailed[10] exponential distribution*

$$P\left(W > t\right) = \rho\mathrm{e}^{-(\mu-\lambda)t} \tag{10}$$

In this case also, ρ denotes the server occupation rate. Similarly, the response time distribution which measures the total time a customer stays in the system including his service time has a cumulative distribution.

Theorem 3.9: *Under the condition of theorem 2.8 above, the cumulative distribution of the response time T for the FCFS-M/M/1 queuing system is the exponential distribution*

$$p_T\left(t\right) = \left(\mu - \lambda\right)\mathrm{e}^{-(\mu-\lambda)t} \tag{11}$$

Similarly, for the M/M/C model, the response time distribution exists.

Theorem 3.10: *For the FCFS-M/M/C model with finitely many servers C, given that, the stationery mean arrival and service rates are λ and μ respectively, then the complementary waiting time distribution is light tailed exponential and is given by*

$$P\left(W > t\right) = P_q\mathrm{e}^{-(C\mu-\lambda)t} \tag{12}$$

In this case, P_q is the service distribution of all waiting customers in the system.

---

[8] Priorities arise normally owing to variations in wants, needs etc of customers. Consequently, it is significant to understand modeling in this sphere.

[9] Priorities are embedded in our nature. Thus modeling must take account of it for effectiveness of approximations.

[10] Its convergence is faster. Thus it attains it limit within a realistic service time.

4. Analytic Applications

Example 4.1:

Suppose we wish to inquire on what to expect in a call center where customers arrive according to a Poisson process at a rate say λ to make calls with service times similarly exponentially distributed at a rate say μ when the arrival rate is strictly less than the service rate. Obviously, we are in the M/M/1 environment and performance measures such as the expected number of customers $E[N]$, expected response time $E[T]$, expected queue length $E[N_q]$ and expected waiting time $E[W]$ can be computed analytically.

Solution 4.1:

From elementary probability, we have

$$E[N] = \sum_{k=0}^{\infty} k p_k \tag{13}$$

That means by (2)

$$E[N] = \sum_{k=0}^{\infty} k(1-\rho)\rho^k = (1-\rho)\rho \frac{\mathrm{d}}{\mathrm{d}\rho} \sum_{k=0}^{\infty} \rho^k = \frac{\rho}{1-\rho} \tag{14}$$

Similarly, the expected response time for the call center can be analyzed. The following theorem due to D.C. Little is significant in analyzing this parameter.

Little's Theorem

Under steady state condition, the average number of customers $E[N]$ in a system is a product of the average at which customers arrive λ and the average time $E[T]$ customers spend in the system. In mathematical sense,

$$E[N] = \lambda E[T] \tag{15}$$

Thus, if we apply (14) on the expected number of customers $E[N]$, the expected response time in the call center $E[T]$ is given by

$$E[T] = \frac{1}{\lambda}\left(\frac{\rho}{1-\rho}\right) = \frac{1}{\mu-\lambda} \tag{16}$$

Also, the expectation of the waiting time $E[W]$ can be computed from that of the response time given that, the expected service distribution is known. Define

$$E[W] = E[T] - E[S] \tag{17}$$

Here, $E[S]$[11] is the limiting expectation that, the call device is busy. This expectation equals to ρ. Consequently,

$$E[W] = \frac{1}{\mu-\lambda} - \rho \tag{18}$$

Finally, the number of customers in the queue waiting to make a call can be computed via Little's theorem connecting the expected number in the queue and the expected waiting time. Thus,

$$E[N_q] = \frac{\rho^2}{1-\rho} \tag{19}$$

Example 4.2:

A telephone business center[12] with a single telephone head has two types of customers; customers with urgent call needs (A) and those with normal call needs (B). As a rule, the center gives priority to A over B. Given that, both A and B arrive according to a Poisson process with steady rates λ_1 and λ_2 respectively and that, the service times of both customer classes is exponentially distributed with a constant mean $1/\mu$, what is the expected number of customers $E[N]$ in the system and the expected response time $E[T]$ when the priority is preemptive and non-preemptive.

---

[11]This expectation is non-zero for a system in state.

[12]This type of problems forms the essence of modeling under light traffic in queuing systems. In addition, they are the initiating problems in modeling queuing systems generally.

Solution 4.2:

Suppose that $\sum_{i=1}^{2} \rho_i < 1$. If this condition holds then, analysis is possible.

Now, denote by N_1 and N_2 the number of type A and B customers in the telephone call center respectively. Let T_1 and T_2 be the respective response times for the two types of customers. Intuitively, the response time of A is independent of B. Thus, for type A customers, we have

$$E[T_1] = \frac{1}{\mu - \lambda_1} \tag{20}$$

Similarly, the expected number of customers in the system is equal to

$$E[N_1] = \frac{\rho_1}{1 - \rho_1} \tag{21}$$

For customer type B, we claim that, the service distribution of all types of customers is exponential. Consequently, the distribution of customers in the system will not depend on the service arrangement and so, the joint expected number of customers $E[N_1 + N_2]$ in the telephone center is given by

$$E[N_1 + N_2] = \frac{\sum_{i=1}^{2} \rho_i}{1 - \sum_{i=1}^{2} \rho_i} \tag{22}$$

Furthermore, the expected number of type B customers $E[N_2]$ is the difference between the joint expectation (22) and that of the first category. Thus,

$$E[N_2] = \frac{\rho_1 + \rho_2}{1 - \rho_1 - \rho_2} - \frac{\rho_1}{1 - \rho_1} = \frac{\rho_2}{(1 - \rho_1)(1 - \rho_1 - \rho_2)} \tag{23}$$

Finally, under pre-emptive priority the expected response time $E[T]$ is given by

$$E[T_2] = \frac{1}{\lambda_2} E[N_2] = \frac{\dfrac{1}{\mu}}{(1 - \rho_2)(1 - \rho_1 - \rho_2)} \tag{24}$$

Example 4.3

A manufacturing system has finitely many servers C. If customers arrive according to a Poisson process at a stationery rate say λ and are served negative exponentially on the average $2/\mu_i$. What is the expected queue length[13] and the expected waiting time in the system.

Solution 4.3

Let $E[N_q]$ be the expected number of customers who upon arrival, has to waiting before service and $E[W]$ be their expected waiting time in the system. We can extract for instance the limiting distribution of $E[N_q]$ from the distribution given in (3). The extraction process will give

$$E[N_q] = \sum_{k=0}^{\infty} k p_k - \sum_{k=0}^{C} C p_C = \sum_{k=0}^{\infty} (k - C) p_{k-C} \tag{25}$$

where p_k is as defined in (3). After substitution and rearranging, we will have

$$E[N_q] = \sum_{k=0}^{\infty} (k - C) p_k = p_0 \frac{(C\rho)^C}{C!} \sum_{k=C}^{\infty} (k - C) \rho^{k-C} = p_0 \frac{(C\rho)^C}{C!} \frac{\rho}{(1 - \rho)^2} = \frac{(C\rho)^C}{C!} \frac{\rho}{(1 - \rho)} \tag{26}$$

Thus, the expected waiting customers is given by

$$E[N_q] = \frac{(C\rho)^C}{C!} \frac{\rho}{(1 - \rho)} \tag{27}$$

---

[13]That is, the number of customers in the waiting.

And by dividing (26) by the arrival rate λ, the expected waiting time of customers in the system follows. This is consequence of Little's theorem. Thus,

$$E[W] = \frac{E[N_q]}{\lambda} = \frac{(C\rho)^C}{C!} \frac{\rho}{\lambda(1-\rho)} \qquad (28)$$

Similarly, that of the number of customers and the response time can also be computed. We left it to the reader as an exercise!!

Example 4.4

In a certain river, customers arrival follow a Poisson process at a rate of λ to take service with service times exponentially distributed at a rate of μ. What is the expected number of customers $E[N]$ in the system and the mean response time $E[T]$.

Solution 4.4

Since it is unimaginable to comprehend a lost customer in a river, we can model it as an M/M/∞. Similarly, it can be idealized that the mean number of services is the same as the sum of all small portions of the river being put to use out of its area[14].

Now, remember that for the M/M/∞, the limiting distribution is a Poisson distribution with parameter ρ. Consequently, $E[N]$ is the mean of a Poisson random variable. In this case, $E[N]$ is simple the server occupation ρ. Similarly, the other expectation can be seen in the light of Little's theorem.

5. Realistic Applications

Example 5.1

A local council wishes to determine the size and staffing of a local maternity hospital. Historically, it was gathered that, there is the average of 10 births per day in the ward to which most women are discharged after 2 days from the ward. Though, in one year, about 5 percent of the women admitted take 10 days for different complications. What is the average length of stay in the ward[15].

Solution 5.1

Here, the council is interested at how long a given admit will stay, possibly for reconstruction and re-staffing. In this case, Little's theorem provides the solution.

Now, denote by $E[L]$ the expected length of stay of admits in the ward. Intuitively, this period depends on the type of admit. If we let λ_1 and λ_2 to represent the arrival rate of admit type 1 and admit type 2 respectively corresponding to the response times T_1 and T_2 then by Little's theorem, we have

$$E[L] = \lambda_1 E[T_1] + \lambda_2 E[T_2]$$
$$= 0.95 \times 2 + 0.05 \times 10$$
$$= 2.4 \text{ days}$$

Example 5.2

A telecom company wishes to upgrade her system capacity. On records, calls arrive in the existing system at a rate of 500 per minutes and each call resides for an average of 5 minutes. If the system presently has 20 service points, what is the stationery distribution of lost calls in this system.

Solution 5.2

In telecommunications modeling, the server occupation rate is commonly known as the average system load. Since, the interest here is on lost customers, the Erlang-B model is most suitable in this computation. Consequently, using

$$P_b = \frac{\dfrac{A^C}{C!}}{\displaystyle\sum_{i=0}^{C} \dfrac{A^i}{i!}} \qquad (29)$$

---

[14]With good modeling practice and estimation, the problem could similarly be modeled as the M/M/C/C queuing system. A student of modeling may wish to compute that and compare the two results.

[15]Problems like this and many others prove the justifications of modeling in the light of social and economic systems for effective management.

Where C is the number of servers and A is the average system load, we have[16]

$$P_b = \frac{\dfrac{(2500)^{20}}{20!}}{\displaystyle\sum_{i=0}^{20} \dfrac{(2500)^i}{i!}}$$

Example 5.3

The department of Mathematics of a University X has a single printer connected to the local area network (LAN) of the department. Recently, it is observed that, the delay time of processing printing jobs is on the increase and there is the need to reduce it. Show by analytic modeling how this problem can be tackled out.

Solution 5.3

This is a decision making problem. Suppose printing jobs arrive the printer as Poisson process independently at a rate of λ and are served exponentially at a rate μ_1 by the problematic printer. There are 2 possible ways to solve this delay problem:

1) A new printer with service rate $a\mu_1$ where $a > 1$ will be bought to replace the existing printer.

Under this modeling, let $E[T_1]$ and $E[T_2]$ are the expected response time of the existing and the new printer respectively. Then, it is clear for this M/M/1 model that,

$$E[T_2] = \frac{1}{a\mu_1 - \lambda} < \frac{1}{\mu_1 - \lambda} = E[T_1], \text{ since } a > 1.$$

Thus,

$$E[T_2] < E[T_1].$$

2) If the department is in no income state then, n-old printers[17] ($n > 1$) similar to the existing one be put to usage side by side with the existing printer.

Under this modeling, the arrival rate for the M/M/C queuing system here is partitioned to be on average, λ/n. Let $E[T]$ denotes the expected response time of the printing arrangement. Obviously,

$$E[T] = \frac{1}{n}E[T_1] = \frac{1}{n(\mu_1 - \lambda)} < \frac{1}{(\mu_1 - \lambda)} = E[T_1], \text{ since } n > 1.$$

Thus, $E[T] < E[T_1]$.

Example 5.4

An M/G/2 queuing system has one-exponential and one general server. Given that, the general server is regularly varying at infinity index η and the system is conditioned such that, the general server is available for use only if the exponential server is busy, what is the steady state expectation of the number of customer in the system. Also, Compute the expected response time $E[T]$ if the arrival rate is 0.5μ, where μ is the exponential service rate.

Solution 5.4

Denote by $B(t)$ the service time distribution of customers served by the regularly varying server with a mean of β. If the arrival and the exponential service rates are λ and μ respectively then, in steady state, the stability condition for this system is $\lambda < \mu + 1/\beta$ holds. Under this condition, we can argue that if $\lambda < \mu + 1/\beta$, it is trivial that is $\lambda < \mu$. This is analogous to keeping a customer with an infinite service time on the regularly varying server. Consequently, the M/G/2 model in this case behaves like the M/M/1. Thus, the limiting expectation for the number of customers in the system is given by (20). Now, given that, $\rho = 0.5$ and $E[T]$ is the expected response time, it is obvious that, $E[T] = 2/\mu$.

Example 5.5

A bank has 13 counters. Customers arrive as a Poisson process at a rate of 3 per minute and stay in a single queue. If each counter staff needs on average 4 minutes to deal with a customer. What is the steady state probability that all the counter staff are idle.

---

[16]The student is expected to complete this as a free lunch.

[17]This may hold only if the existing printer is not the first to be used in the department. Otherwise, the first option suffices.

Solution 5.5

We can easily compute this distribution if we model the bank as an M/M/C queue. In this case, $C = 13$, $\lambda = 3$, $\mu = 1/4$, $\lambda/\mu = 12$ and $\rho = \lambda/C\mu = 12/13$. Applying the idle probability distribution for the M/M/C model given by

$$p_0 = \left(\sum_{k=0}^{C-1} \frac{a^k}{k!} + \frac{a^C}{C!} \frac{1}{1-\rho} \right)^{-1} \tag{30}$$

We have

$$p_0 = \left(\sum_{k=0}^{12} \frac{12^k}{k!} + \frac{12^{13}}{12!} \right)^{-1}.$$

Exercises:

1) In a gas station with a single pump, cars arrive according to a Poisson process at an average of 20 cars per hour to receive service. The time required for service is exponential at a mean rate of 3 minutes. Given that, a car may refuse to enter the station with probability $q_n = n/4$, determine;
a) The stationary distribution of cars in the station.
b) The mean number of cars at the station.
c) The mean response time of cars remaining in the station.
2) A repair man fixes broken computers. The repair time is deemed exponentially distributed with a mean of 30 minutes. Broken computers arrive at his repair shop according to a Poisson stream with an average of 10 broken computers per day (he works for 8 hours in a day).
a) What is the fraction of time that the repair man has no work to do?
b) How many computers are on average, at his repair shop?
c) What is the mean response time for a computer to be repaired?
3) Consider two parallel machines A and B having a common buffer where jobs arrive according to a Poisson process with rate β. The processing times (service times) are exponentially distributed with means $1/\mu_1$ for A and $1/\mu_2$ for B, where $(\mu_2 < \mu_1)$. The service discipline is first come first serve and during idle state, an arrival is assigned the faster machine. Assuming that, $\dfrac{\lambda}{\mu_1 + \mu_2} < 1$, determine;
a) Determine the distribution of the number of jobs in the system.
b) Use this information to derive the mean number of jobs in the system.
c) Decide when it is better not to use the slower machine at all.
4) A computer consists of 3 processors. Their main task is to execute jobs from users. These jobs arrive according to a Poisson process with rate 15 jobs per minute. The execution time[18] is exponentially distributed with mean of 10 seconds. Given that, when a processor completes a job and there are no other jobs waiting to be executed, the processor starts to execute continuous-time maintenance jobs that are exponentially distributed with a mean of 5 seconds. But as soon as a main job arrives, the processor interrupts the execution of the maintenance job and starts to execute the main job. The execution of the maintenance job will be resumed later (at the point where it was interrupted).
a) What is the expected number of processors busy with executing jobs from users.
b) How many maintenance jobs are on average completed per minute.
c) What is the probability that a job from a user has to wait.
d) Determine the mean waiting time of a job from a user.

6. Conclusion

We demonstrate modeling in light traffic queuing systems for business owners with little appreciation for Mathematics. The aim is to encourage the use of the well grounded theories of the subject to areas where application will lead to improvements and better service provision. The functional limit theorems selected and applied here are some of the well known theorems in queuing theory. As for the proofs and more general discussions on this subject, there are vast amount of literature covering this topic on queuing theory today.

---

[18]Execution time here means the service time.

Acknowledgements

We are grateful to all the literature sources used in this note. Also is to the anonymous referees.

References

[1]   Medhi, J. (2003) Stochastic Models in Queuing Theory. Academic Press. An Imprint of Elsevier Science (USA).

[2]   Franken, P., Koonig, D., Arndt, U. and Schmidt, V. (1982) Queues and Point Processes. Akademie-Verlag Publication, Germany.

[3]   Federgruen, A. and Tijms, H.C. (1980) Computation of the Stationary Distribution of the Queue Size in M/G/1 with Variable Service Rate. *Journal of Applied Probability*, **17**, 515-522. http://dx.doi.org/10.2307/3213040

[4]   Hoksad, P. (1978) Approximation for the M/G/m Queue. *Journal of Operation Research*, **26**, 511-523.

[5]   Hoksad, P. (1979) On the Steady State Solution of the M/G/2 Queue. *Advanced Applied Probability*, **11**, 240-255. http://dx.doi.org/10.2307/1426776

[6]   Smith, J.M. (2002) M/G/C/K Blocking Probability Models and System Performance. *Performance Evaluation*, **52**, 237-267. http://dx.doi.org/10.1016/S0166-5316(02)00190-6

[7]   Tijms, H.C., Van Hoorn, M.H. and Federgruen, A. (1981) Approximation for the Steady State Probabilities in the M/G/C Queue. *Advances in Applied Probability*, **13**, 186-206. http://dx.doi.org/10.2307/1426474

[8]   Abate, A. and Whitt, W. (1994) Asymptotics for M/G/1 Low-Priority Waiting-Time Tail Probabilities. *Queuing Systems*, **25**, 173-233. http://dx.doi.org/10.1023/A:1019104402024

[9]   Leland, W.E., Taqqu, M.S., Willinger, W. and Wilson, D.V. (1994) On the Self-Similar Nature of Ethernet Traffic (Extended Version). *IEEE/ACM Transactions on Networking*, **2**, 1-15. http://dx.doi.org/10.1109/90.282603

[10] Karagiannis, T., Molle, M., Faloutsos, M. and Broido, A. (2008) A Nonstationary View on Poisson Internet Traffic. *Proceedings of IEEE INFOCOM*, 84-89.

7

Multi-Knapsack Model of Collaborative Portfolio Configurations in Multi-Strategy Oriented

Shujuan Luo, Sijun Bai, Suike Li

School of Management, Northwestern Polytechnical University, Xi'an, China
Email: luosj@mail.nwpu.edu.cn, baisj@nwpu.edu.cn, lisk@mail.nwpu.edu.cn

Abstract

Aiming at constructing the multi-knapsack model of collaborative portfolio configurations in multi-strategy oriented, the hybrid evolutionary algorithm was designed based on greedy method, combining with the organization of the multiple strategical guidance and multi-knapsack model. Furthermore, the organizing resource utility and risk management of portfolio were considered. The experiments were conducted on three main technological markets which contain communication, transportation and industry. The results demonstrated that the proposed model and algorithm were feasible and reliable.

Keywords

Multi Knapsack Model, Multi Strategy, Collaborative Portfolio, Evolutionary Algorithm

1. Introduction

In the practical competitions, in order to reduce the operational risk and obtain certain profit space in the new field, many enterprises will choose the diversified development strategy. However, it is proved that numerous organizations did not gain the expected return in the diversification process. The reasons are the mistake of making strategy and the implementation difficulty of diversification strategy. Applying the theory of portfolio management and system method to the implementation of diversification strategy could improve the capability of strategy implementation and achieve the target of the organization diversification strategy. Meantime, the organization will encounter a variety of project opportunities in the expanding process of diversification. Facing numerous alternative projects, how to select and configure them is the urgent problem needed to be solved.

A successful implementation of project cannot do without the guidance of project organization strategy.

Diversification of project portfolio selection and configuration must conform with the organization strategy, at the same time, the multi-strategy requires a specific project and project portfolio selection to achieve. Through the analysis of diversification strategy, our conclusion is that every implementation way, based on the diversification strategy, can be regarded as a project management process, and each new work can be managed and realized as a new project. Therefore, taking the project portfolio allocation concept to realize the diversification of organizational strategy is effective and necessary [1]. The success of diversification of organizational strategy not only depends on the success or failure of individual project, but also depends on the evaluation, selection, configuration and benefits of the project portfolio [2] [3]. This paper is the study of multiple strategic orientation and resource limited project portfolio allocation problem, the goals are: 1) by portfolio allocation to achieve organizational gross income maximum, portfolio risk minimum; 2) to solve the problem of stability and coordination among project portfolios.

2. Assumptions of the Questions of Collaborative Portfolio Configurations in Multi-Strategy Oriented

In reality, the decision of the diversification strategy must consider whether there are good projects, long-term growth prospects and their own ability of resources. For this, considerable theoretical researches and enterprise management practices were made. For example, Fu Jun, one entrepreneur, proposed, when carrying out the projects or target industries which multiple choices would enter, we must carefully analyze the following basis for decision: First is the potential of the selected projects and industry to become bigger and stronger rapidly. If the market capacity of the project industry is limited, too scattered or difficult to expand, it is unfavorable to choose. Second is the sustained development prospect of the selected projects and industry. Although some selected projects and industry have the potential to become bigger and stronger and meet the first basis, they may be eliminated because of the lack of the possibility to develop for a long time. Finally, do the new executive projects and the project combination fit the organization's reserve capacity of existing resources? The assumptions of diversification decisions of Founder are as followed. One is the traditional and stable project. Selecting the base project whose income is relatively stable to ensure that it can continue making stable earnings for the group and reduce the adverse impact of the economic fluctuation. The other is a high-tech project. Through the project selection and execution to master the key technology in the new field, set the key technology as a prime mover of the sustained development of the group [4].

This paper built the portfolio allocation model guiding by the pluralistic strategy. The specific problems were briefly described as follows. Assuming the diversification strategy of high-level decisions of the organization decided to achieve the pluralistic strategy by projecting and the project portfolio management. On the basis of allocating the organizational resources, m portfolios were established. According to the demand of the pluralistic strategy, the organizational resource type and quantity which each project portfolio can allocate were limited. The organization prepared n projects. Under certain constraints, these projects were selected to these combinations, achieving the biggest implementation of the strategic goals. In view of the complexity of the combination and allocation of the projects in the diversifying strategy oriented, to simplify the complex problems and build the model consistent with the actual situation of the organization, we need make assumptions on the following problems:

2.1. Resource Sharing

Based on the analysis of organizational diversification strategy classification and resource requirements, and the project portfolio allocation of the sub strategy in the multi-strategy, project resource sharing is widespread and is the prerequisite for scale effect. Project resource sharing illustrates the soft constraints of resources of the project portfolio are the key characteristics which distinguish this from the hard constraints of the traditional knapsack problems knapsack and goods size. The project resource sharing is related to the property of the resources. Non-consumption resources, such as the fixed assets, are relatively more likely to be shared. In addition, it is related to the project similarity of the portfolio. If the similarity degree is high, the possibility of resource sharing is relatively large. The number of the configuration of the project is concerned. The more projects the same combination executes, the higher degree of resource sharing could be.

Since the references are less, the measurement of the total amount of resource sharing is difficult. According to the above analysis, the following relevant assumptions are taken:

i) The sharing degree of the resource k can be expressed as φ_k, means consumable resources cannot be shared. Means resources can be fully shared (such as fixed assets);

ii) The resource which the resource k allocate for the project portfolio j is $C_{k,j}$, the similarity of project portfolio j is μ_j, $\mu_j \in (0,1)$, $\mu_j \approx 0$ means the projects are completely different. $\mu_j \approx 1$ means the projects are almost same but there is no same projects exactly. So $\mu_j \neq 1$.

The implementation number of the project portfolio j is n_j. The sharing coefficient $\theta_{k,j}$ and sharing amount of $\pi_{k,j}$, which belong to the resource k in the project portfolio j, are calculated by:

$$\theta_{k,j} = \varphi_k * \mu * \frac{n_j}{n} \tag{1}$$

$$\pi_{k,j} = \theta_{k,j} * C_{k,j} \tag{2}$$

In the project portfolio allocation model, the resource constraints should be adjusted according to the organizational strategy, project portfolio and the project portfolio resource sharing accounting of the resource characteristics.

2.2. Risk Management

The project portfolio allocation of the multi-strategy oriented need consider the overall portfolio risks. The first step is risk identification, analysis and assessment of individual projects, but the project portfolio risk of the organization cannot be the simple sum of each individual objective risk. The difficulty of the risk management of the organization project combination should also be considered. The difficulty of the project portfolio management is related to the similarity between the projects, which means if the difficulty of the risk management is low, the overall risk will be reduced. At the same time, it is related to the number of the selected project portfolio. The more items the same portfolio executes, the more difficult the risk management of project portfolio is. Meantime, the risk of portfolio configuration will increase.

Ditto, the risk tolerance of project portfolio j is R_j. Assume that the similarity of project portfolio j is μ_j, and the implementation number of project portfolio j is n_j. ω_j means the difficulty of the risk management coefficient and the risk increase of the project portfolio j is ∂_j;

$$\omega_j = \left(\frac{1}{\mu_j} - 2 \right) * \frac{n_j}{n} \tag{3}$$

$$\partial_j = \omega_j * R_j \tag{4}$$

According to the Formula (3) and (4), we can obtain some conclusions. If $\mu_i \in (0, 0.5)$, $\omega_i > 0$, $\partial_i > 0$, the risk amount increases with the difficulty of the risk management. If $\mu_i \in (0.5, 1)$, $\omega_i < 0$, $\partial_i < 0$, the risk amount decreases with the difficulty of the risk management. Assuming that when $\mu_i = 0.5$, the risk management??s difficulty is moderate, the increase of the risk is 0.

3. The Multi-Knapsack Model and Algorithm Design of Portfolio Collaborative Configuration

3.1. The Multi-Knapsack Model of Portfolio Collaborative Configuration

1) the explanations of the symbols

Corresponding to the diversification strategy of the organization, the number of the established configuration project portfolio j is $m, j = 1, 2, \cdots, m$; Project I, which is in the candidate item pool, is waiting to be selected by project portfolio, and the total number of candidate items is n, $i = 1, 2, \cdots, n$; The volume v_i expressed the Strategic value of the project i. $f(x)$ expressed the volume of the total strategy, which was realized by the project portfolio configuration [5]. The decision variables $x_{i,j}$ expressed the decision of the project portfolio j for the project i. When project i was chosen by project portfolio j, $x_{i,j} = 1$. If not $x_{i,j} = 0$; $q_{k,i}$ expressed the requirement of project i for the resource k. r_i expressed the risk of project I; The Constraint formula of resource and risk to the project portfolio can be seen from the (1) to (4).

2) the objective function and the constraint conditions

$$\max f(x) = \sum_{j=1}^{m}\sum_{i=1}^{n} v_i x_{i,j} \tag{5}$$

$$s.t. \sum_{j=1}^{m} x_{i,j} \leq; i = 1, \cdots, n \tag{6}$$

$$\sum_{i=1}^{n} q_{k,i} x_{i,j} - \theta_{k,j} * C_{k,j} \leq C_{k,j}; j = 1, \cdots, m \tag{7}$$

$$\sum_{i=1}^{n} r_i x_{i,j} + \omega_j * R_j \leq R_j; j = 1, \cdots, m \tag{8}$$

$$\theta_{k,j} = \varphi_k * \mu_j * \frac{n_j}{n}; j = 1, \cdots, m \tag{9}$$

$$\omega_j = \left(\frac{1}{\mu_j} - 2\right) * \frac{n_j}{n}; j = 1, \cdots, m \tag{10}$$

$$x_{i,j} = 0 \text{ or } 1; i = 1, \cdots, n, j = 1, \cdots, m \tag{11}$$

$$n_j = \sum_{i=1}^{n}; j = 1, 2, \cdots, m \tag{12}$$

In the formulas, the multi knapsack model of the portfolio allocation (5) is a 0 - 1 integer programming [6]. The all solutions of this model are the matrix space of $m * n$. Each solution of this matrix is used to express the relationship between the project and the project portfolio. If project I was selected by project portfolio j, $x_{i,j} = 1$; If not, $x_{i,j} = 0$. Formula (6) means that each project can only be selected by one project portfolio. Formulas from (7) to (12) respectively expressed the resource capacity of the project portfolio and the risk tolerance boundaries. The resource which the resource k allocate for the project portfolio j is $C_{k,j}$. This multi-knapsack model of project portfolio allocation is a typical NP-complete problem. The complexity of the model calculation is $O(2^{mn})$.

3.2. Greedy Method—Genetic Algorithm Design

The problem of the simple genetic algorithm is that when it faces the larger solution model, it is possible that all the individual values are 0 and Iterative could interrupt. Therefore, the greedy algorithm is consulted and introduced to initialize the groups and repair the unfeasible solution. Combining the SGA with the traditional greedy algorithm, the search speed and the accuracy of the algorithm are improved. Furthermore, it can over- come the problem of being easy to fall into local optimum, which is one disadvantage of the traditional method [7] [8].

1) the genetic coding and fitness function

Treat with the integer coding. For example, coding for the multiple knapsack problem, which contains 20 projects and 3 portfolios. It can be expressed as the following integer set:
$X = \{x_1, x_2, \cdots, x_{20}\}$ $(x_i = l \in \{0,1,2,3\})$. It means the project i is allocated to the portfolio l $(l = 0)$ represents that 3 backpacks do not choose this project). If $x_2 = 1$, it means 2 projects would be selected and allocated in portfolio l. So the search space of this solution will be smaller, and the speed of convergence will be faster.

The fitness function is defined as $fitness(x) = \sum_{i=1}^{n}(x_i = 0) p_i$. In the process of operation, the adaptive values are arranged by ascending. Treat the number of original adaptive value as the new adaptive value and change the scale of the adaptive value. The purpose of doing so is to reduce the probability of the premature convergence or stop of the iteration.

2) the genetic operator and the algorithm flow

First of all, use the completely random method to initialize the population. And then repair the unfeasible solution. Secondly, based on the greedy method to produce an approximate optimal solution, replace the worst individual of the population with it to get an optimization Initial population. The algorithm flow is as shown in **Figure 1**.

4. Case Analysis and Conclusions

4.1. The Description of Case Background

The headquarters of company Z decided to change the position of Chinese Company, and help it become the product research and development, processing, manufacturing and comprehensive service base of the Asia Pacific region. According to its core competence and diversification development strategy, it formulated a diversification strategy. The diversification strategy, namely the 3 technologies: radio frequency technology (HF), optical fibre technology (FO), cable and polymer technology (NF) and 3 application areas, namely the three main markets: Communication, Transportation, industry [9]. Taking one technical field of Z company in three Chinese market ("1 × 3") as the research object to carry out diversification analysis. Transform it into a mathematical model of the multi knapsack problem, which contains 3 backpack (portfolios) and 20 items (alternative projects) and then establish the project portfolio collaborative allocation model in the diversification strategy-oriented. Finally, analyze the dynamic process of allocation and coordination of the project portfolio.

4.2. Basic Data Input

In order to verify the effectiveness of the model and algorithm, this paper combined with the actual situation of Z company, and simulated project portfolio allocation problems. In this simulation, there are 3 portfolio ($m = 3$), 20 items ($n = 20$) and 2 kinds of resources ($k = 2$). Assume that the degree of resource sharing are $\varphi_1 = 0.6$, $\varphi_2 = 0.2$, separately. The relevant data, which has been fuzzy transformed, had shown in **Table 1** and **Table 2**.

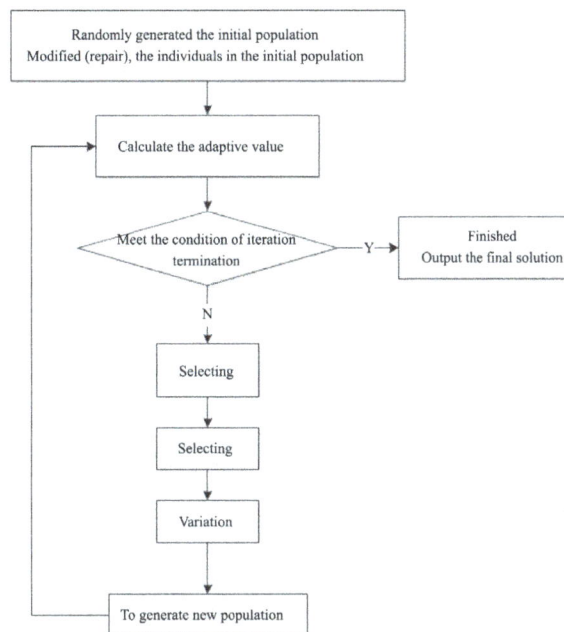

Figure 1. Hybrid genetic algorithm flow chart.

Table 1. The parameters of portfolio configuration.

Project portfolio allocation	$C_{1,j}$	$C_{2,j}$	R_j	μ_j
Portfolio 1 (j = 1)	142	98	0.96	0.6
Portfolio 2 (j = 2)	165	100	0.98	0.4
Portfolio 3 (j = 3)	193	102	0.95	0.7
Total	500	300	-	-

Table 2. The parameters of candidate item.

Project number	v_j	$q_{1,j}$	$q_{2,j}$	r_j
Project 1 ($j = 1$)	47	30	11	0.141
Project 2 ($j = 2$)	57	37	14	0.171
Project 3 ($j = 3$)	63	39	18	0.189
Project 4 ($j = 4$)	50	31	13	0.150
Project 5 ($j = 5$)	55	36	16	0.144
Project 6 ($j = 6$)	51	24	21	0.153
Project 7 ($j = 7$)	60	31	23	0.180
Project 8 ($j = 8$)	58	35	17	0.174
Project 9 ($j = 9$)	44	22	16	0.132
Project 10 ($j = 10$)	43	25	20	0.113
Project 11 ($j = 11$)	50	20	24	0.151
Project 12 ($j = 12$)	51	32	13	0.153
Project 13 ($j = 13$)	54	33	15	0.157
Project 14 ($j = 14$)	58	41	11	0.174
Project 15 ($j = 15$)	57	33	18	0.171
Project 16 ($j = 16$)	50	28	16	0.152
Project 17 ($j = 17$)	60	36	18	0.181
Project 18 ($j = 18$)	59	42	14	0.146
Project 19 ($j = 19$)	74	46	22	0.222
Project 20 ($j = 20$)	55	29	20	0.165

4.3. Experiment Results

The parameters of the hybrid evolutionary algorithm can be set as follows: the number of the population size, *i.e.* individual species, is *Nind* = 30. The maximum number of the evolutionary generation is *Maxgen* = 100. The crossover rate is *cr* = 0.1, and the mutation rate is *mr* = 0.1. The results of the portfolio configurations in multi-strategy oriented, which are presented in the **Table 3**, can be gained through the MATLAB.

As is shown in **Table 3**, the usage of organizing project resource should reduce the total cost of resource. Considering the difficulty of the risk management of the project portfolio, the portfolio risk is the adjustment. The results illustrate that the sum of project is 17, and the organizing strategical value confirmed by three joint projects is 907. It shows that we have proposed the model and method which can fulfill the target which makes the organizing strategical value the most based on the share and rational allocation of resources. We consider the ability to withstand risk, simultaneously. Project optimum population and the trend of average population along with the change of evolutionary generation are shown in **Figure 2**. It demonstrates that the proposed model and algorithm could gain the global best solution of the joint project portfolios. It also validates that the presented method turns out to be an efficient and steady one.

5. Conclusion

The paper has proposed a novel thought based on multi-organization and project portfolio. Strategy implementation and project portfolio management theory have provided the opportunity to organize the multiple strategies and to guide the project portfolio. Combing with the case of company Z, we constructed the multi-knapsack

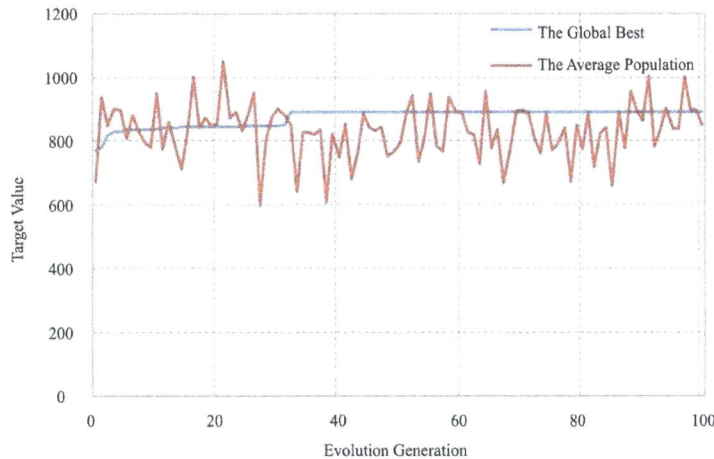

Figure 2. The change trend of the optimal value and the population mean with the evolution generation.

Table 3. The results of project portfolio.

Project portfolio	Selected project	Strategical value	Resource cost by 1	Resource cost by 2	Portfolio risk
Portfolio 1 ($j = 1$)	3, 4, 7, 11, 15	280	128.44	90.12	0.673
Portfolio 2 ($j = 2$)	5, 6, 10, 13, 18	262	140.2	82	0.963
Portfolio 3 ($j = 3$)	1, 9, 12, 14, 16, 17, 20	365	161.258	95.004	0.695
Total cost	-	907	429.898	267.127	-

model of collaborative portfolio configurations in multi-strategy oriented and proposed the hybrid evolutionary algorithm based on the greedy method which succeeds in solving the problem of multiple strategical guidance and project portfolio. The experiment presents the statistical data and shows the feasibility of the proposed model and method.

Acknowledgements

We thank the Editor and the referee for their comments. This paper is supported by National Natural Science Foundation of China under Grant No. 71172123, by Aviation Science Fund under Grant No. 2012ZG53083.

References

[1] Su, L.Y. (2008) The Implementation of Enterprise Diversification Management and Project Management. *Inner Mongolia Science Technology and Economy*, **14**, 278-281.

[2] Li, S.K., Bai, S.J., Guo, Y.T. and Wang, X.B. (2014) A Study on Distributed Collaborative Model of Model Development Based on Logistics. *Industrial Engineering Journal*, **17**, 50-54.

[3] Shen, J. (2009) Analysis on Synergy Effects in Diversification Strategy of Enterprises. *Economy and Management*, **23**, 49-54.

[4] Shi, Y.H. (2005) The Empirical Analysis on the Diversified Profession Selection. *SCI/TECH Information Development and Economy*, **17**, 76-79.

[5] Song, H.S., Fu, R.Y., Xu, R.S. and Song, H.Z. (2009) Hybrid Genetic Algorithm for Multi-Knap Sack Problem. *Computer Engineering and Applications*, **45**, 45-48.

[6] Shi, L., Zhang, Y.H. and Lv, J.H. (2013) Optimization Algorithm of 0-1 Knapsack Problem Based on Absolute greedy and Expected Efficiency. *Application Research of Computers*, **11**, 1-5.

[7] Li, Z.Y., Ma, L. and Zhang, H.Z. (2013) Application of Bat Algorithm in Multi-Objective and Multi-Choice Knapsack Problem. *The Computer Simulation*, **30**, 350-354.

[8]   Zhang, T.N. and Zeng, Q.C. (2003) Business Reengineering and Synergetic Development. *Management Sciences in China*, **16**, 21-24.

[9]   Zhang, X.F. (2011) The Development of HUBER+SUHNER in China. *Intelligent Building and City Information*, **8**, 104-105.

8

Using Malmquist Index Approach to Measure Productivity Change of a Jordanian Company for Plastic Industries

Abbas Al-Refaie, Mohammad D. Al-Tahat, Ruba Najdawi

Department of Industrial Engineering, University of Jordan, Amman, Jordan
Email: abbas.alrefai@ju.edu.jo, altahat@ju.edu.jo

Abstract

Measurement of a production unit-performance is crucial in determining whether it has achieved its objectives or not, and it generates a phase of management process that consists of feedback motivation phases. The purpose of this paper is to analyze the growth potentials of five production machines in a Jordanian company for plastic industries by employing the non-parametric Malmquist productivity index (MPI) over the period from February to July 2014 in both day and night shifts. The productivity change is decomposed into technical efficiency change (TEC) and technological change (TC). Inefficiency values are observed in each period. The percentage of input utilization is determined in all periods. Then, the Malmquist productivity index (MPI) values are calculated for all periods. Finally, comparisons of TEC, TC and MPI are conducted among the five machines and between the day and night shifts for each machine. The MPI results indicate that the needs for internal training, effective operating procedures, and enhancing quality procedures are required to increase the technical efficiency. On the other hand, figuring out more efficient ways of making existing products allowing output to grow at a faster rate than economic inputs, like using new technologies, will increase technological change. In conclusions, Malmquist model analysis shall provide valuable reference information to management when evaluating the progress in the performances of production machines in plastic industry.

Keywords

Malmquist Index, Technical Efficiency, Technological Efficiency, Plastic Industry

1. Introduction

Performance measurement of a production unit is crucial in determining whether it has achieved its objectives or not, improving production efficiency, and dealing with internal or external pressures by monitoring and benchmarking a company's production [1]-[7].

Malmquist productivity index (MPI) proposed is a management tool used to evaluate the productivity progress for multi-inputs and multi-outputs [8]-[12]. The MPI represents Total Factor Productivity (TFP) growth of a Decision Making Unit (DMU) and reflects the increase or decrease in efficiency with progress or regress of the frontier technology over time under multiple inputs and multiple outputs framework [13]-[15]. The TFP index can be used to estimate the productivity change, which is decomposed into efficiency change and technological change. The concept of productivity usually referred to labor productivity, this concept is very much related to TFP, defined as the product of efficiency change (catch-up) and technological change (frontier-shift). If TFP value is greater than one, this indicates a positive TFP growth from period (t) to period ($t + 1$), whereas a value less than one indicates a decrease in TFP growth or performance relative to the previous year. The framework employed in Malmquist can be illustrated in **Figure 1**, where a production frontier representing the efficient level of output (y) that can be produced from a given level of input (x) is constructed. The assumption made is that the frontier can shift over time. The frontier obtained in the current (t) and future ($t + 1$) time periods is labeled accordingly. When inefficiency exists, the relative movement of any given DMU over time will, therefore, depend on both its position relative to the corresponding frontier (technical efficiency) and the position of the frontier itself (technical change). If the inefficiency is ignored, then the productivity growth over time will be unable to distinguish between improvements that derive from a DMU catching up to its own frontier, or those that result from the frontier itself shifting up over time.

The input-based Malmquist productivity change index is formulated as [16]:

$$M_I^{t+1}\left(y^{t+1},x^{t+1},y^t,x^t\right)=\left[\frac{D_I^t\left(y^{t+1},x^{t+1}\right)}{D_I^t\left(y^t,x^t\right)}\times\frac{D_I^{t+1}\left(y^{t+1},x^{t+1}\right)}{D_I^{t+1}\left(y^t,x^t\right)}\right]^{1/2} \tag{1}$$

where M is the productivity of the most recent production point $\left(x^{t+1},y^{t+1}\right)$ using the period $t + 1$ technology relative to the earlier production point $\left(x^t,y^t\right)$ using period t technology, D is input distance functions, the subscript I indicates CCR input-orientation. A value of $M_I^{t+1}\left(y^{t+1},x^{t+1},y^t,x^t\right)$ greater than unity indicates a positive total factor productivity growth between the two periods. Alternatively,

$$M_I^{t+1}\left(y^{t+1},x^{t+1},y^t,x^t\right)=\frac{D_I^{t+1}\left(y^{t+1},x^{t+1}\right)}{D_I^t\left(y^t,x^t\right)}\left[\frac{D_I^t\left(y^{t+1},x^{t+1}\right)}{D_I^{t+1}\left(y^{t+1},x^{t+1}\right)}\times\frac{D_I^t\left(y^t,x^t\right)}{D_I^{t+1}\left(y^t,x^t\right)}\right]^{1/2} \tag{2}$$

In other words, the Malmquist index is

$$\text{Malmquist Index} = \text{Technical Efficiency Change}\times\text{Technological Change} \tag{3}$$

The technical efficiency change (TEC) is given by:

$$\text{TEC}=\frac{D_I^{t+1}\left(y^{t+1},x^{t+1}\right)}{D_I^t\left(y^t,x^t\right)}=\frac{TE\left(t+1\right)}{TE\left(t\right)} \tag{4}$$

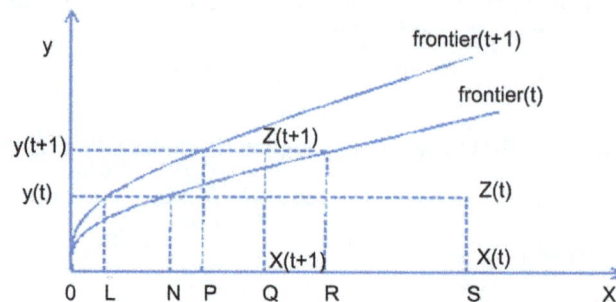

Figure 1. Productivity changes over time.

where $TE(t + 1)$ and $TE(t)$ represent the technical efficiency at period $(t + 1)$ and (t) respectively and can be calculated using the DEA model in Equation (5).

$$\min \theta \tag{5}$$

Subject to:

$$\sum_{j=1}^{n} \lambda_j x_{ij} \leq \theta x_{io} \qquad i = 1, \cdots, m \tag{5a}$$

$$\sum_{j=1}^{n} \lambda_j y_{kj} \geq y_{ko} \qquad k = 1, \cdots, s \tag{5b}$$

$$\lambda_j \geq 0 \qquad j = 1, \cdots, n \tag{5c}$$

where θ represents the technical efficiency score of unit DMU_o and λ_j represents the dual variables that identify the benchmarks for inefficient units. Also, the technological change (TC) is formulated as:

$$TC = \left[\frac{D_I^t \left(y^{t+1}, x^{t+1} \right)}{D_I^{t+1} \left(y^{t+1}, x^{t+1} \right)} \times \frac{D_I^t \left(y^t, x^t \right)}{D_I^{t+1} \left(y^t, x^t \right)} \right]^{1/2} = \left[\frac{TE(t) \times IEI^{t+1 \to t}}{IEI^{t \to t+1} \times TE(t+1)} \right]^{1/2} \tag{6}$$

where $IEI^{t+1 \to t}$ and $IEI^{t \to t+1}$ represent the intertemporal efficiency indices between $t + 1$ and t and are calculated respectively as:

$$IEI^{t+1 \to t} = \min \theta \tag{7}$$

Subject to:

$$\sum_{j=1}^{n} \lambda_j x^{t+1}{}_{ij} \leq \theta x^t{}_{io} \qquad i = 1, \cdots, m \tag{7a}$$

$$\sum_{j=1}^{n} \lambda_j y^t{}_{kj} \geq y^{t+1}{}_{ko} \qquad k = 1, \cdots, s \tag{7b}$$

$$\lambda_j \geq 0 \qquad j = 1, \cdots, n \tag{7c}$$

Also,

$$IEI^{t \to t+1} = \min \theta \tag{8}$$

Subject to:

$$\sum_{j=1}^{n} \lambda_j x^t{}_{ij} \leq \theta x^{t+1}{}_{io} \qquad i = 1, \cdots, m \tag{8a}$$

$$\sum_{j=1}^{n} \lambda_j y^{t+1}{}_{kj} \geq y^t{}_{ko} \qquad k = 1, \cdots, s \tag{8b}$$

$$\lambda_j \geq 0 \qquad j = 1, \cdots, n \tag{8c}$$

In Equation (2), the TFP index is the product of a measure of technical progress (Technological Change) as measured by shifts in the frontier measured at period $t + 1$ and period t (averaged geometrically) and a change in efficiency (Efficiency Change) over the same period. Technical efficiency refers to the ability to use a minimal amount of input to make a given level of output. If an organization fails to achieve an output combination on its production possibility frontier, and fails beneath this frontier, it can be said to be technically inefficient. Over time the level of output an organization is capable of producing will increase due to technological changes that affect the ability to optimally combine inputs and outputs. These technological changes cause the production possibility frontier to shift upward, as more outputs are obtainable from the same level of inputs. Thus, for any organization in an industry, productivity improvements over time (more outputs for the same or lower level of inputs) may be either technical efficiency improvements (catching up with their own frontier) or technological improvements (the frontier is shifting up over time) or both. Considering the Constant return to scale (CRS) and variable return to scale (VRS) models will result in four efficiencies: technical efficiency change (TEC), technological change (TC), pure technical efficiency change (PTEC), scale efficiency change (SEC), where all of which will be combined together in the total factor productivity change (TFPC).

Plastics are one of the most used materials on a volume basis in countries' industrial and commercial life. Plastics are broadly integrated in today's life style and make a major, irreplaceable contribution to virtually all product areas. A Jordanian company specialized in the production of plastic containers and covers used in food, oil and cosmetics production is interested to assess the performance of its production line during a given period of time. The company started its first production using one injection machine. In order to satisfy the growing demand, it has widened its production with more injection and blowing machines. Currently, the production plant has two types of machines: injection and five blowing machines. The company aims to measure and evaluate the productivity change of five production machines of the same type (blowing machines) M1, M2, M3, M4 and M5 in the company over the period from February to July 2014. This paper, therefore, utilizes Malmquist productivity index to assess the total factor productivity change of the five production machines. The results of this research provide valuable feedback to top managers regarding current improvement decisions and suggest guidelines to future planning. The remaining of this paper including the introduction is organized as follows. Section two describes the data collection and application of MPI. Section three conducts MPI analysis. Section four presents the results and discussions of MPI. The last section concludes the paper.

2. Ease of Use

Figure 2 presents samples of the studied plastic products (covers, pets and containers). The production operation for producing several plastic products in a Jordanian company for plastic industries is depicted in **Figure 3**.

The data were obtained from the production report over a period of six months (February-2014 to July-2014) for both day and night shifts for the five blowing machine; (M1-M5). Data includes the planned production in units (PP), defect quantity in units (DQ), and idle time in units (IT) and are selected as inputs, whereas the actual production quantity in units (PQ) is set output for each period. Each month was divided into two periods; each period consists of two weeks where (H1) represents the first half of the month and (H2) represents the second half of the month. Inputs and outputs data are represented in **Tables 1-5** for M1 to M5, respectively. **Table 6** lists the descriptive statistics of the inputs and the output for both day and night shifts.

3. MPI Analysis

The input-based Malmquist productivity change index described in Equation (1) is used to analyze the perfor-

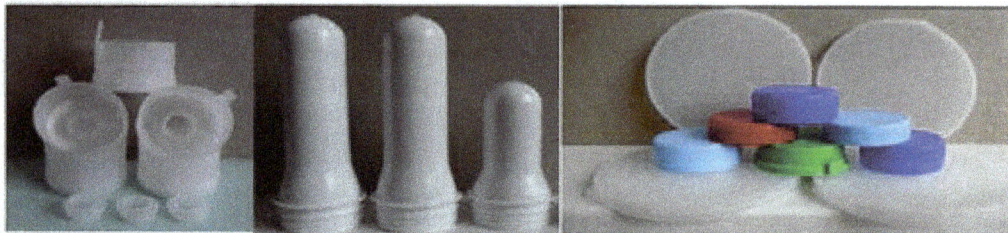

Figure 2. Sample of plastic products.

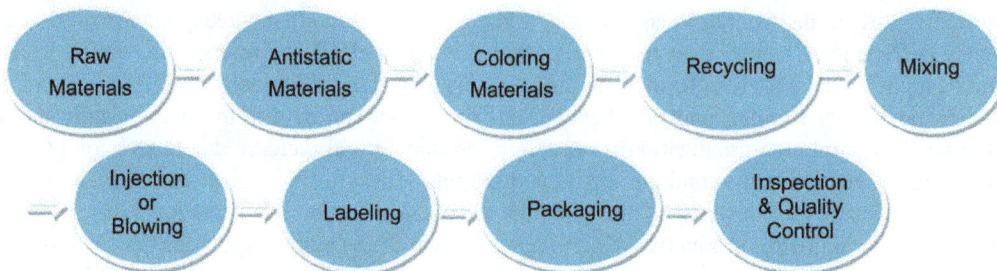

Figure 3. Production operations of plastic products.

Table 1. The inputs and output data for M1.

| Period | Day Shift | | | | Night Shift | | | |
| | Inputs | | | Output | Inputs | | | Output |
	PP (units)	DQ (units)	IT (units)	PQ (units)	PP (units)	DQ (units)	IT (units)	PQ (units)
Feb. H1	24,192	192	1446	22,000	24,192	185	1426	22,300
Feb. H2	24,192	241	8975	14,834	24,192	94	7996	15,731
Mar. H1	24,192	251	5202	18,763	24,192	69	3149	21,419
Mar. H2	24,192	236	8274	15,781	24,192	97	6414	18,359
Apr. H1	20,736	197	2329	18,000	20,736	176	1935	17,221
Apr. H2	27,648	201	569	26,960	27,648	142	51	27,720
May H1	24,192	242	4116	19,984	24,192	120	3420	19,456
May H2	20,736	264	2481	20,042	20,736	53	2081	19,616
Jun. H1	13,824	79	1621	11,955	13,824	32	678	13,500
Jun. H2	10,368	76	3016	7,740	10,368	73	2855	8,318
Jul. H1	15,552	118	3609	12,600	15,552	115	2962	13,350
Jul. H2	22,646	240	3054	19,370	22,464	245	2339	20,750

Table 2. The inputs and output data for M2.

| Period | Day Shift | | | | Night Shift | | | |
| | Inputs | | | Output | Inputs | | | Output |
	PP (units)	DQ (units)	IT (units)	PQ (units)	PP (units)	DQ (units)	IT (units)	PQ (units)
Feb. H1	38,016	201	5105	31,012	38,016	258	2710	33,458
Feb. H2	27,648	151	4778	20,894	25,920	99	4845	20,154
Mar. H1	31,104	203	5817	24,552	31,104	191	2013	28,436
Mar. H2	44,928	514	9736	33,504	41,472	192	7826	33,916
Apr. H1	29,376	328	4770	22,878	29,376	169	4343	23,442
Apr. H2	22,464	293	4491	18,284	22,464	124	4229	19,369
May H1	20,736	183	1417	17,552	17,280	60	23	17,680
May H2	48,384	292	9965	35,496	43,200	87	6906	37,212
Jun. H1	29,376	210	4871	23,393	29,376	122	2498	26,720
Jun. H2	32,832	210	2179	30,880	31,104	173	1826	31,406
Jul. H1	17,280	165	6307	11,721	17,280	222	4182	13,641
Jul. H2	22,464	373	4653	32,441	22,464	393	3785	20,296

mance of the five blowing machines over the period from February to July 2014 in both day and night shift. Firstly, the TE are calculated and presented for the five blowing machines in **Table 7**.

In **Table 7** the TE value (= 0.9771) of the first half of February (Feb. H1) for M1 at day work (M1d1) is calculated using Equation (5) as follows:

$$TE_{M1d1} = \min \theta$$

Table 3. The inputs and output data for M3.

| Period | Day Shift | | | | Night Shift | | | |
| | Inputs | | | Output | Inputs | | | Output |
	PP (units)	DQ (units)	IT (units)	PQ (units)	PP (units)	DQ (units)	IT (units)	PQ (units)
Feb. H1	24,192	189	976	22,516	24,192	178	30	23,400
Feb. H2	24,192	224	284	23,200	24,192	166	30	23,320
Mar. H1	21,492	711	2748	22,004	21,492	825	1298	22,664
Mar. H2	27,648	909	10,385	22,870	27,648	787	9710	23,508
Apr. H1	21,492	822	2054	22,368	24,192	833	3762	22,814
Apr. H2	29,376	923	871	28,442	27,648	414	548	27,496
May H1	8640	288	2358	7320	8640	90	3295	6960
May H2	24,192	716	20,555	17,440	24,192	173	12,759	21,388
Jun. H1	13,824	594	4233	11,320	13,824	118	5230	11,040
Jun. H2	21,492	346	7305	18,074	24,192	159	4524	20,690
Jul. H1	10,368	241	6753	7440	10,368	384	5572	8291
Jul. H2	19,008	601	9266	12,879	19,008	564	9963	13,207

Table 4. The inputs and output data for M4.

| Period | Day Shift | | | | Night Shift | | | |
| | Inputs | | | Output | Inputs | | | Output |
	PP (units)	DQ (units)	IT (units)	PQ (units)	PP (units)	DQ (units)	IT (units)	PQ (units)
Feb. H1	24,192	695	6811	20,288	24,192	516	2731	23,109
Feb. H2	24,192	1008	3080	20,136	24,192	668	345	21,072
Mar. H1	20,736	527	8987	14,464	20,736	560	5036	17,602
Mar. H2	29,376	1069	2786	26,828	29,376	583	1248	28,271
Apr. H1	24,192	991	2882	22,108	24,192	653	1762	22,775
Apr. H2	27,648	983	5475	24,288	27,648	419	5700	24,545
May H1	24,192	800	1098	23,320	24,192	367	919	23,496
May H2	29,376	824	5051	26,356	29,376	274	1982	28,160
Jun. H1	20,736	526	6115	15,595	20,736	228	3454	18,721
Jun. H2	20,736	284	7224	16,131	20,736	226	4793	19,180
Jul. H1	19,008	470	2738	17,006	19,008	604	4457	18,137
Jul. H2	9,612	48	2204	4480	5184	68	17	5112

Subject to:

$$-24{,}192\lambda_1 - 38{,}016\lambda_2 - 24{,}192\lambda_3 - 24{,}192\lambda_4 - 24{,}192\lambda_5 + 24{,}192\theta \geq 0$$

$$-192\lambda_1 - 201\lambda_2 - 189\lambda_3 - 695\lambda_4 - 202\lambda_5 + 192\theta \geq 0$$

$$-1446\lambda_1 - 5105\lambda_2 - 976\lambda_3 - 6811\lambda_4 - 11{,}025\lambda_5 + 1446\theta \geq 0$$

Table 5. The inputs and output data for M5.

Period	Day Shift				Night Shift			
	Inputs			Output	Inputs			Output
	PP (units)	DQ (units)	IT (units)	PQ (units)	PP (units)	DQ (units)	IT (units)	PQ (units)
Feb. H1	24,192	202	11,025	14,997	24,192	119	7857	17,742
Feb. H2	24,192	542	6693	18,568	24,192	372	4979	20,319
Mar. H1	24,192	316	4240	19,254	24,192	208	1981	22,418
Mar. H2	29,376	902	4469	24,869	29,376	247	2688	27,520
Apr. H1	19,008	425	3417	16,365	19,008	360	6354	14,754
Apr. H2	22,464	282	5252	17,645	22,464	101	4230	19,065
May H1	20,736	361	4986	16,908	20,736	119	4554	17,971
May H2	29,376	243	5540	25,712	29,376	69	4118	27,663
Jun. H1	8,640	60	2726	6,134	8,640	31	1444	7463
Jun. H2	27,648	409	3644	25,709	27,648	206	2617	25,677
Jul. H1	22,464	286	7471	15,969	22,464	444	4681	19,616
Jul. H2	22,464	278	4002	20,143	22,464	357	3265	19,651

Table 6. Descriptive statistics of inputs and output.

Statistics item	Maximum	Minimum	Mean	Standard deviation
PP (units)	48,384	5184	23,392	7147.60
DQ (units)	1069	31	338	256.58
IT (units)	20,555	17	4274	3073.62
PQ (units)	37,212	4480	20,151	6648.86

Table 7. The TE values for machines M1-M5.

Period	TE/Day Shift					TE/Night Shift				
	M1	M2	M3	M4	M5	M1	M2	M3	M4	M5
Feb. H1	0.9771	1.0000	1.0000	0.9010	0.6661	0.9530	0.9661	1.0000	0.9876	1.0000
Feb. H2	0.6394	1.0000	1.0000	0.8679	0.8003	0.8320	1.0000	1.0000	0.9036	0.8713
Mar. H1	0.9580	1.0000	1.0000	0.7849	0.9662	1.0000	1.0000	1.0000	0.9088	1.0000
Mar. H2	1.0000	1.0000	0.9298	1.0000	0.9674	1.0000	1.0000	0.8835	1.0000	1.0000
Apr. H1	1.0000	0.8932	1.0000	0.9884	0.9590	1.0000	1.0000	1.0000	1.0000	0.8691
Apr. H2	1.0000	0.8347	0.9929	0.9009	0.8055	1.0000	0.8600	0.9919	0.8855	0.9670
May H1	0.9694	1.0000	0.8789	1.0000	0.9182	0.7860	1.0000	0.7873	0.9493	0.8470
May H2	1.0000	1.0000	0.7459	0.9283	1.0000	1.0000	1.0000	0.9262	1.0000	1.0000
Jun. H1	1.0000	0.9208	0.9469	0.8697	0.8209	1.0000	0.9314	0.8178	0.9245	0.8845
Jun. H2	0.7937	1.0000	0.7943	0.8271	0.9886	0.7946	1.0000	0.8470	0.9161	0.9198
Jul. H1	1.0000	0.8273	0.8087	1.0000	0.8509	1.0000	0.8972	0.8381	1.0000	0.9731
Jul. H2	0.9275	1.0000	0.4692	1.0000	0.8296	1.0000	0.9162	0.7046	1.0000	0.8871

$$22,000\lambda_1 + 31,012\lambda_2 + 22,516\lambda_3 + 20,288\lambda_4 + 14,997\lambda_5 \geq 22,000$$

Similarly, the TE value (= 0.6394) for M1 in Feb. H2 day working (M1d2) is estimated as follows:

$$TE_{M1d2} = \min \theta$$

Subject to:

$$-24,192\lambda_1 - 27,648\lambda_2 - 24,192\lambda_3 - 24,192\lambda_4 - 24,192\lambda_5 + 24,192\theta \geq 0$$

$$-241\lambda_1 - 151\lambda_2 - 224\lambda_3 - 1008\lambda_4 - 542\lambda_5 + 241\theta \geq 0$$

$$-8975\lambda_1 - 4778\lambda_2 - 284\lambda_3 - 3080\lambda_4 - 6693\lambda_5 + 8975\theta \geq 0$$

$$14,834\lambda_1 + 20,894\lambda_2 + 23,200\lambda_3 + 20,136\lambda_4 + 18,568\lambda_5 \geq 14834$$

$$\lambda_j \geq 0, \quad j = 1, \cdots, 5$$

The TE values of the other periods for M1 at day and night shifts are estimated similarly. The TE values at day and night shifts for the other blowing machines (M2-M5) over the period from February-July 2014 are calculated in a similar manner and are also presented in **Table 7**.

Secondly, the $IEI^{t \to t+1}$ and $IEI^{t+1 \to t}$ for the five blowing machines are calculated and then the results are presented in **Table 8**. For example, the $IEI^{Feb.H1 \to Feb.H2}$ (=1.0346) and $IEI^{Feb.H2 \to Feb.H1}$ (=0.6588) for M1d are calculated using the input and output data shown in **Table 9**. Mathematically,

Table 8. The estimated $IEI^{t \to t+1}, IEI^{t+1 \to t}$ values for machines M1-M5.

Period	$IEI^{t \to t+1}, IEI^{t+1 \to t}$ /Day Shift					$IEI^{t \to t+1}, IEI^{t+1 \to t}$ /Night Shift				
	M1	M2	M3	M4	M5	M1	M2	M3	M4	M5
$IEI^{Feb.H1 \to Feb.H2}$	1.0346	1.1758	1.1027	0.8745	0.6810	0.9563	0.9174	1.0034	0.9910	0.8680
$IEI^{Feb.H2 \to Feb.H1}$	0.6588	0.9175	3.5410	0.8943	0.8247	1.1225	1.4087	1.0686	0.9005	0.8683
$IEI^{Feb.H2 \to Mar.H1}$	0.7594	1.1441	10.2020	0.9151	0.8794	0.7270	0.8699	44.5190	3.4980	0.9037
$IEI^{Mar.H1 \to Feb.H2}$	0.8088	0.9712	0.9484	0.7274	0.8299	1.5575	0.9940	0.9719	0.8806	0.9613
$IEI^{Mar.H1 \to Mar.H2}$	1.1435	1.8270	1.0980	0.8308	1.1802	1.7335	1.3665	0.9735	0.8820	1.0604
$IEI^{Mar.H2 \to Mar.H1}$	0.8090	0.9148	0.9094	1.2026	0.9308	0.8474	0.9077	0.9098	1.3794	1.0120
$IEI^{Mar.H2 \to Apr.H1}$	0.7510	0.8547	0.8971	0.9877	0.9232	1.3645	1.2735	0.9029	1.7526	1.1486
$IEI^{Apr.H1 \to Mar.H2}$	1.9150	1.2825	1.1309	1.0007	1.0512	0.8863	0.9649	0.9799	0.9782	0.8082
$IEI^{Apr.H1 \to Apr.H2}$	0.8902	0.7987	0.9482	0.9372	0.8829	0.8283	0.7959	0.9405	0.9390	0.7742
$IEI^{Apr.H2 \to Apr.H1}$	4.5473	0.9289	2.9986	0.9501	0.8976	42.0504	1.1261	3.8818	1.0199	1.3608
$IEI^{Apr.H2 \to MayH1}$	3.0187	0.9389	1.5375	0.9113	0.9086	0.9799	0.8427	0.9720	0.8677	0.8295
$IEI^{MayH1 \to Apr.H2}$	0.8471	0.8681	0.8688	0.9886	0.8362	0.8306	1.5095	0.8035	0.9687	0.8644
$IEI^{MayH1 \to MayH2}$	0.9071	1.5334	0.8766	2.6291	0.8436	0.8461	54.1035	0.8403	1.7995	0.9104
$IEI^{MayH2 \to MayH1}$	1.1169	1.2674	0.7607	0.9548	1.1032	1.2560	1.4516	0.8641	0.9369	1.3606
$IEI^{MayH2 \to Jun.H1}$	1.1176	0.8483	0.8336	1.0375	1.0121	0.9687	1.0139	0.9053	0.9816	0.9643
$IEI^{Jun.H1 \to MayH2}$	1.5024	1.0473	0.8472	0.7781	0.8973	1.2924	1.0268	0.8344	0.9418	0.9112
$IEI^{Jun.H1 \to Jun.H2}$	1.0291	0.8467	0.8706	0.7996	0.7548	2.3239	1.2065	0.7909	0.8941	1.3261
$IEI^{Jun.H2 \to Jun.H1}$	0.8632	1.5399	0.8639	0.8995	1.0752	0.8215	1.0339	0.8758	0.9472	0.9510
$IEI^{Jun.H2 \to Jul.H1}$	0.9538	2.6031	0.8859	0.9258	1.3976	0.9816	3.8161	1.1209	1.0605	2.1769
$IEI^{Jul.H1 \to Jun.H2}$	0.8614	0.7212	0.7630	0.9512	0.7558	0.8502	07818	0.7920	0.9450	0.8648
$IEI^{Jul.H1 \to Jul.H2}$	1.1810	0.7862	0.4969	0.8909	0.6319	1.3707	0.8068	0.8109	0.9676	0.8855
$IEI^{Jul.H2 \to Jul.H1}$	1.4449	1.6898	0.7573	0.8741	1.1953	1.9683	1.1897	0.7345	66.7184	1.3354

Table 9. Inputs and output data for calculation of $IEI^{Feb.H1 \rightarrow Feb.H2}$ and $IEI^{Feb.H2 \rightarrow Feb.H1}$.

	Inputs/Output	M1	M2	M3	M4	M5
Feb. H1	PP	24,192	38,016	24,192	24,192	24,192
	DQ	192	201	189	569	202
	IT	1446	5105	976	6811	11,025
	PQ	22,000	31,012	22,516	20,288	14,997
Feb. H2	PP	24,192	27,648	24,192	24,192	24,192
	DQ	241	151	224	1008	542
	IT	8975	4778	284	3080	6693
	PQ	14,834	20,894	23,200	20,136	18,568

$$IEI^{Feb.H1 \rightarrow Feb.H2} = \min \theta$$

Subject to:

$$-24,192\lambda_1 - 27,648\lambda_2 - 24,192\lambda_3 - 24,192\lambda_4 - 24,192\lambda_5 + 24,192\theta \geq 0$$

$$-241\lambda_1 - 151\lambda_2 - 224\lambda_3 - 1,008\lambda_4 - 542\lambda_5 + 192\theta \geq 0$$

$$-8975\lambda_1 - 4778\lambda_2 - 284\lambda_3 - 3080\lambda_4 - 6693\lambda_5 + 1446\theta \geq 0$$

$$14,834\lambda_1 + 20,894\lambda_2 + 23,200\lambda_3 + 20,136\lambda_4 + 18,568\lambda_5 \geq 22,000$$

$$\lambda_j \geq 0, \quad j = 1, \cdots, 5$$

$$IEI^{Feb.H1 \rightarrow Feb.H2} = \min \theta$$

Subject to:

$$-24,192\lambda_1 - 38,016\lambda_2 - 24,192\lambda_3 - 24,192\lambda_4 - 24,192\lambda_5 + 24,192\theta \geq 0$$

$$-192\lambda_1 - 201\lambda_2 - 189\lambda_3 - 569\lambda_4 - 202\lambda_5 + 241\theta \geq 0$$

$$-1446\lambda_1 - 5105\lambda_2 - 976\lambda_3 - 6811\lambda_4 - 11,025\lambda_5 + 8975\theta \geq 0$$

$$22,000\lambda_1 + 31,012\lambda_2 + 22,516\lambda_3 + 20,288\lambda_4 + 14,997\lambda_5 \geq 14,834$$

$$\lambda_j \geq 0, \quad j = 1, \cdots, 5$$

The remaining $IEI^{t \rightarrow t+1}$ and $IEI^{t+1 \rightarrow t}$ for M1 and the other blowing machines (M2-M5) over the period from February-July 2014 are calculated in a similar manner and then presented in **Table 8** for both day and night shifts.

Utilizing the results displayed in **Table 7** and **Table 8**, the MPI values are calculated as follows. At beginning, the values of technical efficiency change (TEC), which measures the change in efficiency between current (t) and next ($t + 1$) periods, are estimated.

For example, the TEC value between the first half of February (Feb. H1) and the second half of February (Feb. H2) for M1d is calculated by applying Equation (4) as follows:

$$TEC = \frac{TE(\text{Feb. H2})}{TE(\text{Feb. H1})} = \frac{0.6394}{0.9771} = 0.6544$$

The TEC values of M1 for the other periods and the blowing machines from M1-M5 are calculated then presented in **Table 10**. Next, the technological change (TC) values are calculated for all periods. The TC is the development of new products or the development of new technologies that allows methods of production to improve and results in the shifting upwards of the production frontier as more outputs are obtainable from the same level of inputs. The TC includes new production processes, called process innovation, and the discovery of new

Table 10. The calculated TEC, TC and MPI values at day shift.

Machine	M1			M2			M3			M4			M5		
Period	TEC	TC	MPI	TEC	TC	MPI	TEC	TC	MPI	TEC	TC	MPI	TEC	TC	MPI
Feb. H1-Feb. H2	0.6544	0.9864	0.6455	1.0000	0.8833	0.8833	1.0000	1.7920	1.7920	0.9632	1.0304	0.9925	1.2016	1.0039	1.2063
Feb. H2-Mar. H1	1.4983	0.8431	1.2632	1.0000	0.9214	0.9214	1.0000	0.3049	0.3049	0.9044	0.9375	0.8478	1.2072	0.8842	1.0674
Mar. H1-Mar. H2	1.0438	0.8233	0.8593	1.0000	0.7076	0.7076	0.9298	0.9438	0.8775	1.2740	1.0659	1.3580	1.0012	0.8875	0.8886
Mar. H2-Apr. H1	1.0000	1.5969	1.5969	0.8932	1.2961	1.1577	1.0755	1.0826	1.1644	0.9884	1.0124	1.0007	0.9913	1.0718	1.0624
Apr. H1-Apr. H2	1.0000	2.2601	2.2601	0.9345	1.1156	1.0425	0.9929	1.7846	1.7720	0.9115	1.0546	0.9613	0.8400	1.1001	0.9241
Apr. H2-May H1	0.9694	0.5381	0.5216	1.1980	0.8785	1.0524	0.8852	0.7990	0.7073	1.1101	0.9886	1.0973	1.1399	0.8986	1.0242
May H1-May H2	1.0316	1.0925	1.1270	1.0000	0.9092	0.9092	0.8486	1.0112	0.8582	0.9283	0.6255	0.5806	1.0891	1.0958	1.1934
May H2-Jun. H1	1.0000	1.1594	1.1594	0.9208	1.1579	1.0662	1.2695	0.8948	1.1359	0.9369	0.8947	0.8383	0.8209	1.0392	0.8531
Jun. H1-Jun. H2	0.7937	1.0280	0.8160	1.0860	1.2941	1.4054	0.8389	1.0876	0.9124	0.9511	1.0876	1.0344	1.2043	1.0876	1.3098
Jun. H2-Jul. H1	1.2599	0.8467	1.0667	0.8273	0.5787	0.4788	1.0181	0.9197	0.9363	1.2090	0.9219	1.1146	0.8606	0.7927	0.6822
Jul. H1-Jul. H2	0.9275	1.1486	1.0653	1.2087	1.3335	1.6118	0.5802	1.6208	0.9403	1.0000	0.9905	0.9905	0.9751	1.3928	1.3581
Geo. Avg.	0.9953	1.0485	1.0435	1.0000	0.9764	0.9764	0.9335	1.0142	0.9468	1.0095	0.9552	0.9643	1.0202	1.0121	1.0325
Std. dev.	0.2200	0.4620	0.4794	0.1190	0.2508	0.3086	0.1713	0.4524	0.4328	0.1257	0.1280	0.1938	0.1485	0.1617	0.2060
CV	0.2211	0.4407	0.4594	0.1190	0.2568	0.3161	0.1835	0.4461	0.4572	0.1245	0.1340	0.2010	0.1456	0.1597	0.1995

products called product innovation. For illustration, the TC value (= 0.9864) between the first half of February (Feb. H1) and the second half of February (Feb. H2) for M1d is calculated by applying Equation (6) as the following:

$$TC = \left[\frac{TE(\text{Feb. H1}) \times IEI^{Feb.H2 \rightarrow Feb.H1}}{IEI^{Feb.H1 \rightarrow Feb.H2} \times TE(\text{Feb. H2})} \right]^{1/2} = \left[\frac{0.9771 \times 0.6588}{1.0346 \times 0.6394} \right]^{1/2} = 0.9864$$

The calculated TC values for the blowing machines from M1-M5 are calculated in a similar manner and are also presented in **Table 10**. Finally, the MPI is used to measure the productivity change of a DMU over time and is calculated by the multiplication of TEC and TC of the same period. For example, the MPI value (= 0.6544) between the first half of February (Feb. H1) and the second half of February (Feb. H2) for M1d is calculated using Equation (2) as follows:

$$MPI = TEC \times TC = 0.6544 \times 0.9864 = 0.6455$$

The MPI values for all of the blowing machines at day shift from (M1-M5) are calculated in a similar manner and are displayed in **Table 10**. Moreover, the TC, TEC and MPI values for all of the blowing machines (M1-M5) are calculated using the data of night shift and then shown in **Table 11**.

4. Results and Discussions of MPI

4.1. Results and Discussion of TEC, TC and MPI

From **Table 10** and **Table 11**, which present the results of TEC, TC and MPI for the five blowing machines in day and night shifts, the following results are obtained:

In both tables, the coefficient of variation (CV), which equals geometric average divided by standard deviation, is larger than (5%) for TC, TEC and MPI in all machines in day shift. This result indicates that the dispersion is significant and there is a trend in TC, TEC, and MPI. For illustration, the minimum value of TEC for M1 during the whole period in the day shift was (0.6544), the maximum value for it during the whole period was (1.4983), the standard deviation was (0.2200) and the coefficient of variation was larger than (5%) which means

Table 11. The estimated TEC, TC and MPI values at night shift.

Machine	M1			M2			M3			M4			M5		
Period	TEC	TC	MPI	TEC	TC	MPI	TEC	TC	MPI	TEC	TC	MPI	TEC	TC	MPI
Feb. H1-Feb. H2	0.8730	1.1596	1.0123	1.0351	1.2180	1.2607	1.0000	1.0320	1.0320	0.9150	0.9966	0.9119	0.8713	1.0715	0.9336
Feb. H2-Mar. H1	1.2020	1.3351	1.6047	1.0000	1.0690	1.0690	1.0000	0.1478	0.1478	1.0058	0.5003	0.5032	1.1477	0.9627	1.1049
Mar. H1-Mar. H2	1.0000	0.6992	0.6992	1.0000	0.8150	0.8150	0.8835	1.0285	0.9087	1.1003	1.1922	1.3118	1.0000	0.9769	0.9769
Mar. H2-Apr. H1	1.0000	0.8060	0.8060	1.0000	0.8704	0.8704	1.1319	0.9792	1.1084	1.0000	0.7471	0.7471	0.8691	0.8998	0.7820
Apr. H1-Apr. H2	1.0000	0.7125	0.7125	0.8600	1.2826	1.1031	0.9919	2.0398	2.0233	0.8855	1.1075	0.9807	1.1127	1.2569	1.3985
Apr. H2-May H1	0.7860	1.0384	0.8162	1.1628	1.2411	1.4432	0.7937	1.0205	0.8100	1.0720	1.0205	1.0940	0.8760	1.0907	0.9554
May H1-May H2	1.2722	1.0802	1.3742	1.0000	0.1638	0.1638	1.1764	0.9349	1.0998	1.0535	0.7030	0.7406	1.1806	1.1251	1.3283
May H2-Jun. H1	1.0000	1.4094	1.4094	0.9314	1.0428	0.9712	0.8829	1.0217	0.9021	0.9245	1.0187	0.9418	0.8845	1.0336	0.9142
Jun. H1-Jun. H2	0.7946	0.6670	0.5300	1.0736	0.8934	0.9592	1.0358	1.0339	1.0709	0.9909	1.0339	1.0245	1.0399	0.8304	0.8636
Jun. H2-Jul. H1	1.2586	0.8296	1.0441	0.8972	0.4779	0.4287	0.9894	0.8450	0.8361	1.0916	0.9035	0.9863	1.0579	0.6128	0.6483
Jul. H1-Jul. H2	1.0000	1.1983	1.1983	1.0212	1.2017	1.2271	0.8407	1.0380	0.8727	1.0000	0.8304	0.8304	0.9117	1.2861	1.1725
Geo. Avg.	1.0044	0.9618	0.9660	0.9952	0.8300	0.8260	0.9687	0.8898	0.8620	1.0011	1.0977	1.0990	0.9892	0.9950	0.9842
Std. dev.	0.1681	0.2651	0.3430	0.0827	0.3492	0.3698	0.1176	0.4290	0.4353	0.0725	0.2022	0.2114	0.1192	0.1915	0.2260
CV	0.1674	0.2757	0.3550	0.0831	0.4207	0.4477	0.1214	0.4821	0.5050	0.0724	0.1842	0.1924	0.1205	0.1925	0.2297

that the dispersion is significant and there is a trend in TEC for M1 in day shift.

In **Table 10**, the M1 has the largest geometric average of MPI (1.0435) with a growth of 4.35% among the five machines in the day shift. This productivity increase is entirely attributed to technological change growth of 4.85% (1 - 1.0485), because the mean technical efficiency regresses by 0.47% (1 - 0.9953) over the whole period. M2 has a geometric average MPI decrease of 2.36% over the same period, this productivity decrease was entirely attributed to technological change regress of 2.36%, while the mean technical efficiency change held constant.

M3 corresponds to the lowest geometric average of MPI over the five blowing machines, it performed the worst with aggregate decrease of 5.32% over this period in the day shift; this productivity decrease stems from the poor performance in technical efficiency change with a regress of 6.65%, while the technological change had a growth of 1.42%. M4 has also geometric average of MPI decrease of 3.57%, over the same period, this productivity decrease is attributed to technological change regress of 4.48%, while the technical efficiency change had a growth of 0.95%. Finally M5 had a large geometric average MPI growth of 3.25% but this growth was lower than the one for M1, this productivity increase was attributed almost equally to both technical efficiency change and technological change of 2.02% and 3.25%, respectively. In **Table 11**, M4 has the largest geometric average of MPI with a growth of 9.9%, among the five machines in the night shift. This productivity increase is attributed to a growth of both technical efficiency change and technological change of 0.11% and 9.9%, respectively. However, M2 has the lowest geometric average of MPI over the five blowing machines it performed the worst with aggregate decrease of 17.4% over the same period in the night shift; this productivity decrease stemmed from the poor performance of both technical efficiency change with a regress of 0.48%, and technological change regress of 17%. M1 has geometric average MPI regress of 3.4% over the same period; this productivity decrease is attributed to technological change regress of 3.82%, while there is a mean technical efficiency growth of 0.44%. M3 has also geometric average of MPI decrease of 13.8% over the same period, this productivity decrease is attributed to both technical efficiency change regress of 3.13% and technological change regress of 11.02%. Finally, M5 has a geometric average of MPI decrease attributed to both technical efficiency change and technological change regress of 1.08% and 0.5%, respectively.

4.2. Comparison of Results between Day and Night Shifts

Figure 4 and **Figure 5** represent the geometric average value of TEC for the five blowing machines in the day

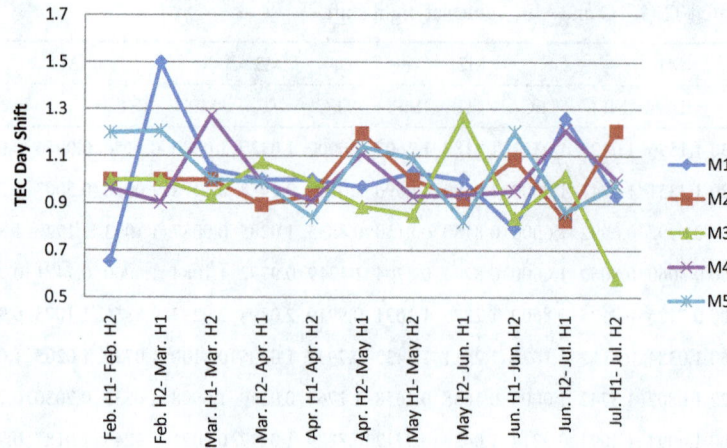

Figure 4. The comparison of TEC values for blowing Machines at day shift.

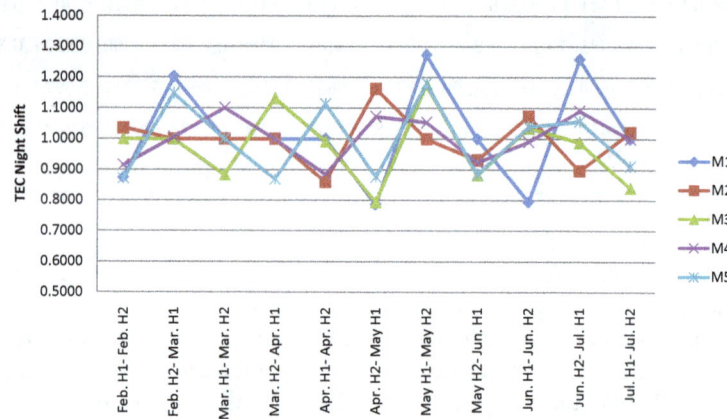

Figure 5. The comparison of TEC values for blowing Machines at night shift.

and night shifts over the period from February-July 2014, respectively. It is seen in **Figure 4** that M5 has the largest geometric average value of TEC (=1.0202) over the period February-July 2014, followed by M4 and M2 with values of 1.009 and 1.0000, respectively. Hence, M2 and M4 are considered efficient. However, M1 and M3 have average values, which are less than one, of 0.9953 and 0.9335, respectively. For the night shift, the geometric average values of TEC in **Figure 5** for M1 and M4 are 1.0044 and 1.0011, respectively. However, average values of TEC for M2, M3, and M5 are 0.9952, 0.9687, and 0.9892, respectively. Consequently, M1 and M4 are considered efficient, whereas M2, M3, and M5 are considered inefficient.

In **Figure 6** for the day shift, the geometric average values of TC for M1 (1.0485), M3 (1.0142), and M5 (=1.0121) are greater than one. However, the average values of TC for M1 (0.9764) and M4 (0.9552) are smaller than one. For the night shift as shown in **Figure 7**, only M4 has a geometric average value of TC (1.0977) larger than one over the period February-July 2014. In contrast, average values of TC for M5 (0.9950), M1 (0.9618), M3 (0.8898), and M2 (0.8300) are smaller than one. Finally, the geometric average value of MPI for the five blowing machines at day and night shifts are depicted in **Figure 8** and **Figure 9**, respectively.

Table 12 represents the values for TEC, TC and MPI progress or regress for the five blowing machines over the period (February-July 2014) for both day and night shifts, the number between two parentheses represent the values of TEC, TC and MPI for the night shift while the other numbers represent the values of TEC, TC and MPI for the day shift. The results in **Table 12** provide valuable feedback to production/planning managers in setting proactive/corrective actions and improvement plans. Finally, comparisons are conducted between MPI values for each of the blowing machines (M-M5) at day and night shifts as shown in **Figure 10**. The differences of the MPI values between day and night shifts are displayed in **Table 13** for all machines.

In **Table 13** for M1, it is noticed that there are significant MPI differences between day and night shifts in all

Table 12. The calculated TEC, TC and MPI values for the five blowing machines at day (night) shift.

Period / Machine		Feb. H1-Feb. H2	Feb. H2-Mar. H1	Mar. H1-Mar. H2	Mar. H2-Apr. H1	Apr. H1-Apr. H2	Apr. H2-May H1	May H1-May H2	May H2-Jun. H1	Jun. H1-Jun. H2	Jun. H2-Jul. H1	Jul. H1-Jul. H2
M1	TEC	−34.56% (−12.7%)	49.83% (20.20%)	4.38% (0%)	0% (0%)	0% (0%)	−3.06% (−21.4%)	3.16% (27.22%)	0% (0%)	−20.63% (−20.54%)	25.99% (25.86%)	−7.25% (0%)
	TC	−1.36% (15.96%)	−15.69% (33.51%)	−17.67% (−30.08%)	59.69% (−19.4%)	126.01% (−28.75%)	−46.19% (3.84%)	9.25% (8.02%)	15.94% (40.94%)	2.8% (−33.3%)	−15.33% (−17.04%)	14.86% (19.83%)
	MPI	−35.45% (1.23%)	26.32% (60.47%)	−14.07% (−30.08%)	59.69% (−19.4%)	126.01% (−28.75%)	−47.84% (−18.38%)	12.7% (37.42%)	15.94% (40.94%)	−18.4% (−47%)	6.67% (4.41%)	6.53% (19.83%)
M2	TEC	0% (3.51%)	0% (0%)	0% (0%)	−10.68% (0%)	−6.55% (−14%)	19.80% (16.28%)	0% (0%)	−7.92% (6.86%)	8.6% (7.36%)	−17.27% (−10.28%)	20.87% (2.12%)
	TC	−11.67% (21.80%)	−7.86% (6.90%)	−29.24% (−18.5%)	29.61% (−12.96%)	11.56% (28.26%)	−12.15% (24.11%)	−9.08% (−83.62%)	15.79% (4.28%)	29.41% (−10.66%)	−42.13% (−52.21%)	33.35% (20.17%)
	MPI	−11.67% (26.07%)	−7.86% (6.90%)	−29.24% (−18.5%)	15.77% (−12.96%)	4.25% (10.31%)	5.24% (44.32%)	−9.08% (−83.62%)	6.62% (−2.88%)	40.54% (−4.08%)	−52.12 (−57.13%)	61.18% (22.71%)
M3	TEC	0% (0%)	0% (0%)	−7.02% (−11.65%)	7.55% (13.19%)	−0.71% (−0.81%)	−11.48% (−20.63%)	−15.14% (17.64%)	26.95% (−11.71%)	−16.11% (3.58%)	1.81% (−1.06%)	−41.98% (−15.93%)
	TC	79.20% (3.20%)	−69.51% (−85.22%)	−5.62% (2.85%)	8.26% (2.08%)	78.46% (103.98%)	−20.1% (2.05%)	1.12% (6.51%)	−10.52% (2.17%)	8.76% (3.39%)	8.03% (−15.5%)	62.08% (3.80%)
	MPI	79.20% (3.20%)	−69.51% (−85.22%)	−12.25% (−9.13%)	16.44% (10.84%)	77.20% (103.98%)	29.27% (−19%)	−14.18% (9.98%)	13.59% (−9.79%)	−8.76% (7.09%)	−6.37% (−16.39%)	−5.97% (−12.73%)
M4	TEC	−3.68% (−8.5%)	−9.56% (0.58%)	27.40% (10.03%)	−1.16% (0%)	−8.85% (−11.45%)	11.01% (7.20%)	−7.17% (5.35%)	−6.31% (−7.55%)	−4.89% (−0.91%)	20.90% (9.16%)	0% (0%)
	TC	3.04% (−0.34%)	−6.25% (−49.97%)	6.59% (19.22%)	1.24% (−25.29%)	5.46% (10.75%)	−1.14% (2.05%)	−37.45% (−29.7%)	−10.53% (1.87%)	8.76% (3.39%)	−7.81% (−9.65%)	−0.95% (−16.96%)
	MPI	−0.75% (−8.81%)	−15.22% (−49.68%)	35.80% (31.18%)	0.07% (−25.29%)	−3.87% (−1.93%)	9.73% (9.40%)	−41.94% (−25.94%)	−16.17% (−5.82%)	3.44% (2.45%)	11.46% (−1.37%)	0.95% (−16.96%)
M5	TEC	20.16% (−12.87%)	20.72% (14.77%)	0.12% (0%)	−0.87% (−13.09%)	−16% (11.27%)	13.99% (−12.40%)	8.91% (18.06%)	−17.91% (−11.55%)	20.43% (3.99%)	−13.94 (5.79%)	−2.49% (−8.83%)
	TC	0.39% (7.15%)	−11.58% (−3.73%)	−11.25% (−2.31%)	7.18% (−10.02%)	10.01% (25.69%)	−10.14% (9.07%)	9.58% (12.51%)	3.92% (3.36%)	8.76% (−16.96%)	−20.73% (−38.72%)	39.28% (28.61%)
	MPI	20.63% (−6.64%)	6.74% (10.49%)	−11.14% (−2.31%)	6.24% (−21.8%)	−7.59% (39.85%)	2.42% (−4.46%)	19.34% (32.83%)	−14.69% (−8.58%)	30.98% (−13.64%)	−31.78% (−35.18%)	35.81% (17.25%)

Table 13. The MPI differences between day and night shifts.

Period	M1	M2	M3	M4	M5
Feb. H1-Feb. H2	−0.3668	−0.3774	0.7600	0.0806	0.2727
Feb. H2-Mar. H1	−0.3415	−0.1476	0.1571	0.3446	−0.0375
Mar. H1-Mar. H2	0.1601	−0.1074	−0.0312	0.0462	−0.0883
Mar. H2-Apr. H1	0.7909	0.2873	0.0560	0.2536	0.2804
Apr. H1-Apr. H2	1.5476	−0.0606	−0.2513	−0.0194	−0.4744
Apr. H2-May H1	−0.2946	−0.3908	−0.1027	0.0033	0.0688
May H1-May H2	−0.2472	0.7454	−0.2416	−0.1600	−0.1349
May H2-Jun. H1	−0.2500	0.0950	0.2338	−0.1035	−0.0611
Jun. H1-Jun. H2	0.2860	0.4462	−0.1585	0.0099	0.4462
Jun. H2-Jul. H1	0.0226	0.0501	0.1002	0.1283	0.0339
Jul. H1-Jul. H2	−0.1330	0.3847	0.0676	0.1601	0.1856
Max dif.	1.5476	0.7454	0.76	0.3446	0.4462
Min dif.	−0.3668	−0.3908	−0.2513	−0.16	−0.4744

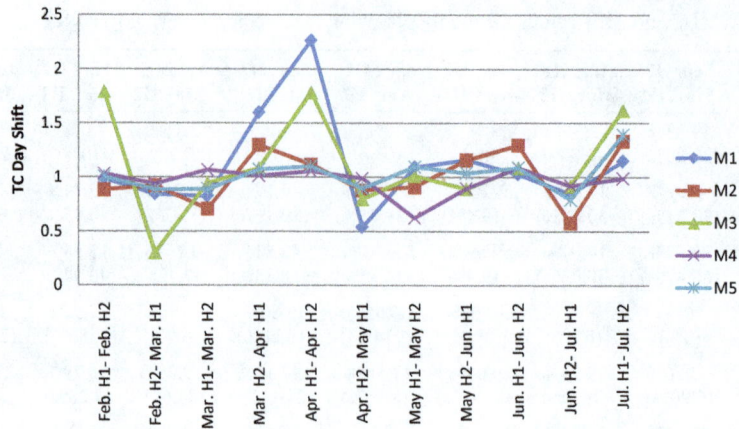

Figure 6. The comparison of TC values for blowing machines at day shift.

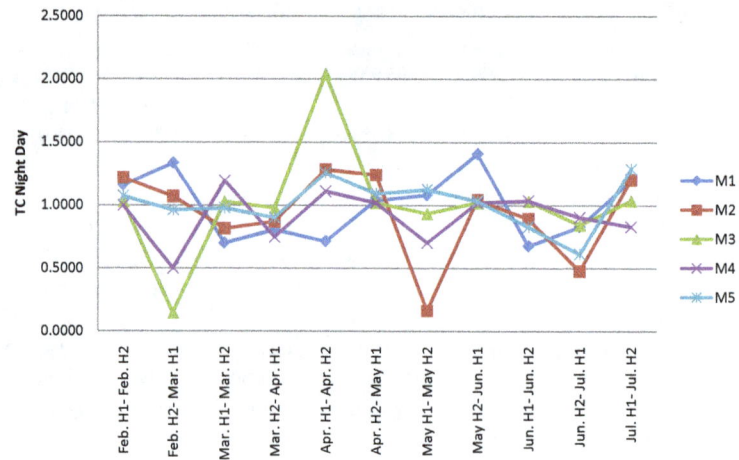

Figure 7. The comparison of TC values for blowing machines at night shift.

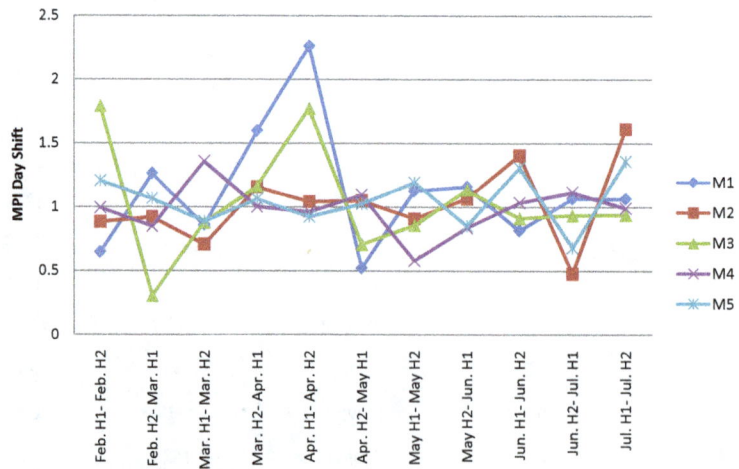

Figure 8. The estimated MPI values at the day shift.

periods; except in Jun. H2-Jul.H1 (0.0226). The largest (smallest) difference is 1.5476 (−1.3668) which corresponds to period Apr. H1-Apr. H2 (Feb. H1-Feb. H2). The MPI values for the day shifts increases from −0.3668 to 1.5476 in Feb. H1-Feb. H2 to Apr. H1-Apr. H2. That is, there exists a regress in the performance of the

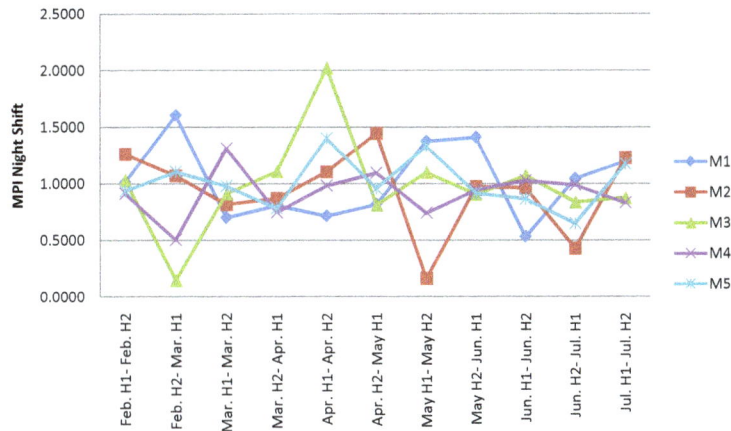

Figure 9. The estimated MPI values at the night shift.

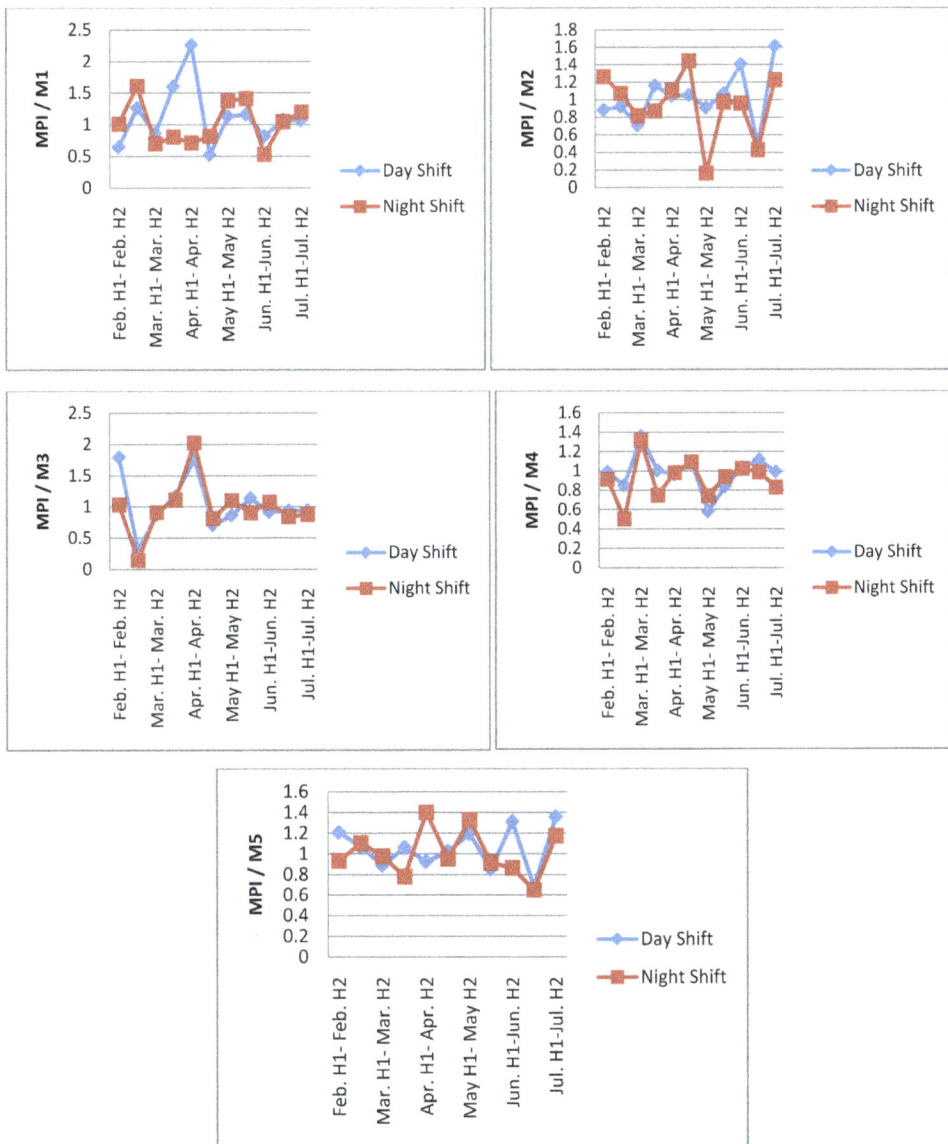

Figure 10. The comparison of MPI values between day and night shifts for each machine.

night shift compared with day shift. Then, the MPI values for the night shift outperform these for the day shift during Apr. H2-May H1 to May H2-Jun. H1. Finally, the difference decreases from Jun. H1-Jun. H2 to Jul. H1-Jul. H2. For M2-M5, the differences can be analyzed similarly. Moreover, slight MPI differences are observed between day and night shifts in most periods for M3 to M5.

4.3. Implications

Technical efficiency change (TEC) refers to the ability to use a minimal amount of planned production, defect quantity and the idle time to make a given level of production quantity. If the company fails to achieve the output combination on its production possibility frontier, and falls beneath this frontier, it is considered technically inefficient. TEC can make use of existing labor, capital, and other economic inputs to produce more output of the same inputs. As more work experience is gained about production, they become more and more efficient. As a result, minor modifications to plant and procedures can contribute to higher levels of productivity.

Further, training new employees and exchanging the experience between experienced employees and newly hired employees has a great influence on productivity improvement. Furthermore, management should revise the hiring policy, incentive programs, and promotion rules to control the employees' turnover rate. Finally, having a reliable quality control system in the company will assure having lower values of defect quantity (DQ). For, the idle time (IT) should be minimized. Interruptions can be caused by confusing or unclear work instructions, incomplete bill of materials, or running out of material. Hence, improving machine reliability and quality using total productive maintenance and quality tools, reducing overproduction and excess inventory, and implementing effective operating procedures help in reducing idle time. On the other hand, technological change (TC) is the development of new products or the development of new technologies that allows methods of production to improve and results in the shifting upwards of the production frontier, as more outputs are obtainable from the same level of inputs. More specifically, technological change includes new production processes, called process innovation and the discovery of new products called product innovation. With process innovation, firms figure out more efficient ways of making existing products allowing output to grow at a faster rate than economic inputs are growing, In the production machines a process innovation entails machines producing more actual production quantity (PQ) at a faster rate than defect quantity (DQ) or idle time in units (IT). Cost of production decline overtime process innovations.

5. Conclusion

This paper assesses the performance of blowing process for plastic industries using Malmquist Index approach during the period from February to July 2014 for both day and night shifts. Five blowing machines are studied. Two primary issues are addressed in the computation of Malmquist indices of productivity growth. The changes in productivity are divided into technical efficiency change (TEC) and technological change (TC). Data include the planned production in units (PP), defect quantity in units (DQ), and idle time in units (IT) and are selected as inputs, whereas the actual production quantity in units (PQ). Inefficiency is observed in each period. The percentage of input utilization is determined in all periods. Then, the Malmquist productivity index (MPI) values are calculated for all periods. Finally, comparisons of TEC, TC and MPI are conducted among the five machines and between the day and night shifts for each machine. It is concluded that: (1) to improve the technical efficiency, a need for internal training, effective operating procedures, and enhancing quality is required, (2) to increase technological change, figuring out more efficient ways of making existing products or using new technologies allowing output to grow at a faster rate than economic inputs is needed, and (3) with the Malmquist productivity index analysis the company is now able to assess the productivity change of the production machines over time. The results of this research will also help decision makers identify the possible causes of decline in productivity within each production machine and guide them in appropriate proactive/corrective plans.

References

[1]   Cooper, W.W., Seiford, L.M. and Kaoru, T. (2000) Data Envelopment Analysis. Kluwer, Boston, MA.

[2]   Charnes, A., Cooper, L.M., Letwin, A.A., *et al.* (1989) Data Envelopment Analysis. Kluwer, Dordtrecht, The Netherlands.

[3]   Al-Refaie, A. (2011) Optimising Correlated QCHs in Robust Design Using Principal Components Analysis and DEA Techniques. *Production Planning and Control*, **22**, 676-689. http://dx.doi.org/10.1080/09537287.2010.526652

[4]   Al-Refaie, A. (2009) Optimizing SMT Performance Using Comparisons of Efficiency between Different Systems Technique in DEA. *IEEE Transactions on Electronic Packaging Manufacturing*, **32**, 256-264. http://dx.doi.org/10.1109/TEPM.2009.2029238

[5]   Al-Refaie, A. (2010) Super-Efficiency DEA Approach for Optimizing Multiple Quality Characteristics in Parameter Design. *International Journal of Artificial Life Research*, **1**, 58-71. http://dx.doi.org/10.4018/jalr.2010040105

[6]   Al-Refaie, A. and Al-Tahat, M. (2011) Solving the Multi-Response Problem in Taguchi Method by Benevolent Formulation in DEA. *Journal of Intelligent Manufacturing*, **22**, 505-521. http://dx.doi.org/10.1007/s10845-009-0312-8

[7]   Al-Refaie, A., Fouad, R., Li, M.-H. and Shurrab, M. (2014) Applying Simulation and DEA to Improve Performance of Emergency Department in a Jordanian Hospital. *Simulation Modeling Practice and Theory*, **41**, 59-72. http://dx.doi.org/10.1016/j.simpat.2013.11.010

[8]   Chen, C.J., Wu, H.L. and Lin, B.W. (2006) Evaluating the Development of High-Tech Industries: Taiwan's Science Park. *Technological Forecasting & Social Change*, **73**, 452-465. http://dx.doi.org/10.1016/j.techfore.2005.04.003

[9]   Coelli, T.J. and Rao, D.S. (2005) Total Factor Productivity Growth in Agriculture: A Malmquist Index Analysis of 93 Countries, 1980-2000. *Agricultural Economics*, **32**, 115-134. http://dx.doi.org/10.1111/j.0169-5150.2004.00018.x

[10]   Odeck, J. (2000) Assessing the Relative Efficiency and Productivity Growth of Vehicle Inspection Services: An Application of DEA and Malmquist Indices. *European Journal of Operational Research*, **126**, 501-514. http://dx.doi.org/10.1016/S0377-2217(99)00305-7

[11]   Worthington, A. (1999) Malmquist Indices of Productivity Change in Australian Financial Services. *Journal of International Financial Markets, Institutions and Money*, **9**, 303-320. http://dx.doi.org/10.1016/S1042-4431(99)00013-X

[12]   Xue, X., Shen, Q., Wang, Y. and Lu, J. (2008) Measuring the Productivity of the Construction Industry in China by Using DEA-Based Malmquist Productivity Indices. *Journal of Construction Engineering and Management*, **134**, 64-71. http://dx.doi.org/10.1061/(ASCE)0733-9364(2008)134:1(64)

[13]   Asmild, M., Paradi, J., Aggarwall, V. and Schaffnit, C. (2004) Combining DEA Window Analysis with the Malmquist Index Approach in a Study of the Canadian Banking Industry. *Journal of Productivity Analysis*, **21**, 67-89. http://dx.doi.org/10.1023/B:PROD.0000012453.91326.ec

[14]   Diskaya, F., Emir, S. and Orhan, N. (2011) Measuring the Technical Efficiency of Telecommunication Sector within Global Crisis: Comparison of G8 Countries and Turkey. *Procedia Social and Behavioral Sciences*, **24**, 206-218. http://dx.doi.org/10.1016/j.sbspro.2011.09.037

[15]   Jia, Y.P. and Liu, R.Z. (2012) Study of the Energy and Environmental Efficiency of the Chinese Economy Based on a DEA Model. *Procedia Environmental Sciences*, **13**, 2256-2263. http://dx.doi.org/10.1016/j.proenv.2012.01.214

[16]   Worthington, A. (2000) Technical Efficiency and Technological Change in Australian Building Societies. *Abacus*, **36**, 180-197. http://dx.doi.org/10.1111/1467-6281.00059

Facility Location Decisions Based on Driving Distances on Spherical Surface

Han Shih

University of Missouri, Columbia, USA
Email: shihh@mizzou.edu

Abstract

Facility location problems are concerned with the location of one or more facilities in a way that optimizes a certain objective such as minimizing transportation cost, providing equitable service to customers, capturing the largest market share, etc. Many facility location decisions involving distance objective functions on Spherical Surface have been approached using algorithmic, meta-heuristic algorithms, branch-and-bound algorithm, approximation algorithms, simulation, heuristic techniques, and decomposition method. These approaches are most based on Euclidean distance or Great circle distance functions. However, if the location points are widely separated, the difference between driving distance, Euclidean distance and Great circle distance may be significant and this may lead to significant variations in the locations of the corresponding optimal source points. This paper presents a framework and algorithm to use driving distances on spherical surface and explores its use as a facility location decision tool and helps companies assess the optimal locations of facilities.

Keywords

Facility Location, Spherical Surface, Euclidean Distance, Great Circle Distance, Clustering, Heuristic Method

1. Introduction

The facility location problem, also known as location analysis or k center problem, is a branch of operations research and computational geometry concerned with the optimal placement of facilities to minimize transportation costs plus facilities costs while considering factors like avoiding placing hazardous materials near housing and competitors' facilities. Poor facility location decisions can result in high transportation costs, inadequate supplies of raw materials and labor, loss of competitive advantage, and financial loss (Reid and Sanders [1]).

The facility location problem on general graphs is NP-hard to solve optimally (Wikipedia [2]). In the past, many facility location decisions involving distance objective functions on Spherical Surface have been approached using algorithmic, metaheuristic algorithms (Brimberg *et al.* [3], branch-and-bound algorithm (Tcha and Lee [4]), heuristic techniques (Francis and White [5], Love *et al.* [6] and Farahani and Masoud [7]), approximation algorithm (Vygen [8] and Shmoys *et al.* [9]), simulation, and decomposition method (Iyigun and Ben-Israel [10]). Ballou [11] presented an exact Center of Gravity method for facility location problems; he used iterative method to solve the facility location problems based on Euclidean distance. Several papers present general problem formulations which are involving the distance functions. The spherical facility location model studied by Katz and Cooper [12] and by Drezner and Wesolowsky [13] is a more realistic model than the Euclidean planar facility location model, especially for large regional facility location problems. Xue [14] has presented a gradient algorithm for solving the spherical facility location problem and proved a global convergence theorem for the algorithm as well as a hull property for the spherical facility location problem. Mwemezi and Huang [15] stated that current practices are dominated by Euclidean and Rectilinear models which are best suited to planar surfaces and presented an alternative distance measurement based on "great circle distance" which represents the shortest path on spherical surface. Sullivan and Peters [16] presented a new location-allocation algorithm that clusters points in a two-dimensional region into mutually exclusive groups, such that the sum of the value of a user-specified objective function calculated for each group is minimized over the whole study region. The squared distance, L-shaped distance and straight line distance functions were selected for the objective function in their example. Bespamyatnikh *et al.* [17] presented efficient algorithms for two problems of facility location, the first to seek a location that *maximizes* a weighted distance function between the facility and the sites, and the second to find a location that *minimizes* the sum (or sum of the squares) of the distances of *k* of the sites from the facility. Levin and Ben-Israel [18] presented a heuristic method to solve large-scale multi facility location problems; the method is using the authors' single facility location method (Levin and Ben-Israel [19]) as a parallel subroutine, and reassigning customers to facilities using the heuristic of *Nearest Center Reclassification*. Rodríguez-Chía and Franco [20] presented a procedure to solve the classical location median problem where the distances are measured with *p*-norms with $p > 2$. The global convergence of the sequence generated by this iterative scheme is proved by considering an approximated problem. Kotian *et al.* [21] developed a procedure that will solve contingent on a necessary and sufficient optimality condition, the planar *k*-centra location problem seeks to find a location that minimizes the sum of the Euclidean distances to the *k* furthest existing facilities. Iyigun and Ben-Israel [10] proposed a probabilistic decomposition method for the K facilities location problem. The problem is relaxed using probabilistic assignments, depending on the distances to the facilities. The probabilities, that decompose the problem into K single-facility location problems, are updated at each iteration together with the facility locations.

These approaches are most based on Euclidean distance or Great circle distance functions. So far, there seem to be no published papers that present the optimal or near optimal facility location decisions involving driving distance objective functions.

2. Motivation

In **Figure 1**, the driving distance, Euclidean distance and Great circle distance between Detroit, MI (42.331427, −83.0457538) and Cleveland, OH (41.4994954, −81.6954088) are calculated as:

Euclidean distance = 90.23 Miles as shown in **Figure 2**.

Great circle distance = 90.251 Miles (SAS Inc. [22], KSU [23] and The Math Forum [24]) as shown in **Figure 3**.

Driving distance = 170 Miles (Google Inc. [25]) as shown in **Figure 4**.

From **Figures 2-4**, it is clear that the Euclidean distance and Great circle distance are not much different, the driving distance are much longer than Euclidean distance and Great circle distance. Moreover, in the U.S, as a whole, road distances between major cities are about 18% greater than the straight-line distances (Love *et al.* [6]). The Euclidean distance is the "ordinary" distance between two points defined as the square root of the sum of the squares of the differences between the corresponding coordinates of the points. In Euclidean three-space (Wikipedia [26] and Wiktionary [27]), the distance between points $\left(x_1, y_1, z_1\right)$ and $\left(x_2, y_2, z_2\right)$ is

$$d = \sqrt{\left(x_2 - x_1\right)^2 + \left(y_2 - y_1\right)^2 + \left(z_2 - z_1\right)^2} \, .$$

Figure 1. Points of Detroit, MI and Cleveland, OH.

Figure 2. Euclidean distance between Detroit, MI and Cleveland, OH.

The great-circle distance is the shortest distance between two points on the surface of a sphere, measured with "Haversine" formula as (Wikipedia [28] and Movable Type Ltd [29]).

Haversine: $a = \sin^2\left(\Delta\varphi/2\right) + \cos\varphi_1 * \cos\varphi_2 * \sin^2\left(\Delta\lambda/2\right)$

Great circle distance: $d = 2 * R * a\tan 2\left(\sqrt{a}, \sqrt{(1-a)}\right)$

where φ is latitude, λ is longitude, R is earth's radius (mean radius = 6371 km);

   $\Delta\varphi$ Is difference between two latitudes radians

   $\Delta\lambda$ Is difference between two longitudes radians

Figure 3. Great circle distance between Detroit, MI and Cleveland, OH.

Figure 4. Driving distance between Detroit, MI and Cleveland, OH.

where $a \tan 2$ is the arctangent function with two arguments. In terms of the standard arctan function, whose range is $(-\pi/2, \pi/2)$, it can be expressed as follows:

$$a\tan 2(y,x) = \begin{cases} \arctan\left(\dfrac{y}{x}\right) & \text{if } x > 0, \\[2mm] \arctan\left(\dfrac{y}{x}\right) + \pi & \text{if } x < 0 \text{ and } y \geq 0, \\[2mm] \arctan\left(\dfrac{y}{x}\right) - \pi & \text{if } x < 0 \text{ and } y < 0, \\[2mm] +\dfrac{\pi}{2} & \text{if } x = 0 \text{ and } y > 0, \\[2mm] -\dfrac{\pi}{2} & \text{if } x = 0 \text{ and } y < 0, \\[2mm] \text{undefined} & \text{if } x = 0 \text{ and } y = 0. \end{cases}$$

Microsoft C# uses Math. Atan2(y, x) function to calculate Atan2(y, x)

Several papers have applied Weiszfeld [30] iterative method to solve spherical facility locations problems involving Euclidean distance and Great circle distance. Katz and Cooper [12] proposed the following closed form iteration formula for solving the spherical facility location problems. There are no any iteration formulas in relation to driving distances.

For Euclidean distance
$$\overline{x}^{k+1} = \frac{\sum_{j=1}^{N} \dfrac{w_j}{\left|\overline{x}^k - \overline{x}_j\right|} \overline{x}_j}{\left|\sum_{j=1}^{N} \dfrac{w_j}{\left|\overline{x}^k - \overline{x}_j\right|} \overline{x}_j\right|}$$

For Great circle distance

$$\overline{x}^{k+1} = \frac{\sum_{j=1}^{N} \left\{ \dfrac{w_j}{\left[\left|\overline{x}^k - \overline{x}_j\right|\left(1 - \left(\dfrac{\left|\overline{x}^k - \overline{x}_j\right|}{2}\right)^2\right)^{1/2}\right]} \right\} \overline{x}_j}{\sum_{j=1}^{N} \left\{ \dfrac{w_j}{\left[\left|\overline{x}^k - \overline{x}_j\right|\left(1 - \left(\dfrac{\left|\overline{x}^k - \overline{x}_j\right|}{2}\right)^2\right)^{1/2}\right]} \right\} \overline{x}_j}$$

So far, there seem to be no published papers that have applied iterative method based on driving distances. Katz and Cooper [12] stated that if points are widely separated, the difference between Euclidean and Great circle distances may be significant, and this may lead to significant variations in the location of the corresponding optimal source points. Aykin [31] stated that the assumption that the earth can be considered as a plane is not appropriate and introduces error if the demand points are spread over a relatively large region.

These facts motivate me to do research and propose an approach for facility location decisions based on driving distances on spherical surface using heuristic technique and helps companies assess the optimal locations of facilities. In particular, the following problem is worth pursuing:

Given:

- The location of each destination in terms of their co-ordinates
- The requirement at each destination
- A set of shipping costs for the region of interest

To determine:

- The optimal number of facilities
- The location of each facility

The optimal facilities locations are those that minimize the driving distances related costs plus facilities costs.

The implementation method proposed in this paper is carried out by using heuristic iterative approach. This paper is organized as follows: Sections 1 is Introduction. Section 2 states the motivation for researching this paper and proposes a framework and algorithm. In Section 3, the facility locations decisions implementation of the proposed framework and algorithm is carried out using a hypothetical case study. In conclusion, Section 4 summarizes this study and outlines its contributions and future research.

3. Implementation Methodology for Facility Location Decisions

3.1. Algorithm

The algorithm of this paper is to develop a method to determine the optimal facility location to minimize the sum of facilities cost and the sum of the volume of goods at a destination multiplied by the transportation rate to ship to the destination multiplied by the Google maps driving distance based on the following assumptions:

1) The good of every destination points can be transported in one time, and the velocity is not changed.
2) The one destination point is only served by one warehouse.
3) The cost is related the length from the warehouse to the destination point, the transport conditions are not considered. Transportation cost is related to the distance only. The transportation cost equals the distances traveled times a fixed price per unit, distance.
4) The warehouse locations are located at populated places (cities/towns).
5) All service facilities are identical.
6) Each destination point wishes to minimize the cost of acquiring the product.
7) The company treats each cluster independently.

The strategy of the algorithm involves the following steps:

Step 1. Generate Google maps driving distance matrix from the set of destination locations' latitudes/longitudes.

Step 2. Perform K means clustering based on the destination locations driving distance matrix to generate K clusters.

Step 3. Calculate starting point of facility location for each cluster using Center of Gravity method and sets as current facility location.

Step 4. Construct the circle whose radius is the maximal Google maps driving distance from current facility location.

Step 5. Query the populated cities within the circle constructed in Step 4.

Step 6. Calculate the cost from each queried city to all destination locations, the queried city with the smallest cost will become new current point.

Step 7. Repeat steps 4 to 6 until no new current point can be found.

Step 8. Randomly select several cities points as current points and perform steps 4 to 6 to show that the local optimal facility location is the global optimal or nearly global optimal facility location.

Step 9. Calculate the total cost for each set of clusters.

Step 10. Perform another set of clustering if desired and repeat Steps 3 to 9 until no clustering is desired.

Step 11. Compare the cost of each set of clusters; select the set of clusters with the minimal total cost and optimal facility locations.

Figure 5 shows the flowchart of Facility location decisions process.

In **Figure 5**, the implementation details are described as follows and a lot of C# programs have been written for the implementation (Source codes are available from the author).

Step 1. A Google maps driving distance matrix is generated as a distance data set and used as input in SAS

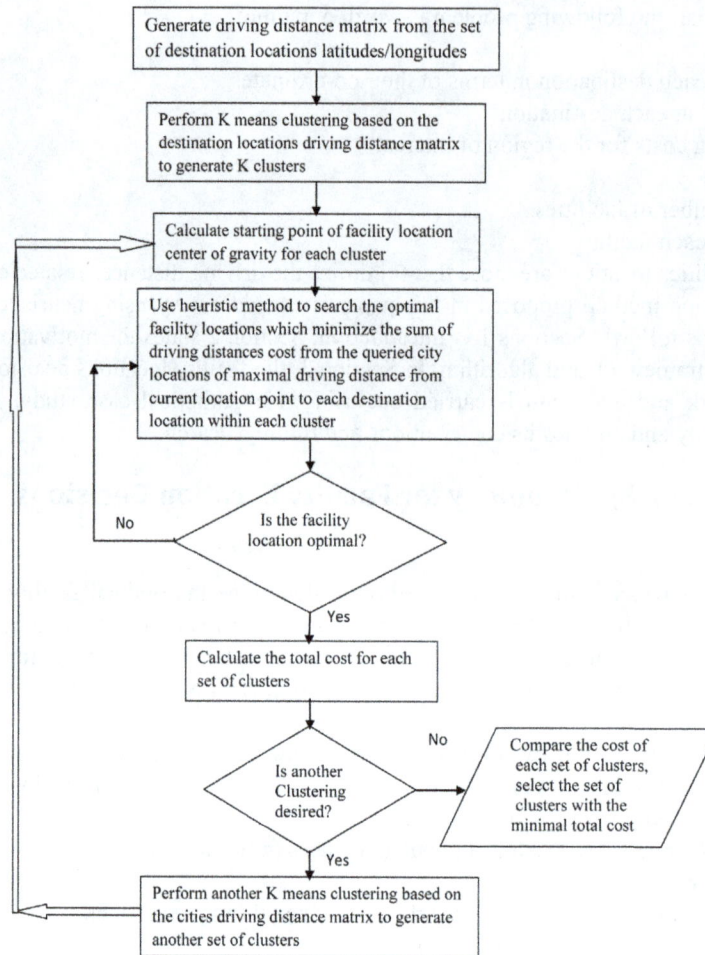

Figure 5. Flowchart of facility location decisions process.

Clustering procedure. A C# program is written using the Google Maps API (Google Inc. [25]) to create a driving distance matrix that will be used as input in SAS Clustering procedure.

Step 2. After driving distance matrix is generated, the next step is to cluster the cities based on Driving distances. This paper will use SAS Clustering procedure (SAS Inc. [32]) to cluster for the cities. TYPE = DISTANCE data set can be used as an input data set to PROC MODECLUS or PROC CLUSTER (SAS Inc. [32]). PROC CLUSTER ignores the upper triangular portion set and assumes that all main diagonal values are zero, even if they are missing. PROC MODECLUS uses the entire distance matrix and does not require the matrix to be symmetric. Since the driving distance matrix data set is not necessary symmetric, the MODECLUS Procedure is used in my paper to cluster the cities.

Step 3. Calculate the initial starting facility location centers for each cluster. Several quantitative techniques (Heizer and Render [33], Reid and Sanders [1] and Pearson [34]) have been developed to assist initial site selection. This paper uses Center of Gravity method (Ballou [11], Geo Midpoint [35], Chen and He [36], Heizer and Render [33] and Kuo and White [37]) to calculate the starting point of facility location. The Center of Gravity method assumes that the cost is directly proportional to distance and volume shipped, inbound and outbound transportation costs are equal, and it does not include special shipping costs for less than full loads. Using latitude and longitude coordinates might be helpful (Heizer and Render [33] and Chase *et al.* [38]) to calculate the initial facility location centers for each cluster. This paper assumes a spherical earth and sea-level points, the following formula is used to perform spherical coordinate conversion from latitude/longitudes to Cartesian coordinates for each destination location. The formula is much more complex for ellipsoidal earth (Ligas and Banasik [39]).

$$x_i = r * \cos\left(radianlat_i\right) * \cos\left(radianlon_i\right) \tag{1}$$

$$y_i = r * \cos\left(radianlat_i\right) * \sin\left(radianlon_i\right) \tag{2}$$

$$z_i = r * \sin\left(radianlat_i\right) \tag{3}$$

where r is the earth's radius.

Next, compute weight volume of good multiple by shipping cost for each location

$$w_i = v_i * r_i \tag{4}$$

where v_i is the volume of good to be shipped to location i, r_i is the shipping rate per unit of good to location i. The initial center of gravity X_C, Y_C and Z_C are calculated as

$$X_C = \sum_i^n x_i w_i \bigg/ \sum_i^n w_i \tag{5}$$

$$Y_C = \sum_i^n y_i w_i \bigg/ \sum_i^n w_i \tag{6}$$

$$Z_C = \sum_i^n z_i w_i \bigg/ \sum_i^n w_i \tag{7}$$

Then, the initial starting Cartesian coordinates are converted to latitude/longitudes points as input for Step 4.

$$lon = a\tan 2\left(Y_C, X_C\right) \tag{8}$$

$$hyp = sqrt\left(X_C * X_C + Y_C * Y_C\right) \tag{9}$$

$$lon = a\tan 2\left(Z_C, hyp\right) \tag{10}$$

Step 4. Calculate the driving distances from the current point calculated in Step 3 to each destination location of each cluster using the Google Maps API (Google Inc. [25]).

Step 5. Search the optimal facility locations. All distances are calculated using the Google maps driving distances. Let the starting point be the current point calculated in Step 3. Use the maximal Google maps driving distance calculated from Step 4 as radius of current point. The cities are queried in a circular pattern around the current point at a maximal driving distance. This paper utilizes a function in Map Suite WinForms Desktop free trial Edition (Thinkgeo LLC [40]) and a populated places database of Natural earth data source (Natural Earth [41]) to query all cities with the maximal driving distance of current location point.

Step 6. Calculate the total cost between each queried city to all destinations locations within each cluster. Compare the cost of each queried city and find the city which has minimal cost. If any of the queried cities has a new minimal cost, then that city will become new current point and go back to repeat steps 4 to 6 until no new smallest cost can be found. Otherwise, If none of the queried cities has a new minimal cost, then the current city becomes the optimal facility location for that cluster.

Step 7. Randomly select several points (Hillier and Lieberman [42]) and perform Steps 5 to 6 to show that the local optimal facility location obtained in Step 6 is the global optimal or nearly global optimal facility location.

Step 8. Calculate the total cost for all clusters in the cluster set.

Step 9. Generate other clusters if desired and repeat steps 3 to 8 to calculate the total cost for all clusters in the cluster set until no clustering is desired.

Step 10. Select the set of clusters which has the minimal total cost, the optimal facilities locations will be the cities found in Step 7.

3.2. Methodology Implementation

Based on the formulas from several papers (Rodriguez-Chia and Valero-Franco [20], Mwemezi and Huang [15], Katz and Cooper [12] and Litwhiler and Aly [43]), this paper formulates the model as follows:

$$\text{Min } TC = \sum_{k=1}^{m}\sum_{i=1}^{n_k} V_{ki} R_{ki} d_{ki} + \sum_{k=1}^{m} Y_k = \sum_{k=1}^{m}\left(\sum_{i=1}^{n_k} V_{ki} R_{ki} d_{ki} + Y_k\right) \tag{11}$$

Subject to: $m > 0$

$$n_k > 0 \quad \forall k, k = 1, \cdots, m$$

$$V_{ki} > 0 \quad \forall k, k = 1, \cdots, m; \forall i, i = 1, \cdots, n_k$$

$$R_{ki} > 0 \quad \forall k, k = 1, \cdots, m; \forall i, i = 1, \cdots, n_k$$

$$d_{ki} > 0 \quad \forall k, k = 1, \cdots, m; \forall i, i = 1, \cdots, n_k$$

$$Y_k > 0 \quad \forall k, k = 1, \cdots, m$$

$$\sum_{i=1}^{k} n_k = \text{total \# of destinations}$$

where TC = Total cost for m clusters

n_k = destination locations for each cluster

V_{ki} = Volume at destination i in cluster k

R_{ki} = Transportation rate to destination i in cluster k

d_{ki} = Driving distance from the facility to be located to destination i in cluster k

Y_k = facility cost at each cluster k

Given the assumptions described in Section 3.1. We search the optimal facility to minimize the sum of facilities cost and the sum of the volume at a destination multiplied by the transportation rate to ship to the destination multiplied by the driving distance for the cluster.

3.3. Hypothetical Case Study

ABC Company has 88 market cities located in Midwest, USA; the company seeks to build several warehouse centers to ship their goods to those markets to serve their customers. Assume that travel originates and ends at the warehouse centers. The coordinates (Latitude/Longitudes) of each market location and volume of goods and transportation rate for each market location are provided. The facility cost is 100,000.00. The goal is to minimize the sum of facilities costs and the sum of transportation costs. **Figure 6** is the graphical map of 88 major market cities.

Appendix A is the coordinates (Latitude/Longitudes) of each market location and volume of goods and transportation rate for each market location. The facility cost is 100,000.00.

Step 1. Converting major cities data to driving distance matrix

Use coordinates (Latitude/Longitudes) provided in **Appendix A** as inputs to create a driving distance matrix that will be used as inputs to MODECLUS clustering procedure in SAS. **Appendix B** is part of Google maps distance matrix of 88 Midwest major market cities.

Step 2. Clustering techniques

Set the driving distance matrix in **Appendix B** as input, the SAS MODECLUS Procedure is used to perform clustering for the cities. **Table 1** is cluster output from SAS code:

Proc modeclus data = work. Driving Data (TYPE = DISTANCE) list m = 1 k = 3; id location; run;

Figure 7 is the map of 14 clusters for the 88 major market cities located in Midwest, USA.

The following steps 3 to 7 are performed for cluster 1.

Step 3. Using equations from (1) to (7), the initial center of gravity for cluster 1 is calculated as in **Table 2** and shown in **Figure 8**.

Using equations (8) to (10), the Latitude and Longitude of Center of gravity is calculated as 41.5995134613, −96.62241872.

Step 4. **Table 3** shows the driving distances from current point to each market city location.

Step 5. The surrounding cities/towns with radius of maximum driving distance (157 miles) from Center of gravity are queried with Map suite tools and Natural populated placed data, the queried results are show in **Table 4** and the map is shown in **Figure 9**.

Step 6. **Table 5** shows the costs from each city to all market cities locations. From **Table 5**, the Depot 15 has the smallest cost, and then Depot 15 (Omaha, NEB) will become new current point. The driving distances from new current point to each market city location are calculated as in **Table 6**. The maximum driving distance is 190 miles.

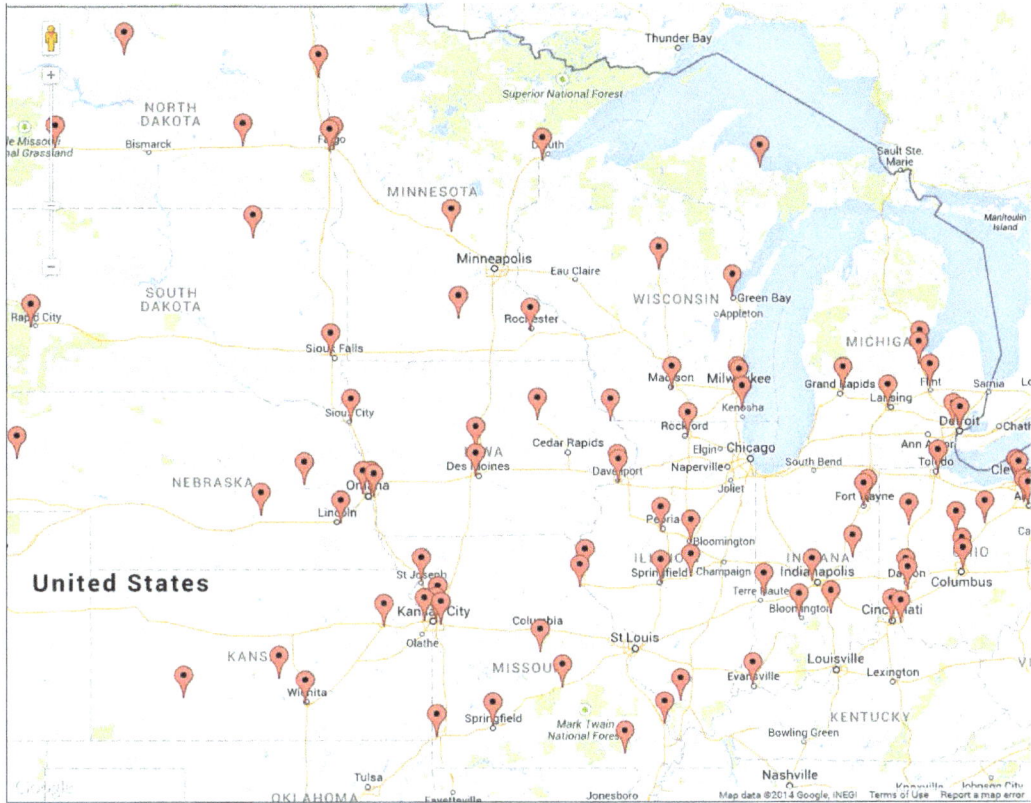

Figure 6. Map of 88 major market cities located in Midwest, USA.

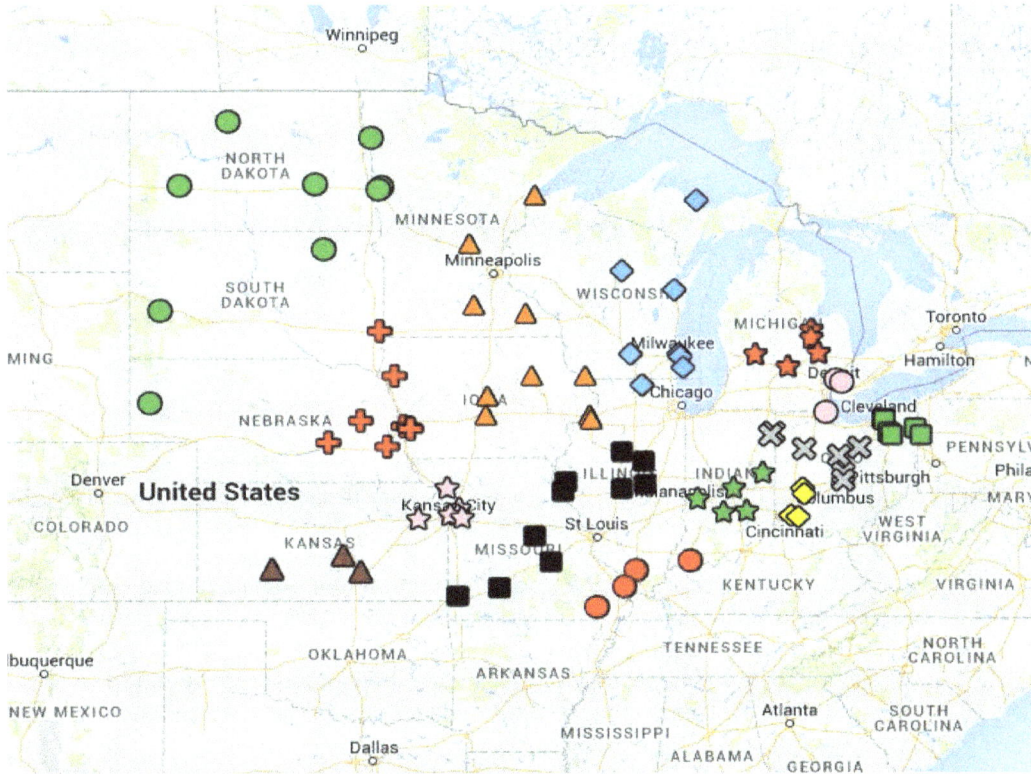

Figure 7. Map of 14 clusters for the 88 major market cities in Midwest, USA.

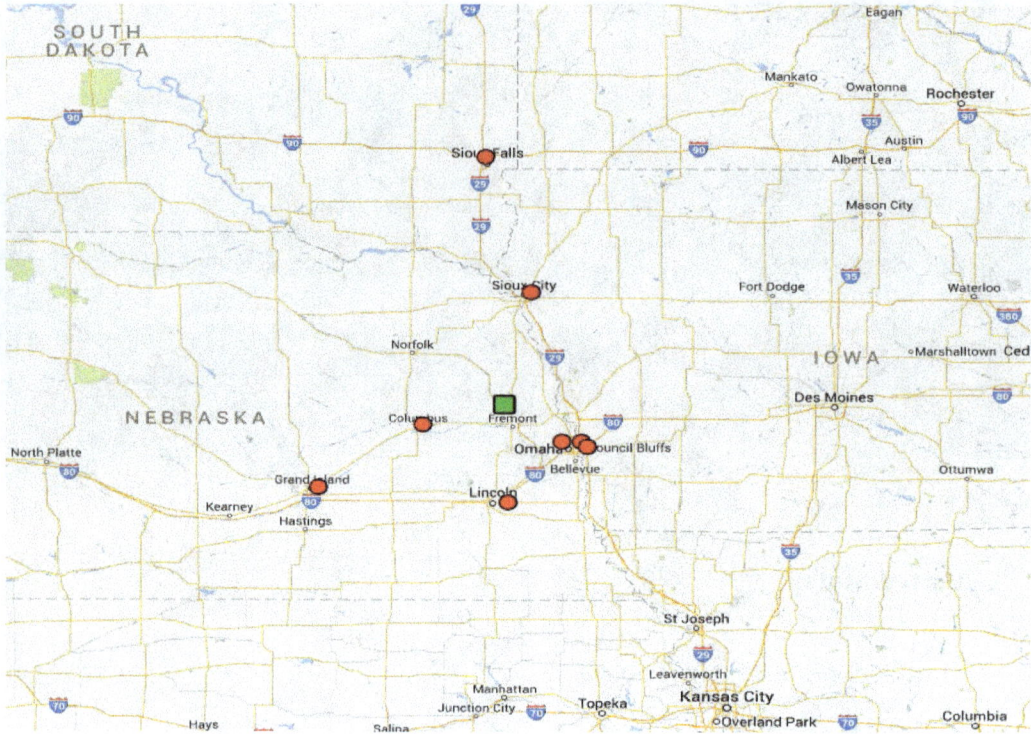

Figure 8. Cluster 1 with initial starting point- center of gravity (Scribner, NE—Green square).

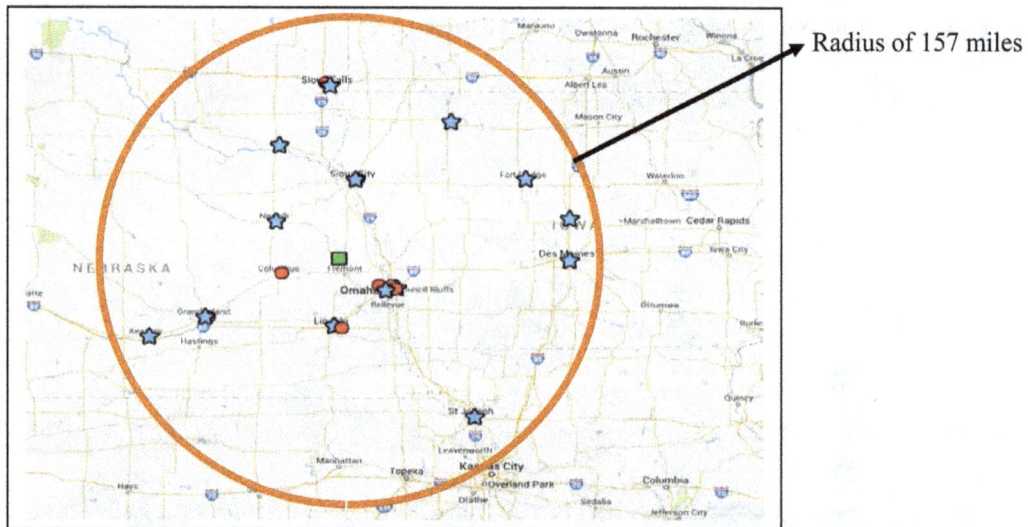

Radius of 157 miles

Figure 9. Cluster1 with Cities (Blue stars) of radius of maximum driving distance of current point (Scribner, NE—Green square).

Repeat Steps 4, 5, & 6. The surrounding cities/towns with radius of maximum driving distance (190 Miles in **Table 6**) from Depot 15 (Omaha, Nebraska, USA) are queried with Map Suite tools and Natural populated placed data, the queried results are show in **Figure 10**.

Table 7 is the queried cities within the radius of maximum driving distance (190 miles) from current point. **Figure 10** is the map with queried cities of radius of maximum driving distance of current point. **Table 8** shows the cost from each city to all market cities locations.

From **Table 8**, use Depot 23 as the current point (Omaha, Neb.) and repeat steps 4 to 6, there is no new smallest cost can be found.

Table 1. Modeclus cluster procedure with m = 1, K = 3.

Sums of Density Estimates Within Neighborhood											
Cluster	Location	Estimated Density	Same Cluster	Other Clusters	Total	Cluster	Location	Estimated Density	Same Cluster	Other Clusters	Total
1	Loc45	0.000179	0.000946	0	0.000946	8	Loc13	0.000361	0.0006687	0	0.000669
	Loc46	0.00084	0.002311	0	0.002311		Loc14	0.000306	0.0007238	0	0.000724
	Loc67	0.000106	0.000179	0.000151	0.00033		Loc15	0.00031	0.0007238	0	0.000724
	Loc1F	0.001495	0.001655	0	0.001655		Loc16	0.000363	0.0006711	0	0.000671
	Loc1G	0.000816	0.002335	0	0.002335	9	Loc19	0.000303	0.0006104	0	0.00061
	Loc1H	0.000299	0.002311	0	0.002311		Loc27	0.000303	0.0006099	0	0.00061
	Loc1I	0.000225	0.000988	0	0.000988		Loc28	0.000217	0.0003028	0.000275	0.000578
	Loc1J	0.000173	0.000524	0	0.000524		Loc29	0.000307	0.0006061	0	0.000606
2	Loc92	0.000744	0.00183	0	0.00183		Loc31	0.000216	0.0006099	0	0.00061
	Loc93	0.001086	0.001489	0	0.001489	10	Loc4	0.000266	0.0005507	0	0.000551
	Loc94	0.000648	0.00183	0	0.00183		Loc32	0.000283	0.0005344	0	0.000534
	Loc95	0.000304	0.001392	0	0.001392		Loc33	0.000268	0.000549	0	0.000549
	Loc1A	0.000744	0.00183	0	0.00183	11	Loc42	0.000167	0.0003166	0	0.000317
	Loc1B	0.000259	0.00183	0	0.00183		Loc43	0.00015	0.0003342	0	0.000334
3	Loc5	0.000541	0.001377	0	0.001377		Loc44	0.000167	0.0003686	0	0.000369
	Loc6	0.000685	0.001234	0	0.001234		Loc47	0.000201	0.0004421	0	0.000442
	Loc8	0.000513	0.001377	0	0.001377		Loc50	0.000228	0.0004151	0	0.000415
	Loc9	0.000693	0.001198	0	0.001198		Loc62	7.64E−05	0.0002726	0	0.000273
	Loc10	0.000354	0.000839	0	0.000839		Loc63	0.000151	0.000318	0	0.000318
	Loc11	0.000325	0.000867	0	0.000867		Loc64	0.000151	0.0002726	0	0.000273
4	Loc41	8.27E−05	0.000259	0	0.000259		Loc65	0.000122	0.0002273	0	0.000227
	Loc51	0.000636	0.001125	0	0.001125		Loc79	0.000214	0.00043	0	0.00043
	Loc52	0.000561	0.0012	0	0.0012	12	Loc66	0.000202	0.0004095	0	0.000409
	Loc53	0.000564	0.001197	0	0.001197		Loc68	9.42E−05	0.0003802	0	0.00038
	Loc54	0.000211	0.000754	0	0.000754		Loc69	7.41E−05	0.0001398	0	0.00014
	Loc55	0.000143	0.000677	0	0.000677		Loc70	0.000209	0.0004018	0	0.000402
	Loc56	0.000116	0.000354	0	0.000354		Loc71	0.0002	0.0004111	0	0.000411
	Loc78	0.000194	0.000776	0	0.000776		Loc72	0.000171	0.0004111	0	0.000411
5	Loc34	0.000363	0.000856	0	0.000856		Loc74	8.65E−05	0.000265	0	0.000265
	Loc35	0.000496	0.000724	0	0.000724		Loc75	9.42E−05	0.0002573	0	0.000257
	Loc36	0.00036	0.000859	0	0.000859		Loc1K	5.33E−05	7.411E−05	0.000173	0.000247
	Loc37	0.000254	0.000507	0	0.000507	13	Loc30	0.000143	0.0001705	0.000216	0.000386

Continued

	Loc						Loc				
	Loc40	0.000143	0.000617	0	0.000617		Loc86	0.00017	0.0003514	0	0.000351
6	Loc1	0.000341	0.000883	0	0.000883		Loc90	0.000208	0.0003068	0	0.000307
	Loc2	0.000473	0.000751	0	0.000751		Loc91	0.000136	0.0003786	0	0.000379
	Loc3	0.00041	0.000786	0	0.000786	14	Loc1C	0.000123	0.0001332	0.000259	0.000392
	Loc12	0.000313	0.000883	0	0.000883		Loc1D	0.000133	0.0002348	0	0.000235
	Loc18	0.000278	0.000684	0	0.000684		Loc1E	0.000112	0.0002558	0	0.000256
	Loc23	0.000275	0.000552	0	0.000552						
	Loc24	0.000274	0.000553	0	0.000553						
7	Loc80	0.000223	0.000611	0	0.000611						
	Loc81	0.000357	0.000619	0	0.000619						
	Loc83	0.000151	0.000417	0	0.000417						
	Loc84	0.000395	0.000611	0	0.000611						
	Loc85	0.000254	0.000752	0	0.000752						
	Loc89	0.000162	0.000405	0	0.000405						
	Loc96	0.000104	0.000158	0.000744	0.000902						
	Loc97	0.000159	0.00032	0	0.00032						
	Loc98	0.000158	0.000317	0	0.000317						
	Loc99	0.000158	0.000262	0	0.000262						

Table 2. Center of Gravity for cluster 1.

City Location	Latitude	Longitude	X Coordinate	Y Coordinate	Z Coordinate	Vi	Ri
IA Loc45	42.505046	−96.38003	−324.327336	−2900.569393	2674.918685	1100	0.03
IA Loc46	41.251631	−95.87447	−304.638558	−2960.828675	2610.434812	400	0.03
SD Loc67	43.577696	−96.80101	−339.639768	−2847.877667	2729.087508	1200	0.03
NE Loc1F	41.292321	−95.93427	−307.536885	−2958.664433	2612.547959	1500	0.04
NE Loc1G	41.296198	−96.11027	−316.604967	−2957.530006	2612.749234	800	0.04
NE Loc1H	40.798632	−96.59023	−343.959573	−2977.202028	2586.820615	1500	0.04
NE Loc1I	41.437838	−97.37192	−380.817694	−2943.427398	2620.09428	1200	0.04
NE Loc1J	40.922826	−98.33087	−433.420604	−2959.825426	2593.310824	500	0.04

City Location	Xi*Vi*Ri	Yi*Vi*Ri	Zi*Vi*Ri	Vi*Ri
IA Loc45	−10,702.80208861	−95,718.78997969	88,272.31661895	33.00
IA Loc46	−3,655.66269894	−35,529.94409800	31,325.21774265	12.00
SD Loc67	−12,227.03164169	−102,523.59600100	98,247.15027041	36.00
NE Loc1F	−18,452.21310819	−177,519.86597229	156,752.87753847	60.00
NE Loc1G	−10,131.35893869	−94,640.96019008	83,607.97548293	32.00
NE Loc1H	−20,637.57437751	−178,632.12169815	155,209.23690756	60.00
NE Loc1I	−18,279.24929086	−141,284.51511257	125,764.52543849	48.00
NE Loc1J	−8,668.41207663	−59,196.50851119	51,866.21648196	20.00
SUM	−102,754.30422113	−885,046.30156298	791,045.51648143	301.00
AVEAGE	X Coordinate	Y Coordinate e	Z Coordinate	
	−341.37642598	−2,940.35316134	2,628.05819429	

Table 3. Driving distances from current point to each market location.

Driving Distance	From Latitude	Longitude	To Latitude	Longitude	City Location
72.6	41.5995134613	−96.62241872	42.505046	−96.380030	IA Loc45
63.6	41.5995134613	−96.62241872	41.251631	−95.874470	IA Loc46
157	41.5995134613	−96.62241872	43.577696	−96.801010	SD Loc67
54.3	41.5995134613	−96.62241872	41.292321	−95.934270	NE Loc1F
43.6	41.5995134613	−96.62241872	41.296198	−96.110270	NE Loc1G
71.6	41.5995134613	−96.62241872	40.798632	−96.590230	NE Loc1H
61.8	41.5995134613	−96.62241872	41.437838	−97.371920	NE Loc1I
125	41.5995134613	−96.62241872	40.922826	−98.330870	NE Loc1J

Table 4. The queried cities within the radius of maximum driving distance from current point Radius of 157 miles.

Latitude	Longitude	City	State	Country
41.5995134613	−96.6224187157			
43.1452850536	−95.1471745217	Spencer	Iowa	USA
42.5068227326	−94.1802567970	Ft. Dodge	Iowa	USA
40.7007055858	−99.0811462827	Kearney	Nebraska	USA
40.9222682892	−98.3579862892	Grand Island	Nebraska	USA
42.5003890168	−96.3999921079	Sioux City	Iowa	USA
41.2622733755	−95.8608002134	Council Bluffs	Iowa	USA
42.0538529654	−93.6197225360	Ames	Iowa	USA
39.7690311880	−94.8463918475	St. Joseph	Missouri	USA
42.0287123810	−97.4335982684	Norfolk	Nebraska	USA
42.8820194693	−97.3924896665	Yankton	South Dakota	USA
40.8199747915	−96.6800008563	Lincoln	Nebraska	USA
43.5499890331	−96.7299978045	Sioux Falls	South Dakota	USA
41.5799800812	−93.6199809181	Des Moines	Iowa	USA
41.2400008332	−96.0099900734	Omaha	Nebraska	USA

Step 7. Randomly select several points (Hillier and Lieberman [42]) and perform Steps 5 to 6 to show that the local optimal facility location obtained in Step 6 is the global optimal or nearly global optimal facility location. **Tables 9-12** show the results of each randomly selected city to all market cities locations. It is concluded that the global optimal facility location is the city found in Step 6. The optimal facility location will be Omaha, Nebraska shown as in **Figure 11**.

Using the same steps for cluster 1, the optimal facility locations for clusters 2 to 14 are calculated and shown as from **Figures 12-24**.

The total cost for these 14 clusters is calculated using equation (11) and shown as follows:

Cluster 1	Cluster 2	Cluster 3	Cluster 4	Cluster 5	Cluster 6	Cluster 7	Cluster 8	Cluster 9	Cluster 10
21,222.6	5,593.8	3,392.8	38,896.4	11,857.5	11,339.2	57,779.0	1,660.4	9,898.2	1,805.6

Cluster 11	Cluster 12	Cluster 13	Cluster 14	# of Facility	Facility cost per unit	Facilities cost	Total Cost
92,088.5	72,057.0	12,892.2	8,700.0	14	100,000.00	1,400,000.00	**1,749,183.2**

Table 5. Cluster1 costs from queried cities to market locations-iteration1.

Center of Gravity and queried cities with Latitudes/Longitudes	Cities ID	Cost
41.5995864980917, −96.6224184226743; Center Of Gravity	Depot = 1	Cost = 23331
43.14528505360, −95.14717452170; Spencer, Iowa, USA	Depot = 2	Cost = 52554.1
42.50682273260, −94.18025679700; Ft. Dodge, Iowa, USA	Depot = 3	Cost = 57508
40.70070558580, −99.08114628270; Kearney, Nebraska, USA	Depot = 4	Cost = 54229
40.92226828920, −98.35798628920; Grand Island, Nebraska, USA	Depot = 5	Cost = 39933.2
42.50038901680, −96.39999210790; Sioux City, Iowa, USA	Depot = 6	Cost = 31472.9
41.26227337550, −95.86080021340; Council Bluffs, Iowa, USA	Depot = 7	Cost = 21732.4
42.05385296540, −93.61972253600; Ames, Iowa, USA	Depot = 8	Cost = 64183
39.76903118800, −94.84639184750; St. Joseph, Missouri, USA	Depot = 9	Cost = 56370
42.02871238100, −97.43359826840; Norfolk, Nebraska, USA	Depot = 10	Cost = 31357.3
42.88201946930, −97.39248966650; Yankton, South Dakota, USA	Depot = 11	Cost = 42640.6
40.81997479150, −96.68000085630; Lincoln, Nebraska, USA	Depot = 12	Cost = 25181
43.54998903310, −96.72999780450; Sioux Falls, South Dakota, USA	Depot = 13	Cost = 49678
41.57998008120, −93.61998091810; Des Moines, Iowa, USA	Depot = 14	Cost = 58027
41.24000083320, −96.00999007340; Omaha, Nebraska, USA	Depot = 15	Cost = 21222.6

Table 6. Driving distances from current point (Omaha, NE) to each market location.

Driving Distance	From Latitude	Longitude	To Latitude	Longitude	City Location
104	41.24000083320	−96.00999007340	42.505046	−96.380030	IA Loc45
11.4	41.24000083320	−96.00999007340	41.251631	−95.874470	IA Loc46
190	41.24000083320	−96.00999007340	43.577696	−96.801010	SD Loc67
9.5	41.24000083320	−96.00999007340	41.292321	−95.934270	NE Loc1F
10.6	41.24000083320	−96.00999007340	41.296198	−96.110270	NE Loc1G
50.1	41.24000083320	−96.00999007340	40.798632	−96.590230	NE Loc1H
83.5	41.24000083320	−96.00999007340	41.437838	−97.371920	NE Loc1I
143	41.24000083320	−96.00999007340	40.922826	−98.330870	NE Loc1J

3.4. Results: Driving Distance vs. Euclidean Distance and Great Circle Distance

With the data in **Table 13**, I use the steps described in Section 3.2 to search the optimal facility location based on driving distance vs. Euclidean distance and Great circle distance. As the results shown in **Figures 25-27**, the optimal facility location based on driving distance is different from the optimal facility location based on Euclidean distance and Great circle distance.

With the data in this case study, I use the steps described in Section 3.2 to search the optimal facility location based on Euclidean distance and Great circle distance. It is found out that the optimal facilities locations for Clusters 1 to 14 are the same when the distances are based on driving distance, Euclidean distance or Great circle distance. For examples, Cluster 3 (see **Figure 13** and **Figure 28**), Cluster 4 (see **Figure 14** and **Figure 29**), Cluster 5 (see **Figure 15** and **Figure 30**) and Cluster 10 (see **Figure 20** and **Figure 31**) have the same optimal facilities locations.

However, the above results reveal that if the driving distance is much longer than Euclidean distance or Great circle distance because there are lakes, rivers between locations (see **Figures 2-4** and **Figure 25**), the optimal facility location based on driving distance will be different from the optimal facility location based on Euclidean

Table 7. The cities within the radius of maximum driving distance from current point (Omaha, NE).

Latitude	Longitude	City	State	Country
41.2400008332	−96.0099900734			
41.0128829133	−92.4148090024	Ottumwa	Iowa	USA
43.1452850536	−95.1471745217	Spencer	Iowa	USA
42.5068227326	−94.1802567970	Ft. Dodge	Iowa	USA
39.1135805169	−94.6301463771	Kansas City	Kansas	USA
38.9597524201	−95.2552299416	Lawrence	Kansas	USA
39.1940275259	−96.5924351418	Manhattan	Kansas	USA
39.0911139110	−94.4152812111	Independence	Missouri	USA
40.7007055858	−99.0811462827	Kearney	Nebraska	USA
40.9222682892	−98.3579862892	Grand Island	Nebraska	USA
42.5003890168	−96.3999921079	Sioux City	Iowa	USA
41.2622733755	−95.8608002134	Council Bluffs	Iowa	USA
42.0538529654	−93.6197225360	Ames	Iowa	USA
38.8246702261	−97.6071794045	Salina	Kansas	USA
39.7690311880	−94.8463918475	St. Joseph	Missouri	USA
42.0287123810	−97.4335982684	Norfolk	Nebraska	USA
42.8820194693	−97.3924896665	Yankton	South Dakota	USA
40.8199747915	−96.6800008563	Lincoln	Nebraska	USA
39.0500053091	−95.6699849871	Topeka	Kansas	USA
39.1070885098	−94.6040942189	Kansas City	Missouri	USA
43.5499890331	−96.7299978045	Sioux Falls	South Dakota	USA
41.5799800812	−93.6199809181	Des Moines	Iowa	USA
41.2400008332	−96.0099900734	Omaha	Nebraska	USA

Table 8. Cluster1 costs from queried cities to market locations—iteration 2.

Current point and queried cities with Latitudes/Longitudes	Cities ID	Cost
41.24000083320, −96.00999007340; Current point	Depot = 1	Cost = 21222.6
41.01288291330, −92.41480900240; Ottumwa, Iowa, USA	Depot = 2	Cost = 83407
43.14528505360, −95.14717452170; Spencer, Iowa, USA	Depot = 3	Cost = 52554.1
42.50682273260, −94.18025679700; Ft. Dodge, Iowa, USA	Depot = 4	Cost = 57508
39.11358051690, −94.63014637710; Kansas City, Kansas, USA	Depot = 5	Cost = 71387
38.95975242010, −95.25522994160; Lawrence, Kansas, USA	Depot = 6	Cost = 76203
39.19402752590, −96.59243514180; Manhattan, Kansas, USA	Depot = 7	Cost = 61467
39.09111391100, −94.41528121110; Independence, Missouri, USA	Depot = 8	Cost = 75485
40.70070558580, −99.08114628270; Kearney, Nebraska, USA	Depot = 9	Cost = 54229
40.92226828920, −98.35798628920; Grand Island, Nebraska, USA	Depot = 10	Cost = 39933.2

Continued

42.50038901680, −96.39999210790; Sioux City, Iowa, USA	Depot = 11	Cost = 31472.9
41.26227337550, −95.86080021340; Council Bluffs, Iowa, USA	Depot = 12	Cost = 21732.4
42.05385296540, −93.61972253600; Ames, Iowa, USA	Depot = 13	Cost = 64183
38.82467022610, −97.60717940450; Salina, Kansas, USA	Depot = 14	Cost = 76040
39.76903118800, −94.84639184750; St. Joseph, Missouri, USA	Depot = 15	Cost = 56370
42.02871238100, −97.43359826840; Norfolk, Nebraska, USA	Depot = 16	Cost = 31357.3
42.88201946930, −97.39248966650; Yankton, South Dakota, USA	Depot = 17	Cost = 42640.6
40.81997479150, −96.68000085630; Lincoln, Nebraska, USA	Depot = 18	Cost = 25181
39.05000530910, −95.66998498710; Topeka, Kansas, USA	Depot = 19	Cost = 66331
39.10708850980, −94.60409421890; Kansas City, Missouri, USA	Depot = 20	Cost = 72022
43.54998903310, −96.72999780450; Sioux Falls, South Dakota, USA	Depot = 21	Cost = 49678
41.57998008120, −93.61998091810; Des Moines, Iowa, USA	Depot = 22	Cost = 58027
41.24000083320, −96.00999007340; Omaha, Nebraska, USA	Depot = 23	Cost = 21222.6

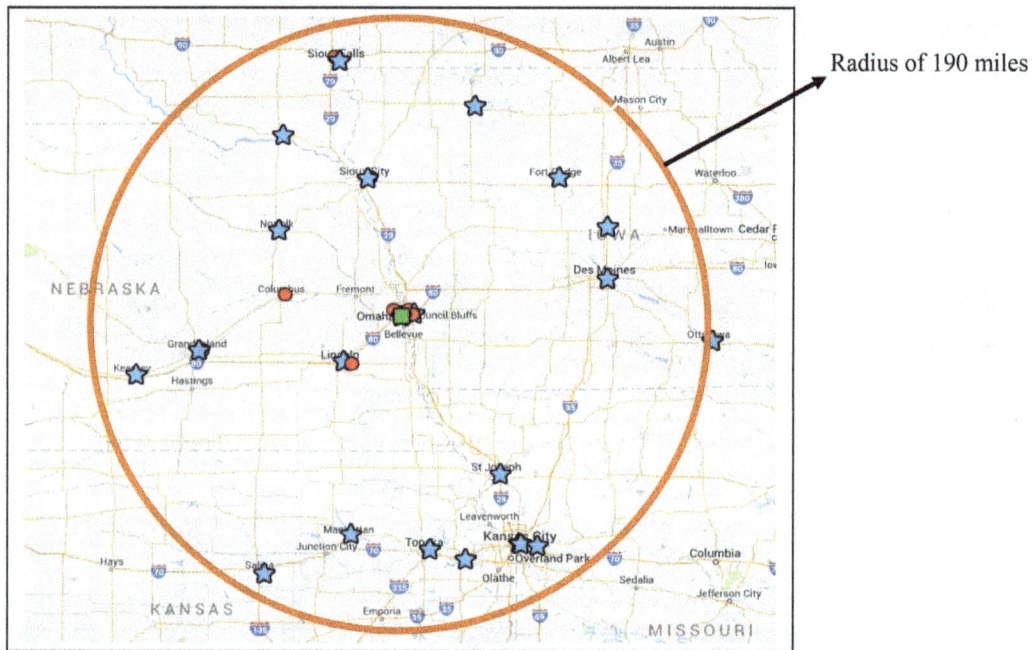

Figure 10. Cluster1 with cities (Blue stars) of radius of maximum driving distance of current point (Omaha, Neb.)

distance and Great circle distance. *The facility location decision based on driving distances is a practical approach. In regard to transportation cost, the driving distances in the presence of geographic barriers should be taken into consideration in facility location decisions.*

3.5. Compare the Total Cost of Different Set of Clusters

Now, this paper uses MODECLUS Procedure to get different set of clustering and select the best set of clusters. **Table 14** is clustering output of 11 clusters from SAS code with $m = 1$, $k = 4$.
Table 15 is clustering output of 9 clusters from SAS code with $m = 1$, $k = 5$.

From equation (11) $TC = \sum_{k=1}^{m}\sum_{i=1}^{n_k} V_{ki}R_{ki}d_{ki} + \sum_{k=1}^{m} Y_k = \sum_{k=1}^{m}\left(\sum_{i=1}^{n_k} V_{ki}R_{ki}d_{ki} + Y_k\right)$

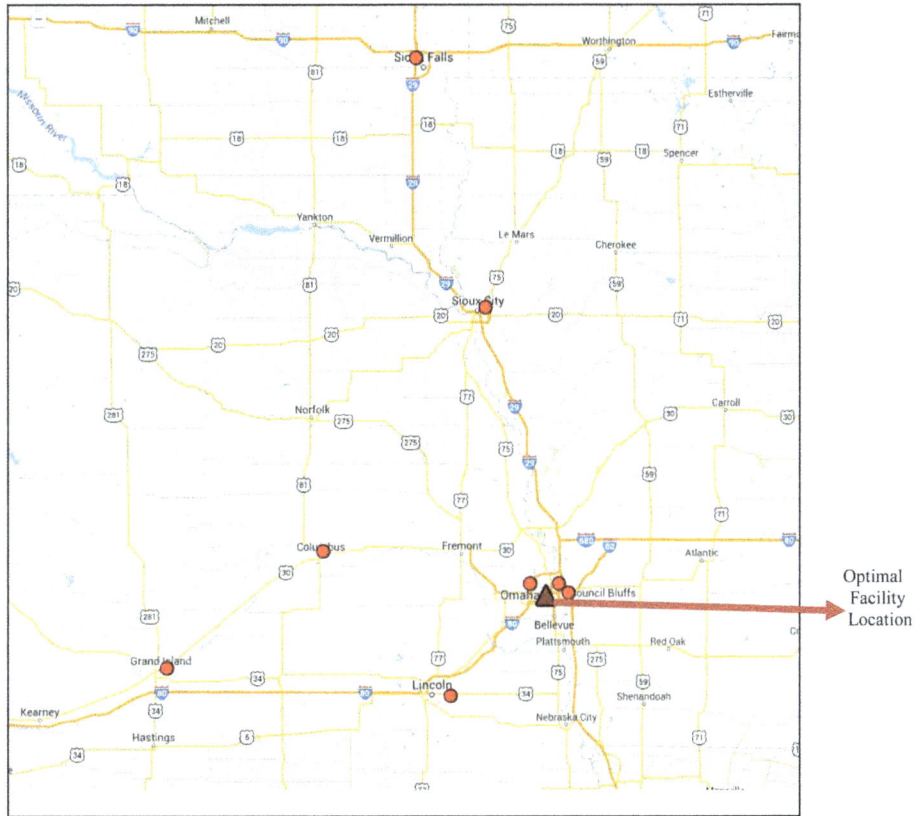

Figure 11. Cluster 1 with optimal facility location (Omaha, Neb.—Brown triangle).

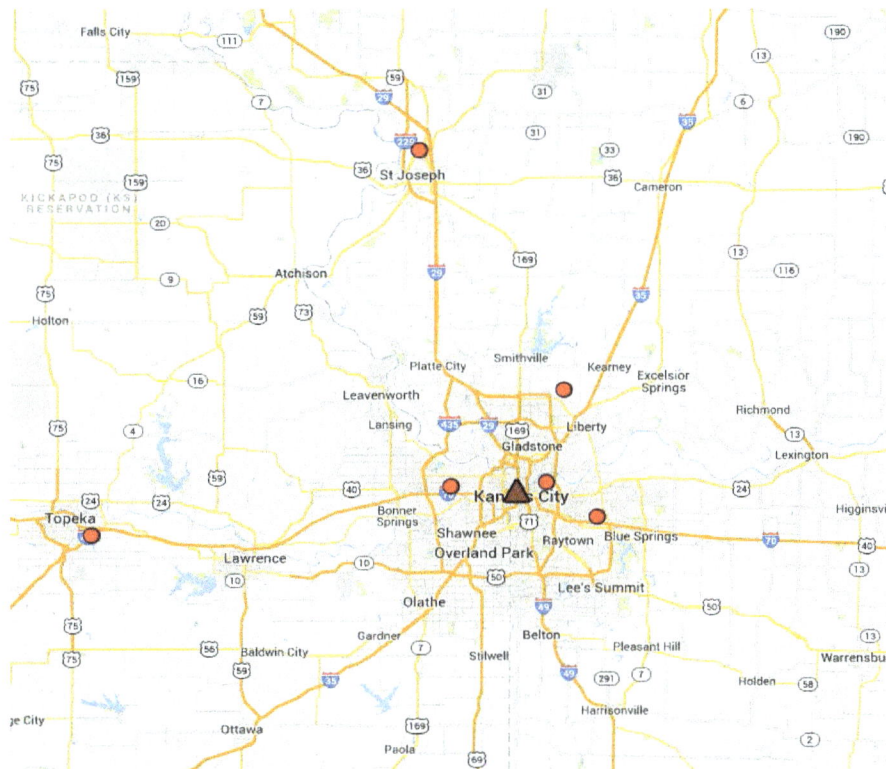

Figure 12. Cluster 2 with optimal facility location (Kansas City, MO—Brown triangle).

Figure 13. Cluster 3 with optimal facility location (Akron, OH—Brown triangle).

Figure 14. Cluster 4 with optimal facility location (Milwaukee, WI—Brown triangle).

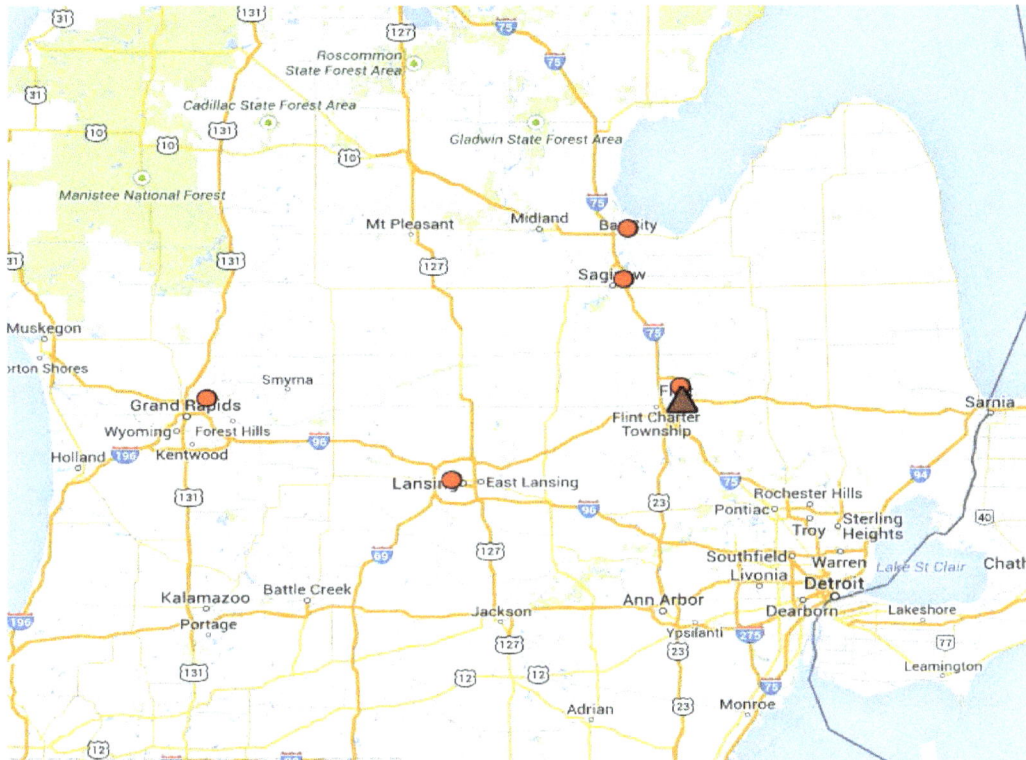

Figure 15. Cluster 5 with optimal facility location (Flint, MI—Brown triangle).

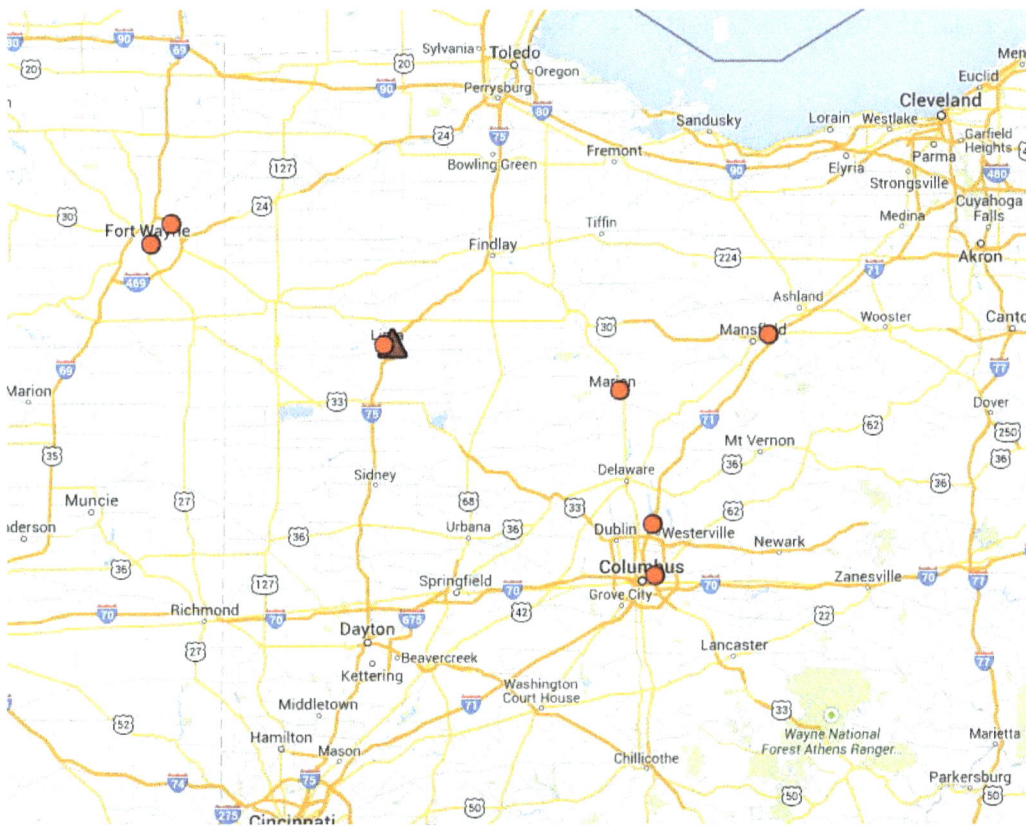

Figure 16. Cluster 6 with optimal facility location (Lima, OH—Brown triangle).

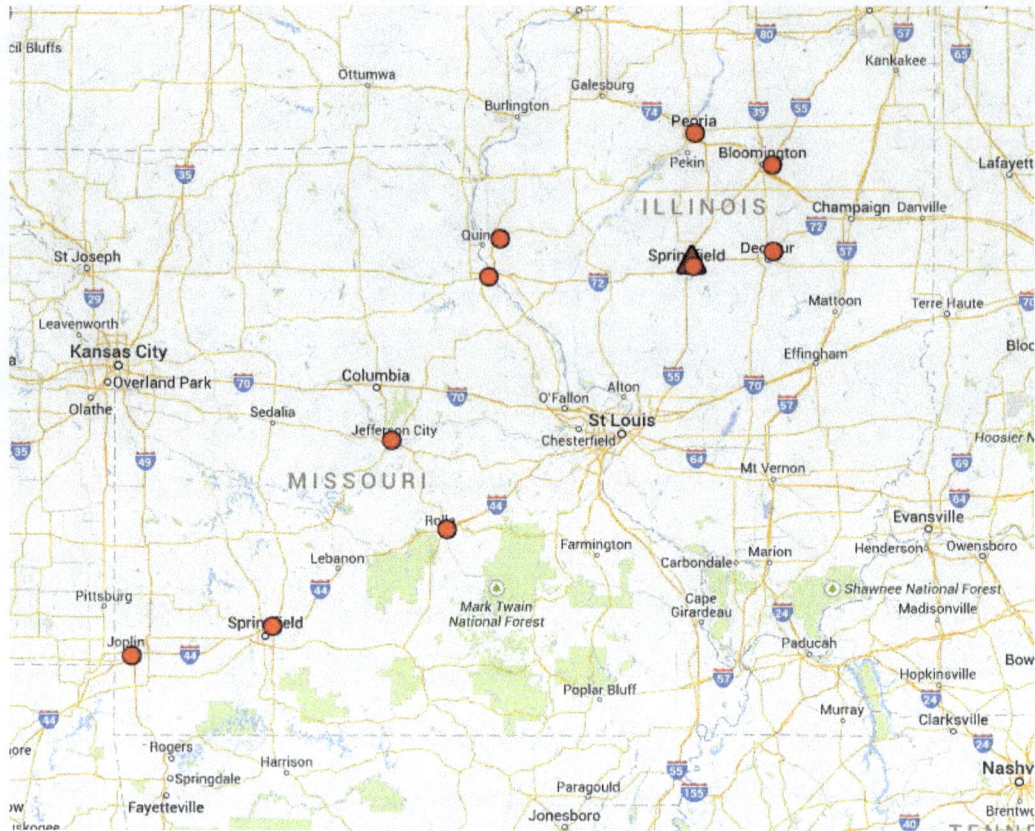

Figure 17. Cluster 7 with optimal facility location (Springfield, IL—Brown triangle).

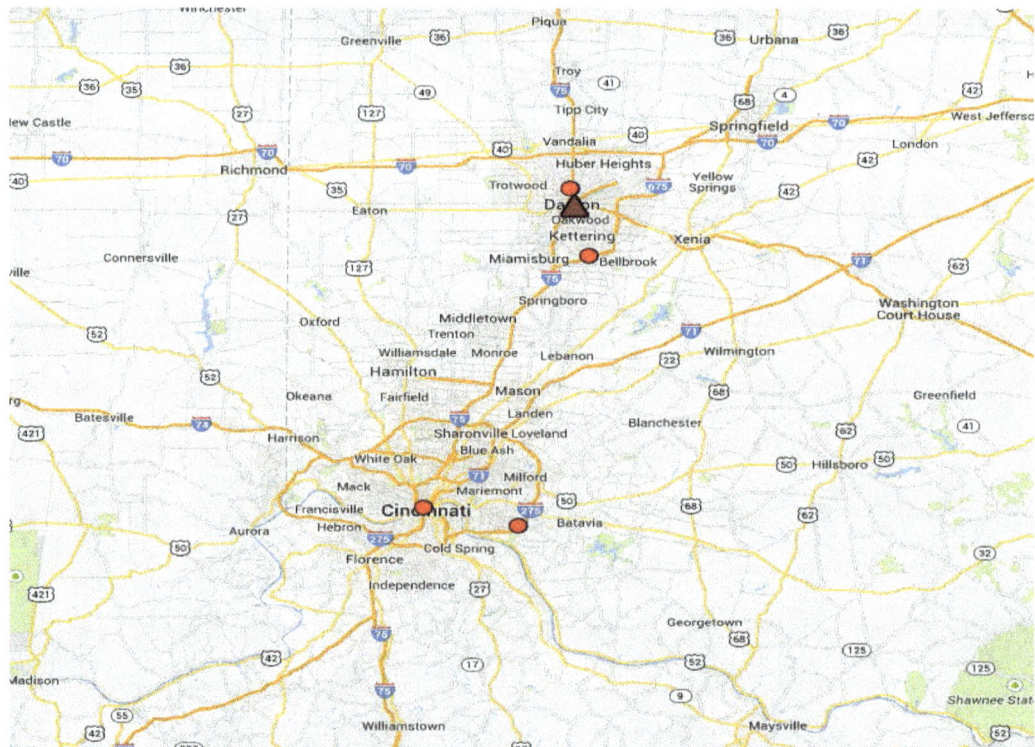

Figure 18. Cluster 8 with optimal facility location (Dayton, OH—Brown triangle).

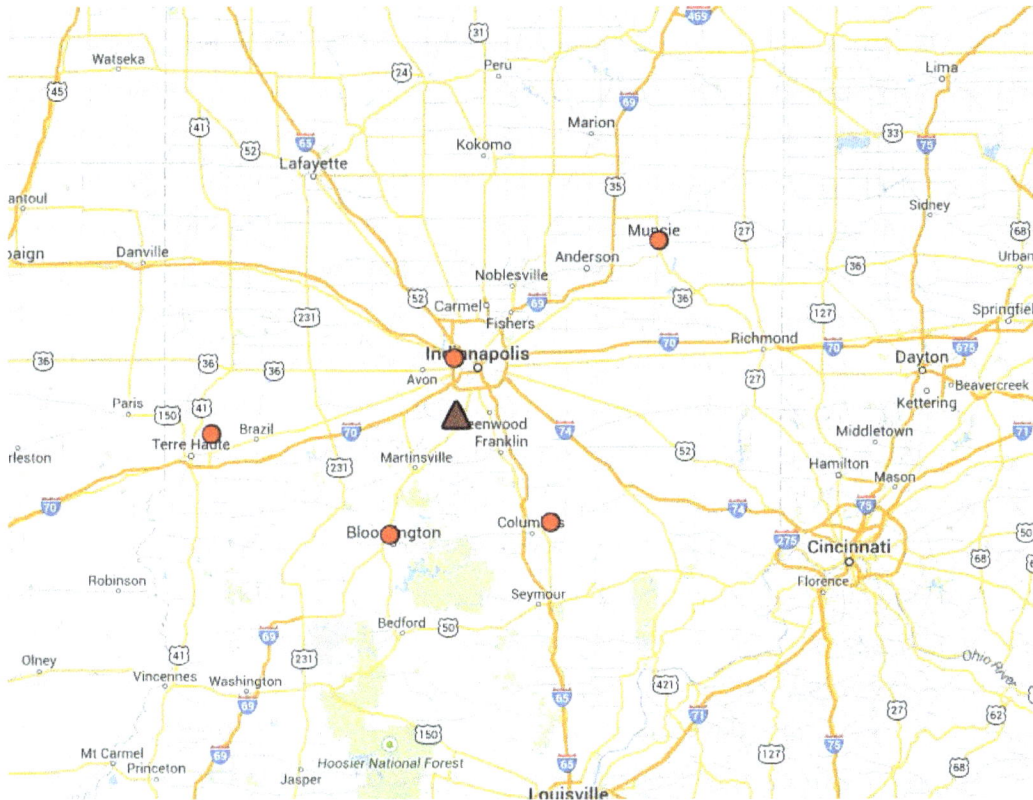

Figure 19. Cluster 9 with optimal facility location (Mooresville, IN—Brown triangle).

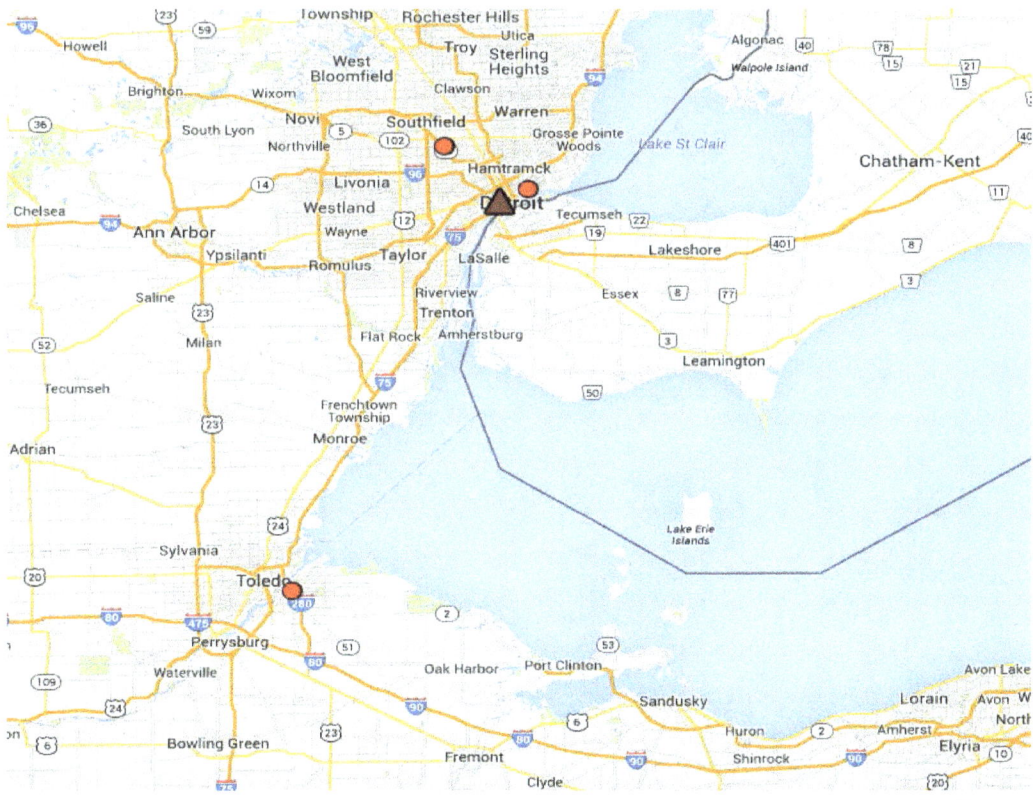

Figure 20. Cluster 10 with optimal facility location (Detroit, MI—Brown triangle).

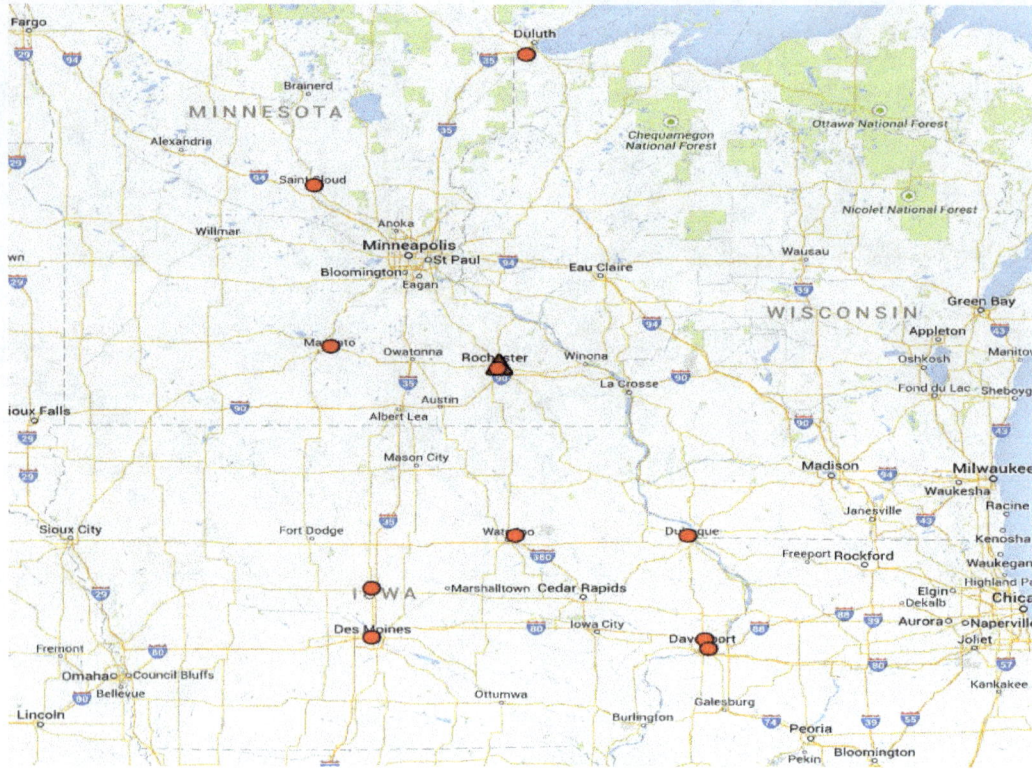

Figure 21. Cluster 11 with optimal location (Rochester, MN—Brown triangle).

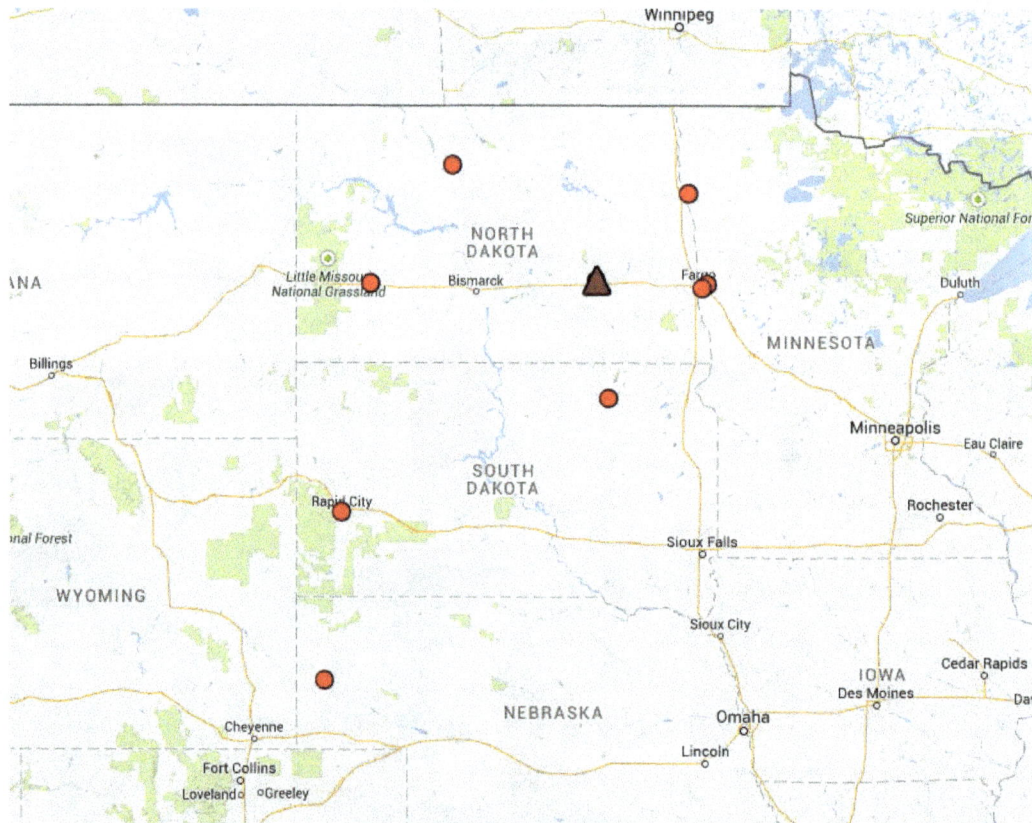

Figure 22. Cluster 12 with optimal facility location (Jamestown, ND—Brown triangle).

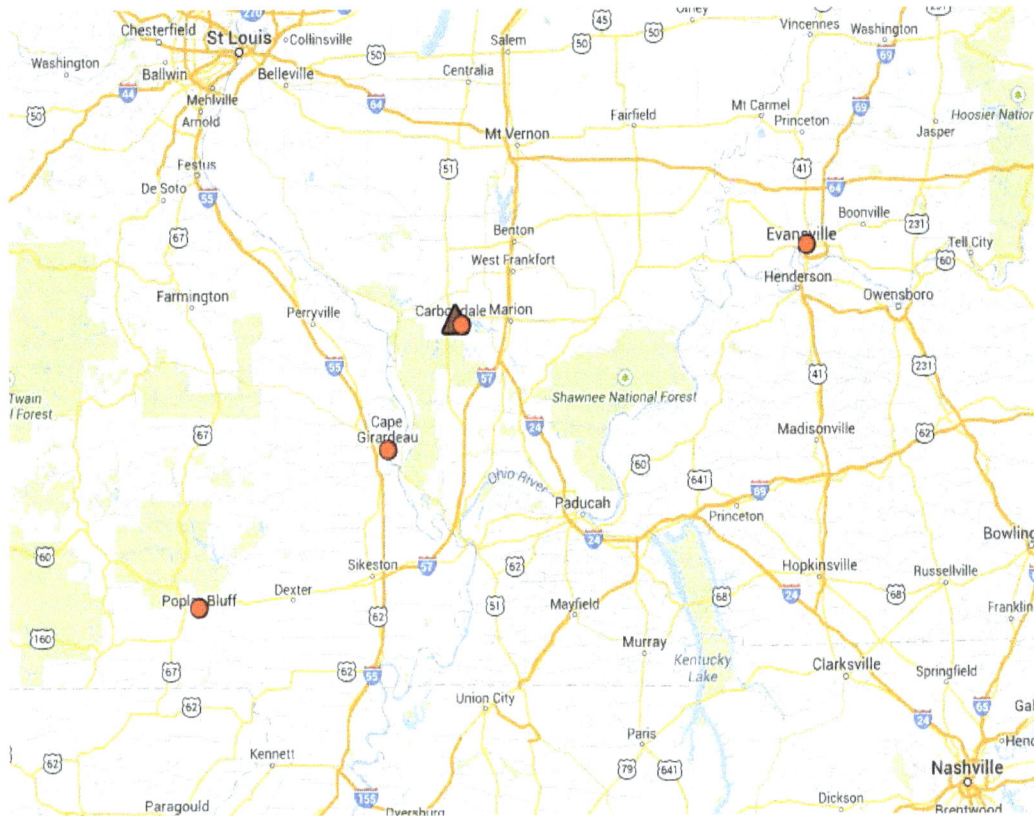

Figure 23. Cluster 13 with optimal location (Carbondale, IL—Brown triangle).

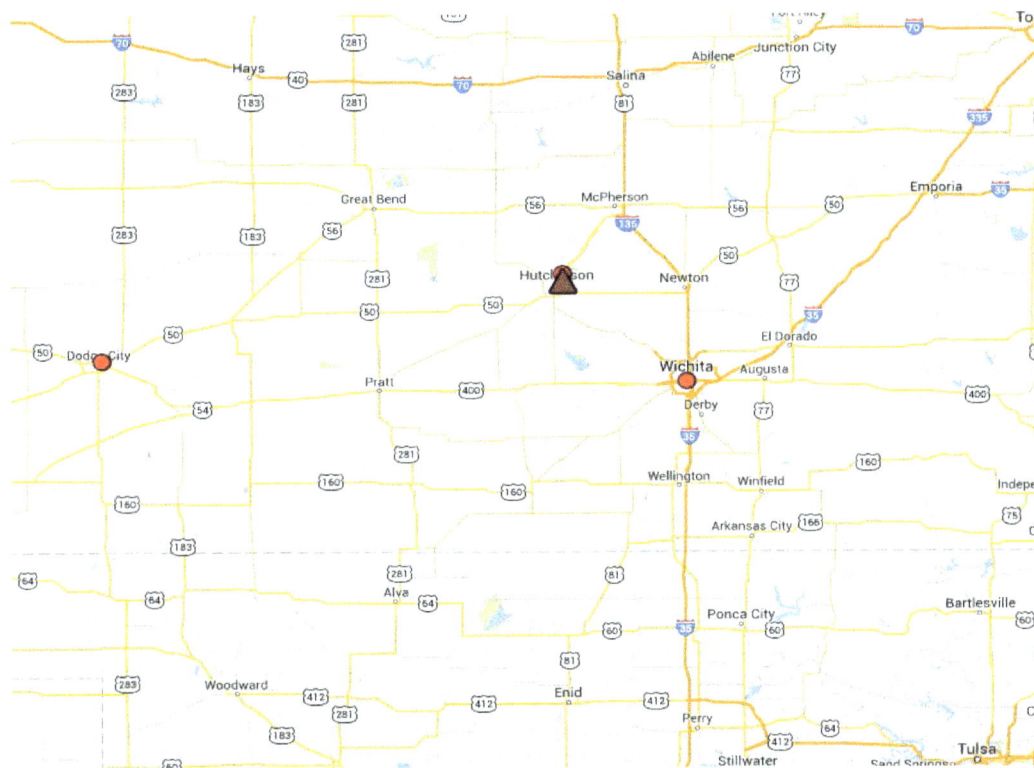

Figure 24. Cluster 14 with optimal facility location (Hutchinson, KS—Brown triangle).

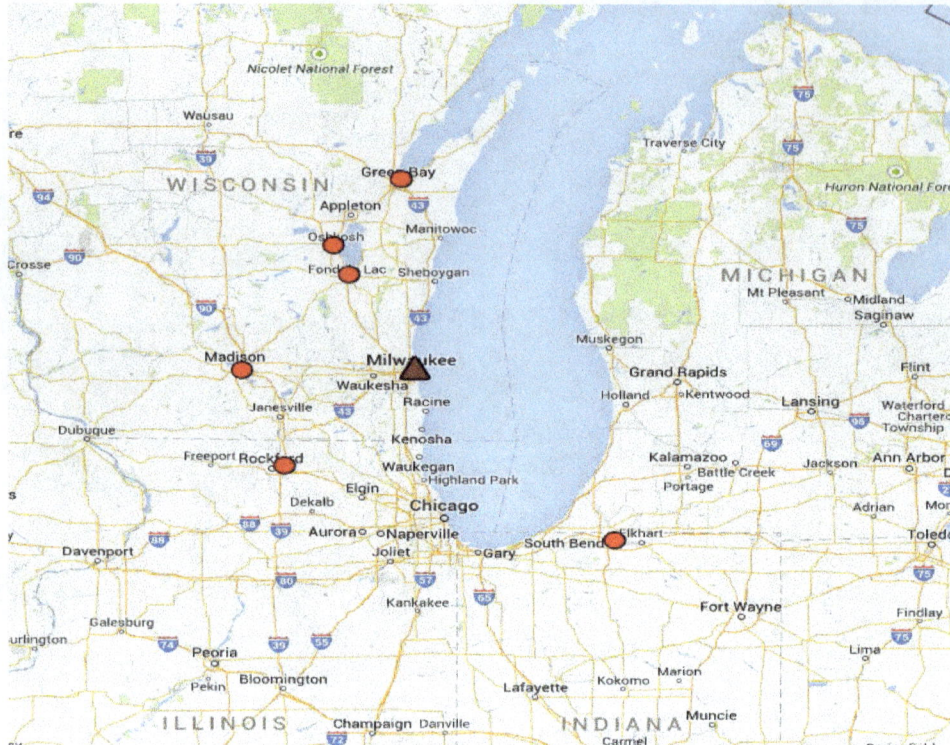

Figure 25. Optimal facility location (Milwaukee, WI) based on driving distance.

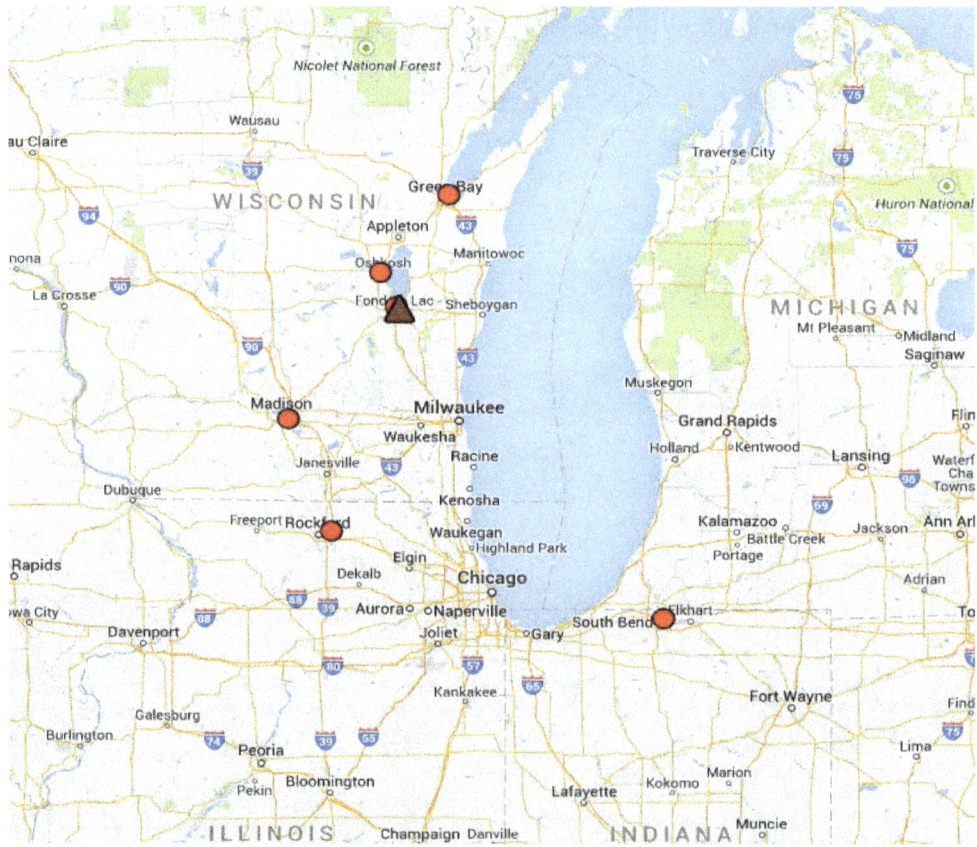

Figure 26. Optimal facility location (Fond du Lac, WI) based on Euclidean distance.

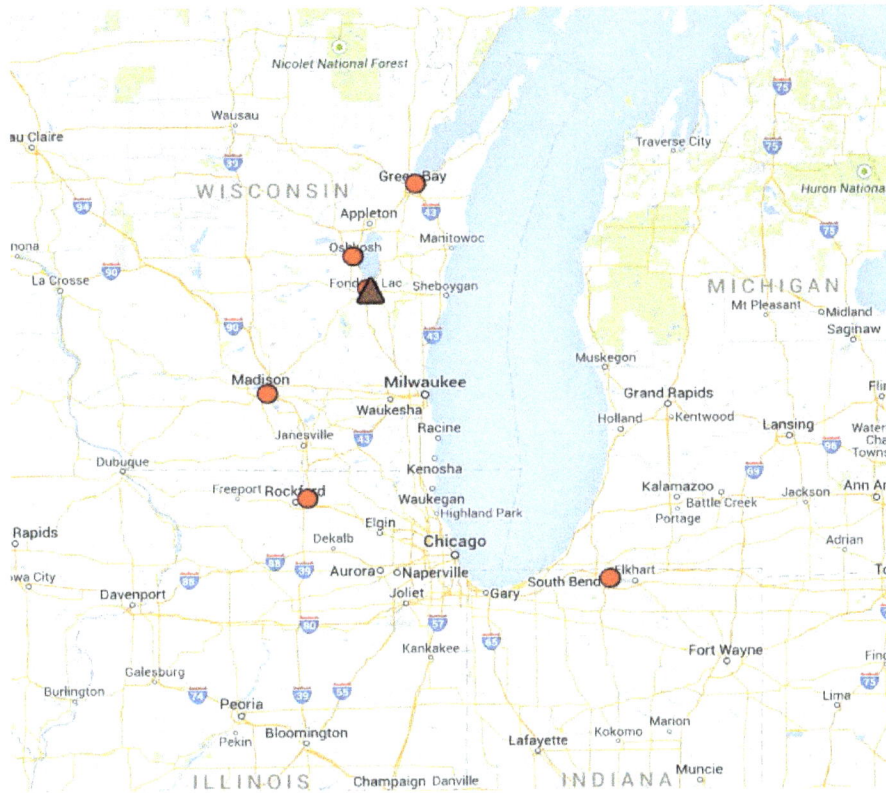

Figure 27. Optimal facility location (Fond du Lac, WI) based on Great circle distance.

Figure 28. Cluster 3 with optimal facility location (Akron, OH—Brown triangle) based on Euclidean distance or Great Circle distance.

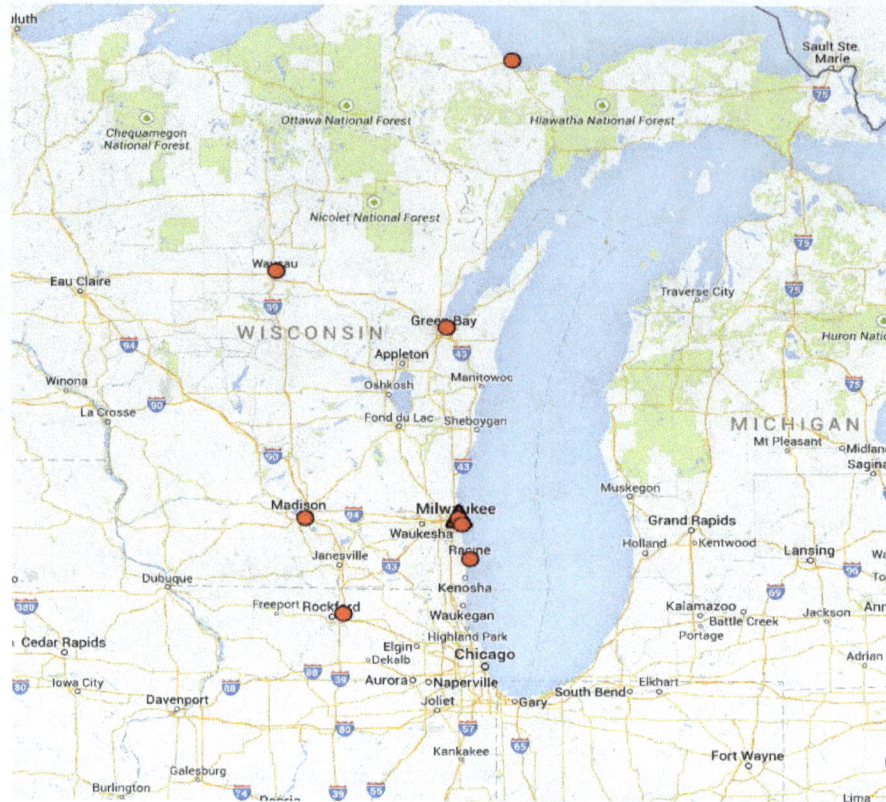

Figure 29. Cluster 4 with optimal facility (Milwaukee, WI—Brown triangle) based on Euclidean distance or Great Circle distance.

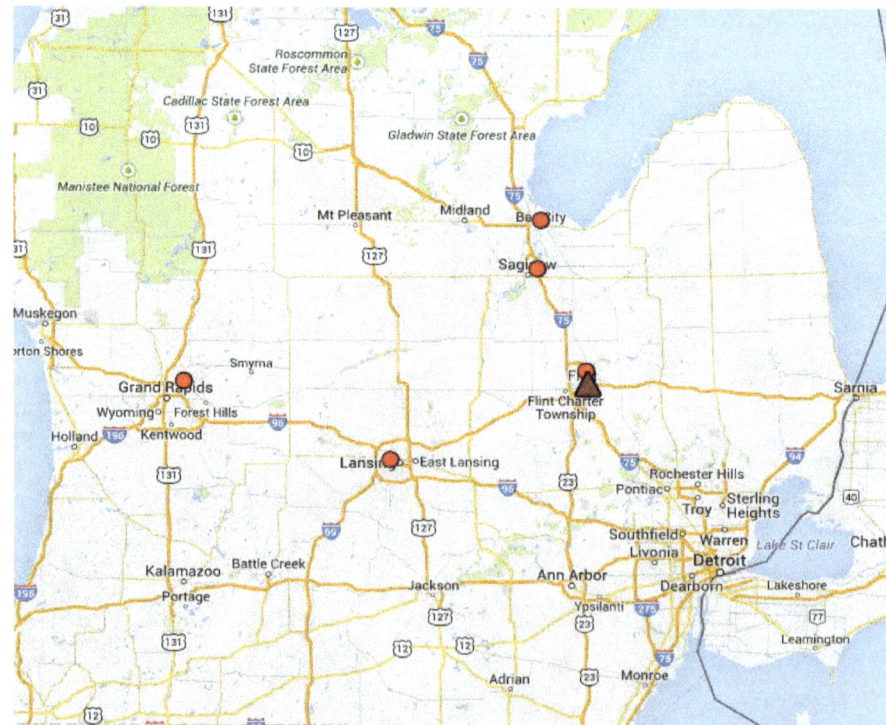

Figure 30. Cluster 5 with optimal facility location (Flint, MI—Brown triangle) based on Euclidean distance or Great Circle distance.

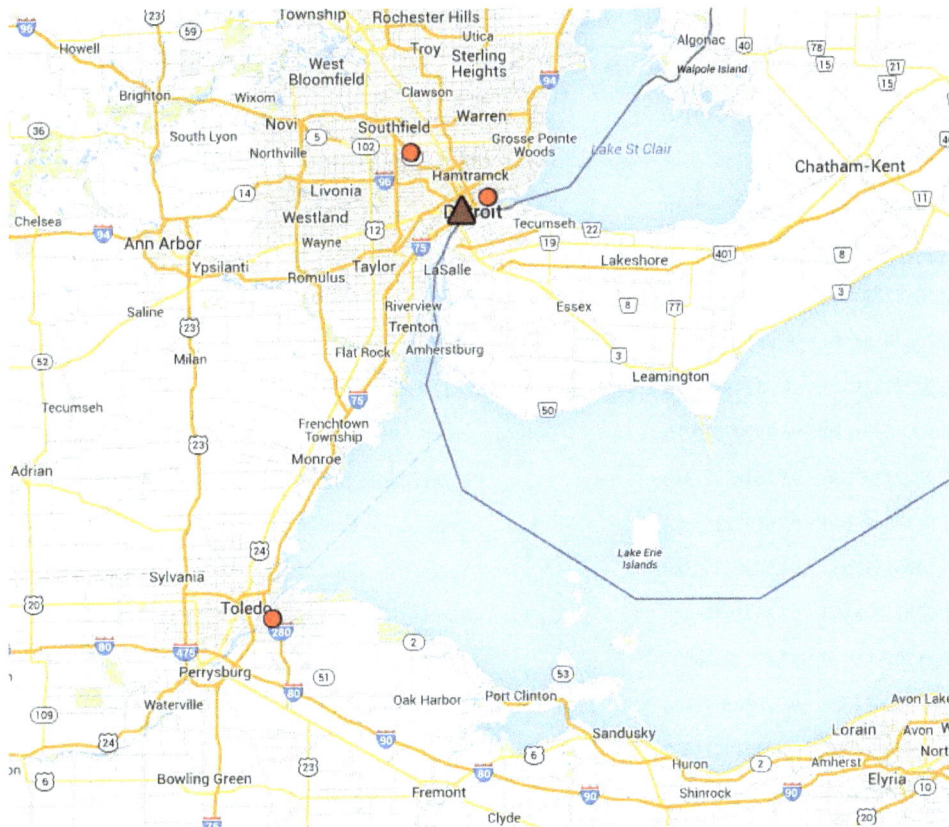

Figure 31. Cluster 10 with optimal facility (Detroit, MI—Brown triangle) based on Euclidean distance or Great Circle distance.

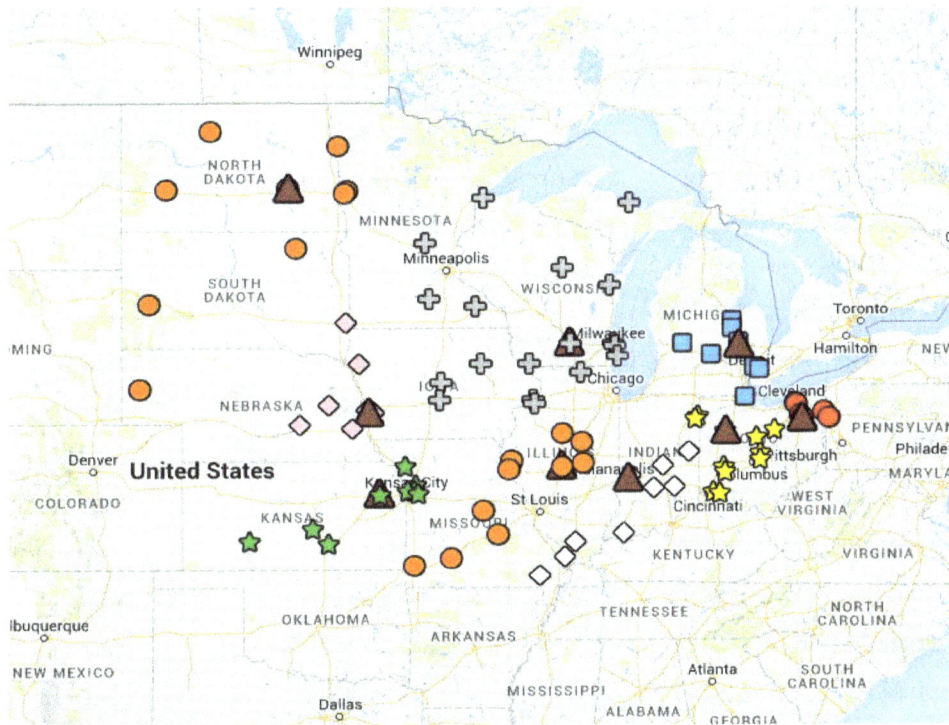

Figure 32. 9 clusters with 9 optimal facilities locations (Brown triangles).

Table 9. Cluster 1 costs from queried cities with Sioux City, Iowa as COG to market locations.

Latitudes/Longitudes of queried cities	Queried cities	Cities ID	Cost
42.50038901680, −96.39999210790	Center Of Gravity(Sioux City, Iowa, USA)	Depot = 1	Cost = 31469.3
44.16362082830, −93.99915674010	Mankato, Minnesota, USA	Depot = 2	Cost = 84665
43.64778668120, −93.36870426610	Albert Lea, Minnesota, USA	Depot = 3	Cost = 87730
43.14528505360, −95.14717452170	Spencer, Iowa, USA	Depot = 4	Cost = 52554.1
42.50682273260, −94.18025679700	Ft. Dodge, Iowa, USA	Depot = 5	Cost = 57508
40.70070558580, −99.08114628270	Kearney, Nebraska, USA	Depot = 6	Cost = 54229
40.92226828920, −98.35798628920	Grand Island, Nebraska, USA	Depot = 7	Cost = 39933.2
42.50038901680, −96.39999210790	Sioux City, Iowa, USA	Depot = 8	Cost = 31469.3
41.26227337550, −95.86080021340	Council Bluffs, Iowa, USA	Depot = 9	Cost = 21822.4
42.05385296540, −93.61972253600	Ames, Ames, Iowa, USA	Depot = 10	Cost = 64243
43.15401837060, −93.20083337580	Mason City, Iowa, USA	Depot = 11	Cost = 83813
42.02871238100, −97.43359826840	Norfolk, Nebraska, USA	Depot = 12	Cost = 31357.3
42.88201946930, −97.39248966650	Yankton, South Dakota, USA	Depot = 13	Cost = 42640.6
44.30676454580, −96.78803043800	Brookings, South Dakota, USA	Depot = 14	Cost = 65267
43.71429424950, −98.02619775570	Mitchell, South Dakota, USA	Depot = 15	Cost = 64908.4
40.81997479150, −96.68000085630	Lincoln, Nebraska, USA	Depot = 16	Cost = 25175
43.54998903310, −96.72999780450	Sioux Falls, South Dakota, USA	Depot = 17	Cost = 49678
41.57998008120, −93.61998091810	Des Moines, Iowa, USA	Depot = 18	Cost = 58147
41.24000083320, −96.00999007340	Omaha, Nebraska, USA	Depot = 19	Cost = 21222.6

Table 10. Cluster 1 costs from queried cities with Council Bluffs, Iowa as COG to market locations.

Latitudes/Longitudes of queried cities	Queried cities	Cities ID	Cost
41.26227337550, −95.86080021340	Center Of Gravity (Council Bluffs, Iowa, USA)	Depot = 1	Cost = 21822.4
41.01288291330, −92.41480900240	Ottumwa, Iowa, USA	Depot = 2	Cost = 83467
43.14528505360, −95.14717452170	Spencer, Iowa, USA	Depot = 3	Cost = 52554.1
42.50682273260, −94.18025679700	Ft. Dodge, Iowa, USA	Depot = 4	Cost = 57508
39.11358051690, −94.63014637710	Kansas City, Kansas, USA	Depot = 5	Cost = 71447
38.95975242010, −95.25522994160	Lawrence, Kansas, USA	Depot = 6	Cost = 74943
39.19402752590, −96.59243514180	Manhattan, Kansas, USA	Depot = 7	Cost = 61455
39.09111391100, −94.41528121110	Independence, Missouri, USA	Depot = 8	Cost = 75605
40.19368227040, −92.58280908380	Kirksville, Missouri, USA	Depot = 9	Cost = 101112
40.70070558580, −99.08114628270	Kearney, Nebraska, USA	Depot = 10	Cost = 54229
40.92226828920, −98.35798628920	Grand Island, Nebraska, USA	Depot = 11	Cost = 39933.2
42.50038901680, −96.39999210790	Sioux City, Iowa, USA	Depot = 12	Cost = 31469.3
41.26227337550, −95.86080021340	Council Bluffs, Iowa, USA	Depot = 13	Cost = 21822.4

Continued

42.05385296540, −93.61972253600	Ames, Iowa, USA	Depot = 14	Cost = 64243
43.15401837060, −93.20083337580	Mason City, Iowa, USA	Depot = 15	Cost = 83813
39.76903118800, −94.84639184750	St. Joseph, Missouri, USA	Depot = 16	Cost = 56430
42.02871238100, −97.43359826840	Norfolk, Nebraska, USA	Depot = 17	Cost = 31357.3
42.88201946930, −97.39248966650	Yankton, South Dakota, USA	Depot = 18	Cost = 42640.6
40.81997479150, −96.68000085630	Lincoln, Nebraska, USA	Depot = 19	Cost = 25175
39.05000530910, −95.66998498710	Topeka, Kansas, USA	Depot = 20	Cost = 65715
39.10708850980, −94.60409421890	Kansas City, Missouri, USA	Depot = 21	Cost = 72082
43.54998903310, −96.72999780450	Sioux Falls, South Dakota, USA	Depot = 22	Cost = 49678
41.57998008120, −93.61998091810	Des Moines, Iowa, USA	Depot = 23	Cost = 58147
41.24000083320, −96.00999007340	Omaha, Nebraska, USA	Depot = 24	Cost = 21222.6

Table 11. Cluster 1 costs from queried cities with Norfolk, Nebraska as COG to market locations.

Latitudes/Longitudes of queried cities	Queried cities	Cities ID	Cost
42.02871238100, −97.43359826840	Center Of Gravity (Norfolk, Nebraska, USA)	Depot = 1	Cost = 31357.3
43.14528505360, −95.14717452170	Spencer, Iowa, USA	Depot = 2	Cost = 52554.1
42.50682273260, −94.18025679700	Ft. Dodge, Iowa, USA	Depot = 3	Cost = 57508
40.70070558580, −99.08114628270	Kearney, Nebraska, USA	Depot = 4	Cost = 54229
40.92226828920, −98.35798628920	Grand Island, Nebraska, USA	Depot = 5	Cost = 39933.2
42.50038901680, −96.39999210790	Sioux City, Iowa, USA	Depot = 6	Cost = 31469.3
41.26227337550, −95.86080021340	Council Bluffs, Iowa, USA	Depot = 7	Cost = 21822.4
42.02871238100, −97.43359826840	Norfolk, Nebraska, USA	Depot = 8	Cost = 31357.3
41.13628623360, −100.77050053100	North Platte, Nebraska, USA	Depot = 9	Cost = 82006
42.88201946930, −97.39248966650	Yankton, South Dakota, USA	Depot = 10	Cost = 42640.6
44.30676454580, −96.78803043800	Brookings, South Dakota, USA	Depot = 11	Cost = 65267
43.71429424950, −98.02619775570	Mitchell, South Dakota, USA	Depot = 12	Cost = 64908.4
40.81997479150, −96.68000085630	Lincoln, Nebraska, USA	Depot = 13	Cost = 25175
43.54998903310, −96.72999780450	Sioux Falls, South Dakota, USA	Depot = 14	Cost = 49678
41.24000083320, −96.00999007340	Omaha, Nebraska, USA	Depot = 15	Cost = 21222.6

Table 16 is the summary of total cost for each set of clusters. From the result shown in **Table 16**, it is concluded that the set of 9 clusters with 9 optimal facilities locations are selected shown as in **Figure 32**.

4. Conclusions and Future Research

Facility location decisions play an important role in the strategic planning and design of logistics/supply chain network. Well-planned location decisions enable the efficient flow of materials through the distribution system,

Table 12. Cluster 1 costs from queried cities with Yankton, South Dakota as COG to market locations.

Latitudes/Longitudes of queried cities	Queried cities	Cities ID	Cost
42.88201946930, −97.39248966650	Center Of Gravity (Yankton, South Dakota, USA)	Depot = 1	Cost = 42,640.6
44.16362082830, −93.99915674010	Mankato, Minnesota, USA	Depot = 2	Cost = 84,665
43.64778668120, −93.36870426610	Albert Lea, Minnesota, USA	Depot = 3	Cost = 87,730
45.12188275050, −95.04330489270	Willmar, Minnesota, USA	Depot = 4	Cost = 92,692
43.14528505360, −95.14717452170	Spencer, Iowa, USA	Depot = 5	Cost = 52,554.1
42.50682273260, −94.18025679700	Ft. Dodge, Iowa, USA	Depot = 6	Cost = 57,508
40.70070558580, −99.08114628270	Kearney, Nebraska, USA	Depot = 7	Cost = 54,229
40.92226828920, −98.35798628920	Grand Island, Nebraska, USA	Depot = 8	Cost = 39,933.2
42.50038901680, −96.39999210790	Sioux City, Iowa, USA	Depot = 9	Cost = 31,469.3
41.26227337550, −95.86080021340	Council Bluffs, Iowa, USA	Depot = 10	Cost = 21,822.4
42.05385296540, −93.61972253600	Ames, Iowa, USA	Depot = 11	Cost = 64,243
43.15401837060, −93.20083337580	Mason City, Iowa, USA	Depot = 12	Cost = 83,813
42.02871238100, −97.43359826840	Norfolk, Nebraska, USA	Depot = 13	Cost = 31,357.3
41.13628623360, −100.77050053100	North Platte, Nebraska, USA	Depot = 14	Cost = 82,006
42.88201946930, −97.39248966650	Yankton, South Dakota, USA	Depot = 15	Cost = 42,640.6
44.30676454580, −96.78803043800	Brookings, South Dakota, USA	Depot = 16	Cost = 65,267
43.71429424950, −98.02619775570	Mitchell, South Dakota, USA	Depot = 17	Cost = 64,908.4
45.46511761380, −98.48640222350	Aberdeen, South Dakota, USA	Depot = 18	Cost = 105,103
40.81997479150, −96.68000085630	Lincoln, Nebraska, USA	Depot = 19	Cost = 25,175
43.54998903310, −96.72999780450	Sioux Falls, South Dakota, USA	Depot = 20	Cost = 49,678
41.57998008120, −93.61998091810	Des Moines, Iowa, USA	Depot = 21	Cost = 58,147
41.24000083320, −96.00999007340	Omaha, Nebraska, USA	Depot = 22	Cost = 21,222.6
44.36833701480, −100.35055200400	Pierre, Pierre, USA	Depot = 23	Cost = 108,595

Table 13. Six locations with volume and transportation rate.

Location	Latitude	Longitude	Volume	Rate
WI Loc1	43.037647	−89.39154	900.00	0.06
WI Loc2	44.530892	−88.04482	1,100.00	0.06
WI Loc3	44.018871	−88.61324	800.00	0.06
WI Loc4	43.785391	−88.48704	900.00	0.06
WI Loc5	42.280019	−89.03347	500.00	0.08
IN Loc6	41.683381	−86.25001	1,500.00	0.07

and lead to decreased costs and improved customer service. This paper has focused on the implementation of facility location decisions based on driving distances on a sphere surface. Two objectives have been achieved in this paper. Given the location of each destination in terms of their coordinates, the requirement at each destination and shipping costs for the region of interest, the proposed methodology in this paper is able to determine the

Table 14. Modeclus cluster procedure with m = 1, K = 4.

| | | Sums of Density Estimates Within Neighborhood | | | | | | | | | |
| | | Estimated | Same | Other | | | | Estimated | Same | Other | |
Cluster	Location	Density	Cluster	Clusters	Total	Cluster	Location	Density	Cluster	Clusters	Total
1	Loc92	0.000878	0.0027064	0	0.002706	7	Loc4	0.000283	0.0006606	0.0003654	0.001026
	Loc93	0.00127	0.0023142	0	0.002314		Loc32	0.000313	0.0009782	0	0.000978
	Loc94	0.00075	0.0028338	0	0.002834		Loc33	0.000348	0.0009438	0	0.000944
	Loc95	0.000389	0.0027064	0	0.002706		Loc34	0.000348	0.0008367	0	0.000837
	Loc1A	0.000687	0.0028973	0	0.002897		Loc35	0.000277	0.0008376	0	0.000838
	Loc1B	0.00031	0.0028338	0	0.002834		Loc36	0.000212	0.000902	0	0.000902
	Loc1C	0.000149	0.0005339	0	0.000534		Loc37	0.000277	0.0008067	0	0.000807
	Loc1D	0.000141	0.0005413	0	0.000541		Loc40	0.000182	0.0008376	0	0.000838
	Loc1E	8.26E−05	0.0005998	0	0.0006	8	Loc80	0.00027	0.0009166	0	0.000917
2	Loc5	0.000571	0.0019546	0	0.001955		Loc81	0.000338	0.0008487	0	0.000849
	Loc6	0.000685	0.0018411	0	0.001841		Loc83	0.000188	0.0007759	0	0.000776
	Loc8	0.00056	0.0019658	0	0.001966		Loc84	0.000285	0.0009017	0	0.000902
	Loc9	0.00071	0.0018154	0	0.001815		Loc85	0.000293	0.0008936	0	0.000894
	Loc10	0.000402	0.0016097	0	0.00161		Loc89	0.000212	0.0006533	0	0.000653
	Loc11	0.000365	0.0016723	0	0.001672		Loc96	0.000137	0.0001684	0.0021472	0.002316
3	Loc1	0.000335	0.0012727	0	0.001273		Loc97	0.000172	0.0005614	0	0.000561
	Loc2	0.000414	0.001194	0	0.001194		Loc98	0.000161	0.0003405	0.0001612	0.000502
	Loc3	0.000454	0.0011543	0	0.001154		Loc99	0.000168	0.0004703	0	0.00047
	Loc12	0.000405	0.0008676	0.00071	0.001578	9	Loc41	8.95E−05	0.0003103	0.0001001	0.00041
	Loc18	0.000365	0.0009998	0	0.001		Loc51	0.000268	0.0008105	0	0.000811
	Loc23	0.00029	0.0006213	0.000283	0.000905		Loc52	0.000285	0.0007934	0	0.000793
	Loc24	0.000256	0.0006557	0.000283	0.000939		Loc53	0.000258	0.0008063	0	0.000806
4	Loc45	0.000238	0.000947	0	0.000947		Loc54	0.000268	0.0008063	0	0.000806
	Loc46	0.000391	0.0012231	0	0.001223		Loc55	0.000185	0.0006787	0	0.000679
	Loc67	0.000158	0.0010277	0.000142	0.00117		Loc56	0.000126	0.0007376	0	0.000738
	Loc1F	0.000398	0.0012163	0	0.001216		Loc78	0.000253	0.0007934	0	0.000793
	Loc1G	0.000438	0.0011764	0	0.001176	10	Loc42	0.000137	0.0003815	0.0003912	0.000773
	Loc1H	0.000387	0.0012271	0	0.001227		Loc43	0.00018	0.000338	0.0003912	0.000729
	Loc1I	0.000272	0.0009818	0	0.000982		Loc44	0.000201	0.0005398	0	0.00054
	Loc1J	0.000157	0.0010973	0	0.001097		Loc47	0.000256	0.000649	0	0.000649
5	Loc13	0.000412	0.0011216	0	0.001122		Loc50	0.000216	0.0004877	0.0002702	0.000758
	Loc14	0.000357	0.0011765	0	0.001177		Loc62	0.0001	0.0004334	0	0.000433

Continued

	Location	Estimated Density	Same Cluster	Other Clusters	Total	Cluster	Location	Estimated Density	Same Cluster	Other Clusters	Total
	Loc15	0.000357	0.0011765	0	0.001177		Loc63	0.000147	0.0004879	0	0.000488
	Loc16	0.000408	0.0011253	0	0.001125		Loc64	0.000142	0.0002914	0.0001578	0.000449
6	Loc19	0.000321	0.0009283	0	0.000928		Loc65	0.000145	0.0003888	0	0.000389
	Loc27	0.000267	0.0006996	0.000412	0.001111		Loc79	0.000231	0.0004727	0.0002702	0.000743
	Loc28	0.000283	0.0003215	0.000647	0.000969	11	Loc66	0.000228	0.0005871	0	0.000587
	Loc29	0.000378	0.0008006	0	0.000801		Loc68	0.00012	0.0006839	0	0.000684
	Loc30	0.000188	0.0007724	0	0.000772		Loc69	6.7E−05	0.0002781	0	0.000278
	Loc31	0.000212	0.0006996	0.000285	0.000985		Loc70	0.000235	0.0005802	0	0.00058
	Loc86	0.000182	0.0005102	0	0.00051		Loc71	0.000131	0.0006839	0	0.000684
	Loc90	0.000161	0.0005308	0	0.000531		Loc72	0.000221	0.0005835	0	0.000583
	Loc91	0.000161	0.000343	0.000161	0.000504		Loc74	9.88E−05	0.0003944	0	0.000394
							Loc75	0.000107	0.0004508	0	0.000451
							Loc1K	5.9E−05	6.704E−05	0.0004289	0.000496

Table 15. Modeclus cluster procedure with m = 1, K = 5.

Sums of Density Estimates Within Neighborhood

Cluster	Location	Estimated Density	Same Cluster	Other Clusters	Total	Cluster	Location	Estimated Density	Same Cluster	Other Clusters	Total
1	Loc5	0.0004	0.0021	0	0.0021	6	Loc19	0.0004	0.0012	0	0.0012
	Loc6	0.0005	0.002	0	0.002		Loc27	0.0003	0.0009	0.0003	0.0013
	Loc8	0.0006	0.0019	0	0.0019		Loc28	0.0003	0.0007	0.0007	0.0014
	Loc9	0.0005	0.0015	0.0005	0.002		Loc29	0.0003	0.0013	0	0.0013
	Loc10	0.0005	0.002	0	0.002		Loc30	0.0002	0.0008	0	0.0008
	Loc11	0.0004	0.0021	0	0.0021		Loc31	0.0002	0.0008	0.0003	0.0011
2	Loc92	0.0004	0.002	0	0.002		Loc86	0.0002	0.0005	0.0003	0.0008
	Loc93	0.0005	0.0019	0	0.0019		Loc90	0.0002	0.0005	0.0002	0.0007
	Loc94	0.0005	0.0018	0	0.0018		Loc91	0.0001	0.0003	0.0003	0.0007
	Loc95	0.0004	0.002	0	0.002	7	Loc41	1E−04	0.0008	0	0.0008
	Loc1A	0.0005	0.0018	0	0.0018		Loc42	0.0002	0.0005	0.0006	0.001
	Loc1B	0.0004	0.002	0	0.002		Loc43	0.0002	0.0004	0.0006	0.001
	Loc1C	0.0002	0.0006	0.0002	0.0007		Loc44	0.0002	0.0008	0	0.0008
	Loc1D	0.0001	0.0011	0	0.0011		Loc47	0.0003	0.001	0	0.001
	Loc1E	1E−04	0.0006	0.0002	0.0008		Loc50	0.0002	0.0008	0.0003	0.0011
3	Loc1	0.0004	0.0017	0	0.0017		Loc51	0.0003	0.0012	0	0.0012
	Loc2	0.0003	0.0017	0	0.0017		Loc52	0.0003	0.0012	0	0.0012
	Loc3	0.0005	0.0016	0	0.0016		Loc53	0.0003	0.0012	0	0.0012

Continued

	Loc						Loc				
	Loc12	0.0005	0.0008	0.0011	0.0019		Loc54	0.0003	0.0012	0	0.0012
	Loc13	0.0003	0.001	0.0003	0.0014		Loc55	0.0002	0.0011	0	0.0011
	Loc14	0.0003	0.0011	0.0003	0.0014		Loc56	0.0001	0.0011	0	0.0011
	Loc15	0.0004	0.0014	0	0.0014		Loc62	0.0001	0.0006	0.0002	0.0008
	Loc16	0.0004	0.0014	0	0.0014		Loc63	0.0002	0.0009	0	0.0009
	Loc18	0.0004	0.0014	0	0.0014		Loc64	0.0001	0.0003	0.0004	0.0008
	Loc23	0.0003	0.0007	0.0006	0.0013		Loc65	0.0002	0.0005	0.0002	0.0006
	Loc24	0.0003	0.0007	0.0006	0.0013		Loc78	0.0003	0.0012	0	0.0012
4	Loc4	0.0003	0.0006	0.0009	0.0015		Loc79	0.0002	0.0008	0.0003	0.0011
	Loc32	0.0003	0.0014	0	0.0014	8	Loc80	0.0003	0.0008	0.0002	0.001
	Loc33	0.0003	0.0013	0	0.0013		Loc81	0.0002	0.0008	0.0002	0.0011
	Loc34	0.0004	0.0012	0	0.0012		Loc83	0.0002	0.001	0	0.001
	Loc35	0.0003	0.0014	0	0.0014		Loc84	0.0003	0.0008	0.0002	0.001
	Loc36	0.0002	0.0014	0	0.0014		Loc85	0.0003	0.001	0	0.001
	Loc37	0.0003	0.0013	0	0.0013		Loc89	0.0002	0.001	0	0.001
	Loc40	0.0002	0.001	0.0003	0.0013		Loc96	0.0002	0.0002	0.0014	0.0016
5	Loc45	0.0003	0.0012	0	0.0012		Loc97	0.0002	0.0008	0	0.0008
	Loc46	0.0003	0.0013	0	0.0013		Loc98	0.0002	0.0006	0.0001	0.0007
	Loc67	0.0002	0.0009	0.0001	0.0011		Loc99	0.0002	0.0005	0.0004	0.001
	Loc1F	0.0003	0.0013	0	0.0013	9	Loc66	0.0002	0.0005	0.0002	0.0006
	Loc1G	0.0004	0.0013	0	0.0013		Loc68	0.0001	0.0005	0.0002	0.0007
	Loc1H	0.0003	0.0013	0	0.0013		Loc69	8E−05	0.0003	0.0002	0.0005
	Loc1I	0.0003	0.0012	0	0.0012		Loc70	0.0002	0.0005	0.0002	0.0007
	Loc1J	0.0002	0.0014	0	0.0014		Loc71	0.0001	0.0006	0	0.0006
							Loc72	0.0002	0.0006	0	0.0006
							Loc74	1E−04	0.0005	0	0.0005
							Loc75	0.0001	0.0006	0	0.0006
							Loc1K	7E−05	8E−05	0.0009	0.0009

Table 16. Summary of total cost for each set of clusters.

14 clusters:

Cluster 1	Cluster 2	Cluster 3	Cluster 4	Cluster 5	Cluster 6	Cluster 7	Cluster 8	Cluster 9	Cluster 10
21,222.6	5,593.8	3,392.8	38,896.4	11,857.5	11,339.2	57,779.0	1,660.4	9,898.2	1,805.6

Cluster 11	Cluster 12	Cluster 13	Cluster 14	# of Facility	Facility cost per unit	Facilities cost	Total Cost
92,088.5	72,057.0	12,892.2	8,700.0	14	100,000.00	1,400,000.00	1,749,183.2

11 clusters:

Cluster 1	Cluster 2	Cluster 3	Cluster 4	Cluster 5	Cluster 6	Cluster 7	Cluster 8	Cluster 9	Cluster 10
46,194.0	3,392.8	11,339.2	21,222.6	1,660.4	46,711.6	19,995.0	57,779.0	38,896.4	92,088.5

Cluster 11	# of Facility	Facility cost per unit	Facilities cost	Total Cost
72,057.0	11	100,000.00	1,100,000.00	1,511,336.5

9 clusters:

Cluster 1	Cluster 2	Cluster 3	Cluster 4	Cluster 5	Cluster 6	Cluster 7	Cluster 8	Cluster 9
3,392.8	46,194.0	18,765.6	19,995.0	21,222.6	46,711.6	180,546.4	57,779.0	72,057.0

# of Facility	Facility cost per unit	Facilities cost	Total Cost
9	100,000.00	900,000.00	1,366,664.0

optimal location of each facility and helps companies assess the locations of facilities. The second object is to establish terminology and an analytical framework for locating optimal facility in perspective. In regard to transportation cost, the driving distance in the presence of geographic barriers should be taken into consideration in facility location decisions. The proposed method in this paper has shown very promising results. The potential benefits of my work include:

• The driving distance measure can be applied to transportation problems, travelling sales person problems, and etc., in Operations Research (OR) study.

• This proposed method can be extended to facility location decisions based on Euclidean distance and Great circle distance functions.

• The framework and methodology proposed in this paper can be applied to improve the management of logistics/supply chain system decisions and planning to make smooth flow throughout the supply chain.

This paper has focused on the facility locations in the region of Midwest, USA. Markets and competition are increasingly global, further research work could include investigation in the process of international facility locations around the world. Therefore, a global driving distance model within and/or outside the US is needed for further work. In addition to qualitative analysis, some factors affecting locations need to be considered, such as labor costs and availability, including wages, productivity, attitudes, age, distribution, unionization, and skills, State and local government fiscal policies (including incentives, taxes, unemployment compensation), proximity to customers, population distribution, quality-of-life issues (education, cost of living, health care, sports, cultural activities, housing, entertainment, religious facilities, etc.).

This paper assumes a spherical earth and uses equation (1) to (7) to calculate initial center of gravity and equation (8) to (10) to convert Cartesian coordinated to latitude/longitudes. Further research could be done on assumption of ellipsoidal earth. Also, further research work can be suggested to apply this proposed methodology on capacity facility location decisions problems.

Geographic Information Systems (GIS) also helps in location analysis; further research work can be approached using GIS instead of Google maps.

Finally, since uncertainty associated with future conditions exists in real world, facility location decisions must account for the inherent uncertainty. This is an area worthy of additional research too.

Acknowledgements

The author is very grateful to the editor and anonymous referees for their valuable comments and suggestions on the earlier version of the manuscript.

References

[1]   Reid, R.D. and Sanders, N.R. (2013) Operations Management. 5th Edition, John Wiley & Sons, Hoboken.

[2] Wikipedia (2014) Facility Location Problem. http://en.wikipedia.org/wiki/Facility_location_problem

[3] Brimberg, J., Hansen, P., Mladenovic, N. and Salhi, S. (2008) A Survey of Solution Methods for the Continuous Location-Allocation Problem. *International Journal of Operations Research*, **5**, 1-12.

[4] Tcha, D.W. and Lee, B.I. (1984) A Branch-and-Bound Algorithm for the Multi-Level Uncapacitated Facility Location Problem. *European Journal of Operational Research*, **18**, 35-43. http://dx.doi.org/10.1016/0377-2217(84)90258-3

[5] Francis, R.L. and White J.A. (1974) Facility Layout and Location: An Analytical Approach. Prentice-Hall, Inc., Englewood Cliffs.

[6] Love, R.F., James, J., Morris, G. and Wesolowsky, G.O. (1988) Facilities Location: Models & Methods. North-Holland Publishing Co., New York.

[7] Farahani, R.Z. and Masoud, H. (2009) Facility Location: Concepts, Models, Algorithm and Case Studies, Springer-Verlag Berlin Heidelberg, Germany. http://dx.doi.org/10.1007/978-3-7908-2151-2

[8] Vygen, J. (2004-2005) Approximation Algorithms for Facility Location Problems (Lecture Notes). Research Institute for Discrete Mathematics, University of Bonn, Bonn, Germany.

[9] Shmoys, D.B., Tardos, E. and Aardal, K. (1998) Approximation Algorithms for Facility Location Problems (Extended Abstract). *Proceedings of the 29th Annual ACM Symposium on Theory of Computing*, El Paso, 4-6 May 1997, 1-21.

[10] Iyigun, C. and Ben-Israel, A. (2010) A Generalized Weiszfeld Method for the Multi-Facility Location Problem. *Operations Research Letters*, **38**, 207-214. http://dx.doi.org/10.1016/j.orl.2009.11.005

[11] Ballou, R.H. (2004) Business Logistics/Supply Chain Management: Planning, Organizing and Controlling the Supply chain. 5th Edition, Pearson/Prentice Hall Inc., New Jersey.

[12] Katz, I.N. and Cooper, L. (1980) Optimal Location on a Sphere. *Computers & Mathematics with Applications*, **6**, 175-196. http://dx.doi.org/10.1016/0898-1221(80)90027-9

[13] Drezner, Z. and Wesolowsky, G.O. (1978) Facility Location on a Sphere. *Journal of the Operational Research Society*, **29**, 997-1004. http://dx.doi.org/10.1057/jors.1978.213

[14] Xue, G.L. (1994) A Globally Convergent Algorithm for Facility Location on a Sphere. *Computers & Mathematics with Applications*, **27**, 37-50. http://dx.doi.org/10.1016/0898-1221(94)90109-0

[15] Mwemezi, J. and Huang, Y. (2011) Optimal Facility Location on Spherical Surfaces: Algorithm and Application. *New York Science Journal*, **4**, 21-28.

[16] Sullivan, E. and Peters, N. (1980) A Flexible User-Oriented Location-Allocation Algorithm. *Journal of Environmental Management*, **10**, 181-193.

[17] Bespamyatnikh, S., Kedem, K., Segal M. and Tamir A. (2000) Optimal Facility Location under Various Distance Functions. *International Journal of Computational Geometry & Applications*, **10**, 523-534. http://dx.doi.org/10.1142/S0218195900000292

[18] Levin, Y. and Ben-Israel, A. (2004) A Heuristic Method for Large Scale Multifacility Location Problems. *Computers & Operations Research*, **31**, 257-272. http://dx.doi.org/10.1016/S0305-0548(02)00191-0

[19] Levin, Y. and Ben-Israel, A. (2002) The Newton Bracketing Method for Convex Minimization. *Computational Optimization and Applications*, **21**, 213-229. http://dx.doi.org/10.1023/A:1013768901780

[20] Rodríguez-Chía, A.M. and Valero-Franco, C. (2013) On the Global Convergence of a Generalized Iterative Procedure for the Minisum Location Problem with ℓ_p Distances for $p > 2$. *Springer and Mathematical Optimization Society, Mathematical Programming, Series A*, **137**, 477-502.

[21] Kotian, S.R., Bonilla, C. and Hale, T.S. (2008) The Planar *k*-Central Location Problem. *The Open Industrial and Manufacturing Engineering Journal*, **1**, 42-49. http://dx.doi.org/10.2174/1874152500801010042

[22] SAS Inc. (2014). http://support.sas.com/documentation/cdl/en/lefunctionsref/67398/HTML/default/viewer.htm#n1korpfg2e18lon1nwpow9qijdxe.htm

[23] Ksu. http://www.math.ksu.edu/~dbski/writings/haversine.pdf

[24] The Math Forum. (1994-2013). http://mathforum.org/library/drmath/view/51756.html

[25] Google Inc. (2014). https://www.google.com/maps/dir/

[26] Wikipedia (2014). http://en.wikipedia.org/wiki/Euclidean_distance

[27] Wiktionary (2014). http://en.wiktionary.org/wiki/Euclidean_distance

[28] Wikipedia (2014). http://en.wikipedia.org/wiki/Great-circle_distance

[29] Movable Type Ltd. (2014). http://www.movable-type.co.uk/scripts/latlong.html

[30] Weiszfeld, E. (1937) Sur le point par lequel la somme des distances de n points donnVes est minimum. *Tohoku Mathematical Journal*, **43**, 355-386.

[31] Aykin, T. (1984) Some Aspects in the Large Region Location Problems on the Surface of the Earth. PhD Thesis, State University of New York at Buffalo.

[32] SAS Inc. (2014).
http://support.sas.com/documentation/cdl/en/statug/63033/HTML/default/viewer.htm#modeclus_toc.htm

[33] Heizer, J. and Render, B. (2011) Principles of Operations Management. 8th Edition, Prentice Hall, Upper Saddle River.

[34] Pearson (2014).
http://www.prenhall.com/divisions/bp/app/russellcd/PROTECT/CHAPTERS/CHAP09/HEAD06.HTM

[35] Geomidpoint (2014). http://www.geomidpoint.com/calculation.html

[36] Chen, Z.X. and He, W. (2010) Study and Application of Center-of-Gravity on the Location Selection of Distribution Center, Logistics Systems and Intelligent Management. *International Conference on Logistics Systems and Intelligent Management*, Harbin, 9-10 January 2010, 981-984.

[37] Kuo, C.C., Richard, E. and White, R.E. (2004) A Note on the Treatment of the Center-of-Gravity Method in Operations Management Textbooks. *Decision Sciences Journal of Innovative Education*, **2**, 219-227.

[38] Chase, R.B., Aquilando, N.J. and Jacobs, F.R. (2006) Operations Management for Competitive Advantage. 11th Edition, McGraw-Hill/Irwin, New York.

[39] Ligas, M. and Banasik, P. (2011) Conversion between Cartesian and Geodetic Coordinates on a Rotational Ellipsoid by Solving a System of Nonlinear Equations. *Geodesy and Cartography*, **60**, 145-159.
http://dx.doi.org/10.2478/v10277-012-0013-x

[40] ThinkGeo LLC. (2014). http://thinkgeo.com/map-suite-developer-gis/desktop-edition/

[41] Natural Earth (2014). http://www.naturalearthdata.com/downloads/

[42] Hillier, F.S. and Lieberman, G.J. (2009) Introduction to Operations Research. 9th Edition, McGraw-Hill Higher Education, New York.

[43] Litwhiler Jr., D.W. and Aly, A.A. (1979) Large Region Location PROBLEMS. *Computer & Operations Research*, **6**, 1-12. http://dx.doi.org/10.1016/0305-0548(79)90009-1

Appendix A: Coordinates (Latitude/Longitudes) and Volume of Goods and Transportation Rate for Each Market City Location

Location	State	Latitude	Longitude	Volume	Rate
Loc1	OH	39.9718	−82.9688	800.00	0.02
Loc2	OH	40.146873	−82.98178	500.00	0.02
Loc3	OH	40.589747	−83.12173	1000.00	0.02
Loc4	OH	41.642549	−83.5104	900.00	0.02
Loc5	OH	41.480881	−81.80036	700.00	0.02
Loc6	OH	41.437018	−81.73488	400.00	0.02
Loc8	OH	41.089405	−81.53973	1500.00	0.02
Loc9	OH	41.144661	−81.61948	800.00	0.02
Loc10	OH	41.252748	−80.80773	1200.00	0.02
Loc11	OH	41.095658	−80.62652	1100.00	0.02
Loc12	OH	40.773556	−82.47612	1300.00	0.02
Loc13	OH	39.104127	−84.53067	500.00	0.02
Loc14	OH	39.066112	−84.32227	600.00	0.02
Loc15	OH	39.788458	−84.21111	1700.00	0.02
Loc16	OH	39.644609	−84.16931	1300.00	0.02
Loc18	OH	40.739517	−84.14468	1600.00	0.02
Loc19	IN	39.792993	−86.28575	500.00	0.03
Loc23	IN	41.071681	−85.15101	1100.00	0.03
Loc24	IN	41.138599	−85.05941	1700.00	0.03
Loc27	IN	39.232235	−85.86356	1200.00	0.03
Loc28	IN	40.192293	−85.38494	1600.00	0.03
Loc29	IN	39.188246	−86.56779	1300.00	0.03
Loc30	IN	37.974642	−87.57349	500.00	0.03
Loc31	IN	39.535285	−87.35222	1400.00	0.03
Loc32	MI	42.352837	−83.02449	800.00	0.05
Loc33	MI	42.427636	−83.19547	1100.00	0.05
Loc34	MI	43.062076	−83.69728	1600.00	0.05
Loc35	MI	43.431375	−83.93267	500.00	0.05
Loc36	MI	43.607523	−83.9162	1100.00	0.05
Loc37	MI	42.735535	−84.62698	700.00	0.05
Loc40	MI	43.022199	−85.61168	1000.00	0.05
Loc41	MI	46.554402	−87.42223	900.00	0.05
Loc42	IA	42.033235	−93.66741	800.00	0.03
Loc43	IA	41.587039	−93.67356	700.00	0.03

Continued

Loc44	IA	42.513636	−92.32418	1200.00	0.03
Loc45	IA	42.505046	−96.38003	1100.00	0.03
Loc46	IA	41.251631	−95.87447	400.00	0.03
Loc47	IA	42.50963	−90.75329	900.00	0.03
Loc50	IA	41.574581	−90.60303	1300.00	0.03
Loc51	WI	42.985465	−87.89998	500.00	0.06
Loc52	WI	43.037313	−87.93373	1800.00	0.06
Loc53	WI	42.714369	−87.82424	1000.00	0.06
Loc54	WI	43.037647	−89.39154	900.00	0.06
Loc55	WI	44.530892	−88.04482	1100.00	0.06
Loc56	WI	44.958382	−89.6693	700.00	0.06
Loc62	MN	46.674141	−92.2276	1200.00	0.07
Loc63	MN	43.991846	−92.49166	1000.00	0.07
Loc64	MN	44.184909	−94.04244	700.00	0.07
Loc65	MN	45.568516	−94.19251	1,100.00	0.07
Loc66	MN	46.861413	−96.75367	800.00	0.07
Loc67	SD	43.577696	−96.80101	1200.00	0.03
Loc68	SD	45.46932	−98.49646	700.00	0.03
Loc69	SD	44.052788	−103.34302	1300.00	0.03
Loc70	ND	46.812118	−96.839	600.00	0.04
Loc71	ND	47.938898	−97.08401	1100.00	0.04
Loc72	ND	46.906983	−98.72826	1000.00	0.04
Loc74	ND	46.878057	−102.8041	800.00	0.04
Loc75	ND	48.277024	−101.31942	1200.00	0.04
Loc78	IL	42.280019	−89.03347	500.00	0.08
Loc79	IL	41.4903	−90.56956	1500.00	0.08
Loc80	IL	40.678037	−89.62737	1200.00	0.08
Loc81	IL	40.459584	−88.96939	800.00	0.08
Loc83	IL	39.96008	−91.302633	1300.00	0.08
Loc84	IL	39.878041	−88.95637	500.00	0.08
Loc85	IL	39.7715	−89.63612	900.00	0.08
Loc86	IL	37.707763	−89.19246	1100.00	0.08
Loc89	MO	39.695798	−91.40084	500.00	0.03
Loc90	MO	37.291432	−89.54065	1200.00	0.03
Loc91	MO	36.759357	−90.41689	600.00	0.03
Loc92	MO	39.053255	−94.4061	1600.00	0.03

Continued

Loc93	MO	39.125304	−94.53234	1000.00	0.03
Loc94	MO	39.3173	−94.48655	700.00	0.03
Loc95	MO	39.812344	−94.84286	1100.00	0.03
Loc96	MO	37.047161	−94.51124	800.00	0.03
Loc97	MO	38.568287	−92.25329	1500.00	0.03
Loc98	MO	37.94384	−91.77105	1200.00	0.03
Loc99	MO	37.257053	−93.29015	600.00	0.03
Loc1A	KS	39.117009	−94.76614	1100.00	0.05
Loc1B	KS	39.013988	−95.65056	600.00	0.05
Loc1C	KS	37.669067	−97.36204	1500.00	0.05
Loc1D	KS	38.094945	−97.929	1200.00	0.05
Loc1E	KS	37.755267	−100.02632	900.00	0.05
Loc1F	NE	41.292321	−95.93427	1500.00	0.04
Loc1G	NE	41.296198	−96.11027	800.00	0.04
Loc1H	NE	40.798632	−96.59023	1500.00	0.04
Loc1I	NE	41.437838	−97.37192	1200.00	0.04
Loc1J	NE	40.922826	−98.33087	500.00	0.04
Loc1K	NE	41.885553	−103.65241	1000.00	0.04

Appendix B: Google Maps Distance Matrix of 88 Midwest Major Market Cities

Obs	Location	Loc 1	Loc 2	Loc 3	Loc 4	Loc 5	Loc 6	Loc 8	Loc 9	Loc 10	Loc 11	Loc 12	Loc 13	Loc 14	Loc 15	Loc 16	Loc 18
1	Loc1	0	14.4	50	143	136	136	125	121	175	173	67.8	110	111	73.9	80.8	95.5
2	Loc2	14.6	0	36	129	123	123	112	108	162	160	54.9	123	124	84.8	91.7	89.2
3	Loc3	50.1	36.2	0	95.6	112	112	101	96.7	151	149	41.6	143	156	90.4	97.3	57.3
4	Loc4	144	130	98.1	0	104	106	131	123	160	170	99.7	202	211	148	160	80.4
5	Loc5	136	122	111	104	0	5.9	39.8	31.5	63.4	79.6	70.8	244	245	206	213	151
6	Loc6	136	123	111	106	6	0	33.2	24.9	56.8	73	71.3	245	246	206	213	152
7	Loc8	125	112	100	131	40.6	33.2	0	9.9	49.1	52.8	60.3	234	235	196	202	157
8	Loc9	121	107	95.9	122	32	24.6	6.2	0	58.2	62.2	55.8	229	230	191	198	152
9	Loc10	170	156	145	160	63.1	56.5	48.2	59.7	0	22.3	105	278	279	240	247	201
10	Loc11	174	160	149	170	79.5	72.9	52.4	62.2	20.1	0	109	282	283	244	251	205
11	Loc12	67.8	54.5	41.8	99.8	71	71.4	60.4	56.1	110	108	0	176	177	138	145	98.1
12	Loc13	111	124	144	202	245	245	234	230	284	282	177	0	16.2	55.2	47.2	125
13	Loc14	111	124	156	210	245	245	234	230	284	282	177	15.4	0	63.7	55.7	134
14	Loc15	73.5	85.6	91.1	148	207	207	196	192	246	244	139	55	63.7	0	12.5	71.7
15	Loc16	80.2	92.2	97.8	163	213	214	203	198	252	251	146	47	55.7	16.7	0	86.7
16	Loc18	95.2	89.8	57.6	80	153	155	159	155	209	207	99.7	125	134	71	82.4	0
17	Loc19	183	195	201	234	329	317	306	302	356	354	249	121	137	117	133	164
18	Loc23	161	161	130	103	197	199	212	208	253	263	153	180	189	126	138	62
19	Loc24	161	161	130	101	196	197	212	208	251	261	153	184	193	130	142	62.1
20	Loc27	185	198	215	267	319	319	308	304	358	356	251	85.2	100	126	118	196
21	Loc28	147	159	140	181	275	277	249	245	299	297	190	103	114	80.2	96.6	92.1
22	Loc29	226	238	243	275	359	359	348	344	398	396	291	131	146	159	164	222
23	Loc30	327	339	372	395	461	461	450	446	500	498	393	218	227	273	265	343
24	Loc31	257	269	275	306	390	391	380	375	429	428	322	186	201	190	207	254
25	Loc32	204	191	159	60.3	163	164	190	181	218	228	158	263	272	209	221	141
26	Loc33	208	194	162	63.6	166	168	193	185	222	232	162	266	275	212	224	145
27	Loc34	250	236	205	118	222	224	249	241	278	288	228	309	317	255	266	187
28	Loc35	277	263	232	145	249	251	276	268	305	315	255	336	344	282	293	214
29	Loc36	290	276	245	158	262	264	289	281	318	328	267	349	357	295	306	227
30	Loc37	260	246	215	128	232	234	259	251	288	298	238	308	317	255	266	186
31	Loc40	323	309	278	191	295	297	322	314	351	361	301	354	362	300	311	232
32	Loc41	631	617	586	499	603	605	630	622	659	669	609	690	698	636	647	568
33	Loc42	675	687	638	588	681	683	708	700	737	747	660	611	626	609	625	570
34	Loc43	652	664	614	564	658	659	684	676	713	723	637	587	602	585	601	546
35	Loc44	617	629	580	530	623	625	650	642	679	689	603	553	568	551	567	512

Continued

36	Loc45	826	838	788	738	832	833	859	850	887	897	811	761	776	759	776	720
37	Loc46	778	790	740	691	784	786	811	802	839	850	763	713	729	711	728	673
38	Loc47	558	570	482	432	526	527	553	544	581	591	505	493	508	491	507	415
39	Loc50	489	501	452	402	495	497	522	514	551	561	475	424	440	422	439	384
40	Loc51	450	415	383	331	424	426	451	443	480	490	406	388	403	383	400	315
41	Loc52	454	418	386	335	428	430	455	446	483	494	409	391	407	387	403	319
42	Loc53	435	400	368	316	409	411	436	428	465	475	391	373	388	369	385	300
43	Loc54	503	475	444	394	487	489	514	506	543	553	467	441	456	436	453	376
44	Loc55	572	536	504	453	546	547	573	564	601	611	527	509	525	505	521	437
45	Loc56	637	601	569	518	611	613	638	629	666	677	592	574	590	570	586	502
46	Loc62	830	802	771	721	814	816	841	833	870	880	793	767	783	763	780	703

47	Loc63	706	678	647	597	690	692	717	709	746	756	669	644	659	639	656	579
48	Loc64	791	763	731	681	775	776	801	793	830	840	754	728	743	724	740	663
49	Loc65	833	805	773	723	817	818	843	835	872	882	796	770	785	766	782	705
50	Loc66	997	969	937	888	981	983	1008	999	1037	1047	960	934	950	930	946	870
51	Loc67	933	945	874	825	918	919	945	936	973	983	897	868	884	866	883	807
52	Loc68	1059	1031	1000	950	1043	1045	1070	1062	1099	1109	1022	996	1012	992	1008	932
53	Loc69	1268	1280	1216	1167	1260	1261	1287	1278	1315	1325	1239	1203	1218	1201	1217	1149
54	Loc70	1002	975	943	893	986	988	1013	1005	1042	1052	966	940	955	936	952	875
55	Loc71	1080	1052	1020	970	1064	1065	1090	1082	1119	1129	1043	1017	1032	1013	1029	952
56	Loc72	1094	1066	1035	985	1078	1080	1105	1097	1134	1144	1058	1032	1047	1027	1044	967
57	Loc74	1288	1260	1229	1179	1272	1274	1299	1291	1328	1338	1252	1226	1241	1221	1238	1161
58	Loc75	1270	1243	1211	1161	1254	1256	1281	1273	1310	1320	1234	1208	1223	1204	1220	1143
59	Loc78	442	414	382	333	426	427	453	444	481	491	405	379	395	375	391	315
60	Loc79	482	494	444	394	488	489	515	506	543	553	467	417	432	415	431	377
61	Loc80	387	399	405	385	478	480	505	497	534	544	453	323	338	321	337	353
62	Loc81	345	357	362	350	443	445	470	462	499	509	410	280	295	278	294	289
63	Loc83	496	508	513	497	591	592	618	614	646	666	561	431	446	429	445	462
64	Loc84	345	357	363	354	482	484	468	463	517	516	411	280	296	278	295	311
65	Loc85	385	397	403	418	512	513	508	503	567	556	451	320	336	318	335	351
66	Loc86	441	453	458	490	575	575	564	560	614	612	507	332	341	374	379	437
67	Loc89	488	500	505	521	614	616	610	606	670	658	553	423	438	421	437	454
68	Loc90	481	494	499	531	615	616	605	600	654	653	547	373	382	414	420	478
69	Loc91	530	542	570	602	664	664	653	649	703	701	596	421	430	485	468	549
70	Loc92	648	660	666	698	781	782	771	767	821	819	714	577	603	581	598	645

Continued

71	Loc93	660	672	677	709	793	793	782	778	832	830	725	589	614	593	609	657
72	Loc94	673	685	691	713	806	808	796	792	846	844	739	603	628	607	623	670
73	Loc95	679	691	697	713	806	808	802	798	861	850	745	615	630	613	629	645
74	Loc96	704	716	721	753	837	837	826	822	876	874	769	633	652	637	653	701
75	Loc97	548	560	565	597	681	681	670	666	720	718	613	477	502	481	497	545
76	Loc98	525	537	543	575	658	659	648	643	698	696	591	454	474	458	475	522
77	Loc99	631	643	649	681	764	765	754	749	804	802	697	560	580	564	581	628
78	Loc1A	670	682	688	719	803	804	793	788	842	841	735	599	625	603	620	667
79	Loc1B	720	732	738	770	854	854	843	839	893	891	786	650	675	654	670	717
80	Loc1C	854	866	872	903	987	988	977	972	1026	1025	919	783	809	787	804	851
81	Loc1D	875	887	892	924	1008	1008	997	993	1047	1045	940	804	829	808	824	872
82	Loc1E	993	1005	1011	1042	1126	1127	1116	1111	1165	1164	1059	922	948	926	943	990
83	Loc1F	786	798	748	698	791	793	818	810	847	857	771	721	736	719	735	680
84	Loc1G	795	807	757	707	800	802	827	819	856	866	780	730	745	728	744	689
85	Loc1H	812	824	797	747	841	842	867	859	896	906	820	747	762	745	761	729
86	Loc1I	857	869	819	769	862	864	889	881	918	928	842	792	807	790	806	751
87	Loc1J	907	919	890	840	933	935	960	952	989	999	913	842	857	840	856	822
88	Loc1K	1213	1225	1197	1147	1240	1242	1267	1259	1296	1306	1219	1148	1164	1146	1163	1129

10

Mathematical Model and Algorithm for Link Community Detection in Bipartite Networks

Zhenping Li[1], Shihua Zhang[2], Xiangsun Zhang[2]

[1]School of Information, Beijing Wuzi University, Beijing, China
[2]National Center for Mathematics and Interdisciplinary Sciences, Academy of Mathematics and Systems Science, Beijing, China
Email: lizhenping66@163.com, zsh@amss.ac.cn, zxs@amt.ac.cn

Abstract

In the past ten years, community detection in complex networks has attracted more and more attention of researchers. Communities often correspond to functional subunits in the complex systems. In complex network, a node community can be defined as a subgraph induced by a set of nodes, while a link community is a subgraph induced by a set of links. Although most researches pay more attention to identifying node communities in both unipartite and bipartite networks, some researchers have investigated the link community detection problem in unipartite networks. But current research pays little attention to the link community detection problem in bipartite networks. In this paper, we investigate the link community detection problem in bipartite networks, and formulate it into an integer programming model. We proposed a genetic algorithm for partition the bipartite network into overlapping link communities. Simulations are done on both artificial networks and real-world networks. The results show that the bipartite network can be efficiently partitioned into overlapping link communities by the genetic algorithm.

Keywords

Bipartite Network, Link Community, Quantity Function, Integer Programming, Genetic Algorithm

1. Introduction

Many interesting systems can be represented as networks [1]-[4]. The networks are composed of nodes and links, each node represents a unit and each link represents a relation between two nodes. Since some nodes or links may have the same function in complex system. One of the most important topics in the area of networks is the community detection, which is a universal problem in many disciplines such as sociology, computer science and biology [5]-[7].

The communities are dense subgraphs induced by a set of nodes or links. If the community is induced by a set of nodes, we call it node community. If a community is induced by a set of links, we call it link community. When we partition a network into node communities, each node must belong to one or more community, some links might belong to no community. When a network is partition into link communities, each link must belong to one community, and each node might belong to one or more communities. By partition the network into link communities, we can find overlapping node or link.

Although most research paid more attention to node community detection, some researchers have investigated link communities and cliques [8]-[12]. In some real-world networks, a link is more likely to have a unique identity while a node often has multiple functions, so the link communities might be more intuitive and informative than the node communities [13] [14].

Given a unipartite network with M links and N nodes, let $P = \{P_1, \cdots, P_K\}$ be a partition of the links into K subsets. $m_c = |P_c|$ be the number of links in subset P_c, $n_c = \bigcup_{e_{ij} \in P_c} \{i, j\}$ be the number of nodes in subgraph induced by P_c. Ahn [8] defined the partition density D as follows

$$D = \frac{2}{M} \sum_c m_c \frac{m_c - (n_c - 1)}{(n_c - 2)(n_c - 1)}.$$

In [12], the authors proposed another partition density H as follows:

$$H = \frac{2}{M} \sum_c \frac{m_c^2}{n_c (n_c - 1)}$$

Obviously, $0 \leq H \leq 1$.

Given the number of communities, we can partition the unipartite network into link communities by maximize D or H.

Besides unipartite networks, there is another special category of network, where nodes are partitioned into two disjoint subsets, there is no link within the same subset. This type of network is called bipartite network. Some real-world relations are more suitable to be represented as bipartite networks [15], such as plant-animal network, scientific publication network, artistic collaboration network, order-item network, paper-author networks, event-attendee networks and so on.

Some research has paid attention to the node community detection problem of bipartite networks [15] [16]. In [15], the authors proposed a projection-based algorithm for node communities detection in bipartite network. In [17], the authors develop a modified adaptive genetic algorithm (MAGA) to detect the node communities in bipartite network. In [18], the authors propose another bipartite modularity detection method which can detect node overlap community. In [19], the authors proposed a hierarchical divisive heuristic for approximate modularity maximization in bipartite graphs. In [20], the authors proposed an algorithm *Bitector* to mine overlapping communities in large scale sparse bipartite networks. In [21], the authors proposed an approach for detecting overlap node communities in a bipartite network based on dual optimization of modularity. In [22] [23], the authors proposed weighted binary matrix factorization framework to detect overlapping communities in bipartite networks. Although the algorithms above can find node communities in bipartite network, current research activity has paid no attention to the link community detection problem in bipartite networks.

In this paper, we will investigate link communities in bipartite network, define the partition density of link communities in bipartite network, and formulate the link community partition problem of bipartite network into an integer programming model. Then we design a genetic algorithm for detecting link communities in bipartite network and conduct validations on some artificial and real-world bipartite networks. By the model and algorithm, the communities including two sets of nodes in bipartite network can be identified simultaneously.

2. Methods

2.1. Link Community Partition Density of Bipartite Networks

Given a bipartite network $G = (U, V, L)$ with $M = |L|$ links and two node sets U and V, where $U \cap V = \varnothing$, $P = \{P_1, \cdots, P_K\}$ is a partition of the links into K subsets. The number of links in subset P_c is $m_c = |P_c|$. The induced node set from link subset P_c is $\bigcup_{l_{ij} \in P_s} \{i, j\}$ (where l_{ij} represents the link connecting node u_i and

v_j), the number of induced nodes in node set U is $s_c = \left| \left(\bigcup_{l_{ij} \in P_s} \{i, j\} \right) \cap U \right|$, the number of induced nodes in

node set V is $t_c = \left| \left(\bigcup_{l_{ij} \in P_s} \{i, j\} \right) \cap V \right|$. The link density H_c of community c in bipartite network is defined as

follows:

$$H_c = m_c / (s_c t_c).$$

The partition density H is the average of H_c:

$$H = \frac{1}{M} \sum_c m_c \cdot H_c = \frac{1}{M} \sum_c \frac{m_c^2}{s_c t_c}.$$

We can see that the maximum of H is 1 and the minimum value of H is 0. $H = 1$ when each community is a complete bipartite network and $H = 0$ when each community is an empty bipartite graph. Given the number of communities, we can find the optimal link community partition of bipartite network by maximizing the value of H.

2.2. Integer Programming Model for Link Community Detection of Bipartite Network

Given a bipartite network $G = (U, V, L)$ with M links and $p + q = |U \cup V|$ nodes (where $p = |U|, q = |V|$), we assume that the number of link communities is K and find the optimal link community partition by maximizing the partition density H. This problem can be formulated into an integer programming model.

Let $U = \{u_1, u_2, \cdots, u_p\}$, $V = \{v_1, v_2, \cdots, v_q\}$ be two disjoint nodes sets of bipartite network G. $A = \left(a_{i \times j} \right)_{p \times q}$ is the adjacent matrix of the bipartite network, where $a_{ij} = 1$ when node u_i and v_j is connected by link l_{ij}, while $a_{ij} = 0$ otherwise.

We also define binary variables x_{ijk}, y_{ik} and z_{jk} to represent the membership of link l_{ij}, node u_i and node v_j for link community k:

$$x_{ijk} = \begin{cases} 1 & \text{if } l_{ij} \in \text{community } k \\ 0 & \text{otherwise} \end{cases} \quad y_{ik} = \begin{cases} 1 & \text{if } u_i \in \text{community } k \\ 0 & \text{otherwise} \end{cases} \quad z_{jk} = \begin{cases} 1 & \text{if } v_j \in \text{community } k \\ 0 & \text{otherwise} \end{cases}$$

The link community detection problem of bipartite network can be formulated into the following integer programming model—Model 1.

$$\max H = \frac{1}{M} \sum_{k=1}^{K} \frac{\left(\sum_{i=1}^{p} \sum_{j=1}^{q} x_{ijk} \right)^2}{\left(\sum_{i=1}^{p} y_{ik} \right) \left(\sum_{j=1}^{q} z_{jk} \right)} \tag{1}$$

$$s.t. \begin{cases} \sum_{k=1}^{K} x_{ijk} = a_{ij} & i = 1, 2, \cdots, p; j = 1, 2, \cdots, q & (2) \\ x_{ijk} \le y_{ik} & i = 1, 2, \cdots, p; j = 1, 2, \cdots, q; k = 1, 2, \cdots, K & (3) \\ x_{ijk} \le z_{jk} & i = 1, 2, \cdots, p; j = 1, 2, \cdots, q; k = 1, 2, \cdots, K & (4) \\ y_{ik} \le \sum_{j=1}^{q} x_{ijk} & i = 1, 2, \cdots, p, k = 1, 2, \cdots K & (5) \\ z_{jk} \le \sum_{i=1}^{p} x_{ijk} & j = 1, 2, \cdots, q, k = 1, 2, \cdots K & (6) \\ x_{ijk} \in \{0, 1\}; i = 1, 2, \cdots, p; j = 1, 2, \cdots, q, k = 1, 2, \cdots, K & (7) \\ y_{ik} \in \{0, 1\}; i = 1, 2, \cdots, p, k = 1, 2, \cdots, K & (8) \\ z_{jk} \in \{0, 1\}; j = 1, 2, \cdots, q, k = 1, 2, \cdots, K & (9) \end{cases}$$

The objective function (1) is to maximize the link partition density H. Constraint (2) means that every link belongs to one community. If there is no link between node u_i and v_j, then variables $x_{ijk} = 0$ for any community k. Constraints (3) and (4) indicate that if link l_{ij} belong to community k, then its adjacent nodes u_i and v_j must belong to the same community k. Constraint (5) and (6) mean that if a node u_i (or v_j) belongs to community k, then there is at least one link adjacent to node u_i (or v_j) belonging to community k. Constraints (7) (8) (9) indicate that the variables are binary.

Since there are a great many of variables in Model 1, it may have large memory overhead when solving the model directly. To decrease the number of variables used, Model 1 can be expressed by using relationship matrix.

Suppose that $U = \{u_1, u_2, \cdots, u_p\}$, $V = \{v_1, v_2, \cdots, v_q\}$ are two disjoint nodes sets, and $L = \{l_1, l_2, \cdots, l_M\}$ is the link set of bipartite network. Define two incidence matrix RS and RT as follows:

$$RS = \begin{pmatrix} s_{11} & s_{12} & s_{13} & \cdots & s_{1M} \\ s_{21} & s_{22} & s_{23} & \cdots & s_{2M} \\ s_{31} & s_{32} & s_{33} & \cdots & s_{3M} \\ \vdots & \vdots & \vdots & \ddots & \vdots \\ s_{p1} & s_{p2} & s_{p3} & \cdots & s_{pM} \end{pmatrix}$$

where

$$s_{im} = \begin{cases} 1 & \text{if } u_i \text{ is an endpoint of link } l_m \\ 0 & \text{otherwise} \end{cases}$$

$$RT = \begin{pmatrix} t_{11} & t_{12} & t_{13} & \cdots & t_{1M} \\ t_{21} & t_{22} & t_{23} & \cdots & t_{2M} \\ t_{31} & t_{32} & t_{33} & \cdots & t_{3M} \\ \vdots & \vdots & \vdots & \ddots & \vdots \\ t_{q1} & t_{q2} & t_{q3} & \cdots & t_{qM} \end{pmatrix}$$

$$t_{jm} = \begin{cases} 1 & \text{if } v_j \text{ is an endpoint of link } l_m \\ 0 & \text{otherwise} \end{cases}$$

Define the binary variables as follows:

$$x_{mk} = \begin{cases} 1 & \text{if } l_m \in \text{community } k \\ 0 & \text{otherwise} \end{cases}$$

$$y_{ik} = \begin{cases} 1 & \text{if } u_i \in \text{community } k \\ 0 & \text{otherwise} \end{cases}$$

$$z_{jk} = \begin{cases} 1 & \text{if } v_j \in \text{community } k \\ 0 & \text{otherwise} \end{cases}$$

Based on the incidence matrix and the above variables, the link community detection problem of bipartite network can be reformulated into the following integer nonlinear programming model, Model 2.

$$\max H = \frac{1}{M} \sum_{k=1}^{K} \frac{\left(\sum_{m=1}^{M} x_{mk} \right)^2}{\left(\sum_{i=1}^{p} y_{ik} \right) \left(\sum_{j=1}^{q} z_{jk} \right)} \tag{10}$$

$$\sum_{k=1}^{K} x_{mk} = 1 \quad m = 1, 2, \cdots, M \tag{11}$$

$$\sum_{m=1}^{M} s_{im} x_{mk} \leq N y_{ik} \quad i = 1, 2, \cdots, p; k = 1, 2, \cdots, K \tag{12}$$

$$\sum_{m=1}^{M} t_{jm} x_{mk} \leq N z_{jk} \quad j = 1, 2, \cdots, q; k = 1, 2, \cdots, K \tag{13}$$

$$s.t. \begin{cases} y_{ik} \leq \sum_{m=1}^{M} s_{im} x_{mk} \quad i = 1, 2, \cdots, p, k = 1, 2, \cdots K \tag{14} \end{cases}$$

$$z_{jk} \leq \sum_{m=1}^{M} t_{jm} x_{mk} \quad j = 1, 2, \cdots, q, k = 1, 2, \cdots K \tag{15}$$

$$x_{mk} \in \{0, 1\}; m = 1, 2, \cdots, M; k = 1, 2, \cdots, K \tag{16}$$

$$y_{ik} \in \{0, 1\}\}; i = 1, 2, \cdots, p, k = 1, 2, \cdots, K \tag{17}$$

$$z_{jk} \in \{0, 1\}; j = 1, 2, \cdots, q, k = 1, 2, \cdots, K \tag{18}$$

Where N is the number of nodes in the network, $N = p + q$. The objective function (10) is to maximize the link partition density. Constraint (11) means that every link belongs to one community. Constraint (12) (13) mean that, if there is some adjacent links of node u_i (v_j) belonging to community k, then node u_i (v_j) must belong to the same community k. Constraints (14) (15) mean that if node u_i (v_j) belongs to community k, then at least one link adjacent to this node must belong to community k. Constraints (16) (17) (18) indicate that the variables are binary.

In Model 1 and Model 2, since every link can belong to one and only one community, we might obtain the result that a pair of nodes belongs to two communities, but the link between this pair of nodes belongs to only one community. To reduce this drawback, we can revise Model 2 into the following model—Model 3.

$$\max H = \frac{1}{\sum_{k=1}^{K} \sum_{m=1}^{M} x_{mk}} \sum_{k=1}^{K} \frac{\left(\sum_{m=1}^{M} x_{mk} \right)^2}{\left(\sum_{i=1}^{p} y_{ik} \right) \left(\sum_{j=1}^{q} z_{jk} \right)} \tag{10'}$$

$$\sum_{k=1}^{K} x_{mk} \geq 1 \quad m = 1, 2, \cdots, M \tag{11'}$$

$$\sum_{m=1}^{M} s_{im} x_{mk} \leq N y_{ik} \quad i = 1, 2, \cdots, p; k = 1, 2, \cdots, K \tag{12'}$$

$$\sum_{m=1}^{M} t_{jm} x_{mk} \leq N z_{jk} \quad j = 1, 2, \cdots, q; k = 1, 2, \cdots, K \tag{13'}$$

$$s.t. \begin{cases} y_{ik} \leq \sum_{m=1}^{M} s_{im} x_{mk} \quad i = 1, 2, \cdots, p, k = 1, 2, \cdots K \tag{14'} \end{cases}$$

$$z_{jk} \leq \sum_{m=1}^{M} t_{jm} x_{mk} \quad j = 1, 2, \cdots, q, k = 1, 2, \cdots K \tag{15'}$$

$$x_{mk} \in \{0, 1\}; m = 1, 2, \cdots, M; k = 1, 2, \cdots, K \tag{16'}$$

$$y_{ik} \in \{0, 1\}; i = 1, 2, \cdots, p, k = 1, 2, \cdots, K \tag{17'}$$

$$z_{jk} \in \{0, 1\}; j = 1, 2, \cdots, q, k = 1, 2, \cdots, K \tag{18'}$$

In model 3, the constraint (11') means that every link must belong to at least one community.

Using model 3, we can partition the network in **Figure 1** into two communities, and link (3, 10) belongs to two communities. Each community is a complete bipartite subnetwork, and the optimal objective function value is 1.

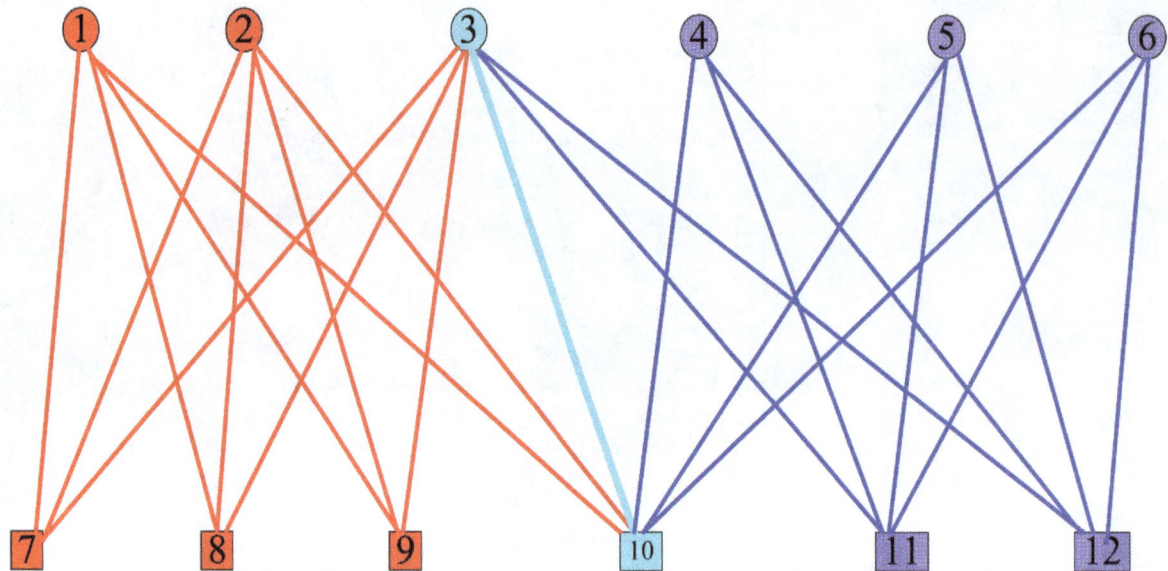

Figure 1. The bipartite network consists of two overlapping communities, each community is a complete bipartite network, they are overlapped by nodes 3,10 and link (3,10). This bipartite network can be partitioned into two communities by model 3, and the objective function value is 1.

2.3. Genetic Algorithm for Link Community Detection of Bipartite Network

Although we can solve Model 2 or Model 3 to partition a bipartite network into link communities for small size of bipartite network. It is difficult to solve the integer programming model for large bipartite networks which might be a NP-hard problem. In addition, most of the algorithms for community detection need some *priori* knowledge about the community structure like the number of communities which is impossible to know in real-life networks. In [12], the authors propose a link community detection algorithm based on the ideas of genetic algorithm and self-organize map (SOM) algorithm, which aims to find the best link community structure by maximizing the network partition density. The algorithm does not need any *priori* knowledge about the number of communities, which makes the algorithm useful in real-life networks. The algorithm outputs the final link community structure and its corresponding overlapping nodes as the result and does not impose further processing on the output. In the following, we will design another genetic algorithm for link community detection of bipartite network.

First of all, we need to design a chromosome representation suitable for the link community detection problem. In our implementation, the chromosome is represented by a matrix $B = (b_{m,c})$, where $m = 1, 2, \cdots, M$, and $c = 1, 2, \cdots, K$. Each element $b_{m,c}$ is the strength with which a link l_m belongs to a community c. Note that $b_{m,c}$ ranges in the interval [0.0, 1.0]. Each link of the bipartite network is subject to the following constraint:

$$\sum_{c=1}^{K} b_{m,c} = 1. \tag{19}$$

Equation (19) represents normalization to 1.0 of link factors of belonging to the communities.

For each chromosome, we design a partition matrix $D = (d_{m,c})$, where $m = 1, 2, \cdots, M$, and $c = 1, 2, \cdots, K$. Each element $d_{m,c}$ is either 0 or 1. Where $d_{m,c} = 1$ if the link l_m is assigned to community c, otherwise, link l_m is not assigned to community c. Matrix D can be calculated from matrix B according to the following equation:

$$d_{m,c} = \begin{cases} 1 & \text{if } b_{m,c} = \max_{1 \leq s \leq K} b_{m,s}. \\ 0 & \text{otherwise} \end{cases} \tag{20}$$

The bipartite network is represented by two incidence matrixes RS and RT, two weighted incidence matrixes ZS and ZT, link adjacent matrix A and weighted link adjacent matrix Q.

$$ZS = \begin{pmatrix} \dfrac{1}{\sqrt{d(u_1)}} & 0 & 0 & \cdots & 0 \\[2ex] 0 & \dfrac{1}{\sqrt{d(u_2)}} & 0 & \cdots & 0 \\[2ex] 0 & 0 & \dfrac{1}{\sqrt{d(u_3)}} & \cdots & 0 \\[2ex] \vdots & \vdots & \vdots & \ddots & \vdots \\[2ex] 0 & 0 & 0 & \cdots & \dfrac{1}{\sqrt{d(u_p)}} \end{pmatrix} \cdot RS$$

$$ZT = \begin{pmatrix} \dfrac{1}{\sqrt{d(v_1)}} & 0 & 0 & \cdots & 0 \\[2ex] 0 & \dfrac{1}{\sqrt{d(v_2)}} & 0 & \cdots & 0 \\[2ex] 0 & 0 & \dfrac{1}{\sqrt{d(v_3)}} & \cdots & 0 \\[2ex] \vdots & \vdots & \vdots & \ddots & \vdots \\[2ex] 0 & 0 & 0 & \cdots & \dfrac{1}{\sqrt{d(v_q)}} \end{pmatrix} \cdot RT$$

where d_{u_i} and d_{v_j} represent the nodes' degree of nodes u_i and v_j, which is the number of links incident to nodes u_i and v_j respectively.

The link adjacent matrix A and the weighted link adjacent matrix Q can be calculated by the following equations:

$$A = (RS)^{\mathrm{T}}(RS) + (RT)^{\mathrm{T}}(RT)$$

$$Q = (ZS)^{\mathrm{T}}(ZS) + (ZT)^{\mathrm{T}}(ZT)$$

The weighted link adjacent matrix Q means the probability for a random walker go from one link to one of its adjacent links across their common node. And this can be regarded as the possibility of two adjacent links belonging to the same community.

2.3.1. The Genetic Algorithm Main Functions

• Input

Input the number of nodes p for node set $|U|$ and q for node set $|V|$ respectively, and the number of links M of the link set $|E|$ in bipartite network, the maximum number of communities K, parameters α, β, θ, where $\alpha \in (0,1), \beta \in (0,1)$.

Input the incident matrixes RS, RT. Calculate the weighted incident matrixes ZS and ZT, the link adjacent matrix A, and the weighted link adjacent matrix Q. Given the number of individuals N, the maximum epochs $epoch_{\max}$, mutation probability $prob_{mutation}$.

• Output

Output the link partition matrix D^* and its fitness value H^*, two nodes set partition matrixes F_1, F_2. Partition the network into communities according to F_1, F_2.

• Initialization: $t = 0$.

Randomly generate initial population $B_1(t), B_2(t), \cdots, B_N(t)$, and give random initial values of D^* and its fitness H^*.

• Step 1. Population Fitness

For every individual $B_1(t), B_2(t), \cdots, B_N(t)$, calculate the partition matrix $D_1(t), D_2(t), \cdots, D_N(t)$, and their fitness value (partition link density value) $H_1(t), H_2(t), \cdots, H_N(t)$.

• Step 2. Population Sorting

Sort $B_1(t), B_2(t), \cdots, B_N(t)$ according to their fitness values in decreasing order. Suppose the sorted chromosomes are $B_1(t), B_2(t), \cdots, B_N(t)$, where $H_1(t) \geq H_2(t) \geq \cdots \geq H_N(t)$.

If $H_1(t) > H^*$, then, $D^* = D_1(t)$, $H^* = H_1(t)$.

If $t = epoch_{\max}$, stop, output D^* and H^*, and calculate the two corresponding node sets partition matrix F_1, F_2. Otherwise, go to Step 3.

• Step 3. Population Crossover

For $i = 1, \cdots, \left\lfloor \dfrac{N}{2} \right\rfloor$, let $B_i(t)$ and $B_{\left\lfloor \frac{N}{2} \right\rfloor + i}(t)$ cross over to produce two temporary individuals (matrixes) $W_i(t)$ and $W_{\left\lfloor \frac{N}{2} \right\rfloor + i}(t)$. If N is an odd number, let $W_N(t) = B_N(t)$.

• Step 4. Population Mutation

Random select $prob_{mutation} N$ temporary individuals (temporary matrices), do mutation operation on each temporary individual.

• Step 5. Population Self Organize Mapping

For each temporary individual, do self organize mapping operation on it.

• Step 6. Population Normalization

For each temporary individual, do normalization on it. Denote the normalized individuals by $B_1(t+1), B_2(t+1), \cdots, B_N(t+1)$. Let $t = t+1$, go to step 1.

2.3.2. Partition Matrix and Fitness Evaluation

For every individual B_i, calculate the partition matrix D_i according to the Formula (20).

For each community s, $1 \leq s \leq K$, let $D_i(:, s)$ be the s-th column of matrix D_i.

Then $E_{i1}(s) = RS \cdot D_i(:, s)$ is a column vector whose element is a non-negative integer. A non-zero element in $E_{i1}(s)$ represents that the corresponding node of the node set $|U|$ belongs to community s. $E_{i2}(s) = RT \cdot D_i(:, s)$ is a column vector whose element is a non-negative integer. A non-zero element in $E_{i2}(s)$ represents that the corresponding node of the node set $|V|$ belongs to community s.

Let $F_{i1}(s)$ and $F_{i2}(s)$ be 0-1 vectors, $f_{i1}(j, s) = 1$ (or $f_{i2}(j, s) = 1$) whenever $e_{i1}(j, s) \geq 1$ (or $e_{i2}(j, s) \geq 1$). $f_{i1}(j, s) = 1$ (or $f_{i2}(j, s) = 1$) means that node u_j (or v_j) belongs to community s. The fitness value of individual B_i is defined by the link partition density of matrix D_i, which can be calculated by the following equation:

$$H_i = \frac{1}{\sum\limits_{s=1}^{K}\sum\limits_{j=1}^{M} D_i(j,s)} \sum_{s=1}^{K} \frac{\left(\sum\limits_{j=1}^{M} D_i(j,s) \right)^2}{\left(\sum\limits_{j=1}^{p} F_{i1}(j,s) \right) \left(\sum\limits_{j=1}^{q} F_{i2}(j,s) \right)}$$

2.3.3. Population Sorting

Sort $B_1(t), B_2(t), \cdots, B_N(t)$ according to their fitness values in decreasing order. Suppose the sorted chromosomes are $B_1(t), B_2(t), \cdots, B_N(t)$, where $H_1(t) \geq H_2(t) \geq \cdots \geq H_N(t)$.

If $H_1(t) > H^*$, then, $D^* = D_1(t)$, $H^* = H_1(t)$.

2.3.4. Population Crossover

For $i = 1, 2, \cdots, \left\lfloor \dfrac{N}{2} \right\rfloor$, do crossover operation on $B_i(t)$ and $B_{\left\lfloor \frac{N}{2} \right\rfloor + i}(t)$ by the following rules: Randomly select a column s, revise the s-th column of $B_{\left\lfloor \frac{N}{2} \right\rfloor+i}(t)$ by the s-th column of $B_i(t)$, and obtain two new temporal individuals $W_i(t)$ and $W_{\left\lfloor \frac{N}{2} \right\rfloor+i}(t)$, where $W_i(t)=B_i(t)$.

In this paper, we revised the s-th column of $B_{\left\lfloor \frac{N}{2} \right\rfloor+i}(t)$ by adding a fraction of the s-th column of $D_i(t)$ (where $D_i(t)$ is the partition matrix corresponding to $B_i(t)$), that is,

$$W_{\left\lfloor \frac{N}{2} \right\rfloor+i}(t)(:,c)=\begin{cases} B_{\left\lfloor \frac{N}{2} \right\rfloor+i}(:,s)+0.1\cdot D_i(:,s) & \text{if } c=s. \\ B_{\left\lfloor \frac{N}{2} \right\rfloor+i}(:,c) & \text{if } c\neq s. \end{cases}$$

2.3.5. Population Mutation

According to the mutation probability $prob_{mutation}$, randomly select $prob_{mutation}\cdot N$ temporal individuals, do mutation operation on each selected individual.

For each selected temporal individual $W_i(t)$, randomly select two parameters j_1, j_2, $1\leq j_1, j_2 \leq M$. There are three mutation rules can be used in this genetic algorithm, *i.e.* exchange the j_1-th row and the j_2-th row in $W_i(t)$, or replace the j_1-th row by the j_2-th row in $W_i(t)$, or replace the elements of the j_1-th row with a randomly selected number in [0.0,1.0]. Three rules lead to no significant difference in this genetic algorithm. In the following simulation, we replace the j_1-th row with the j_2-th row in $W_i(t)$. The other elements in $W_i(t)$ remain unchanged.

2.3.6. Population Self Organizing Map

For every link, find the community it belongs to and calculate its community ID variance. If the community ID variance of a link is larger than a threshold, then increase the weights of this link to its community and the weights of its neighbor links to the same community. If the community ID variance of a link is smaller than the threshold value, then decrease the weights of the link to its community and the weights of its neighbor links to the same community. This process can improve the quality of the partition by eliminating wrongly placed links.

For $i=1,\cdots,N$, do Self Organizing Map (SOM) operations on individual (chromosome) W_i as follows:

• According to temporal matrix W_i, calculate its partition matrix D_i';
• For $j=1,\cdots,M$, do the following operation on link l_j.
• Find the community ID that link l_j belongs to. The community ID corresponds to the maximum element in the j-th row of D_i' (the maximum element must be 1). Suppose the maximum element in the j-th row of D_i' is in the s-th column, which is $D_i'(j,s)$. This means that link l_j belongs to community P_s.

• Calculate the total number $TN(l_j)$ of adjacent links of l_j (including link l_j), and the number of its adjacent links in $TN(l_j)$ belonging to community P_s (denoted by $IN(l_j)$). $TN(l_j)$ is equal to the sum of elements in the j-th row of matrix A, which can be expressed by $TN(l_j)=A(j,:)\cdot I$, where $I=(1,1,\cdots,1)^{\mathrm{T}}$, and $TN(l_j)$ can be calculated by the equation $IN(l_j)=A(j,:)\cdot D_i'(:,s)$.

• Calculate the community ID variance $CV(l_j)$ of link l_j by the following equation.

$$CV(l_j)=\frac{IN(l_j)}{TN(l_j)}$$

• If $CV(l_j)\geq\theta$, then

$$W_i(:,s)=W_i(:,s)+Q(:,j)\cdot\alpha-(I-A(:,j))\cdot\beta$$

Else,

$$W_i\left(:,s\right)=W_i\left(:,s\right)-Q\left(:,j\right)\cdot\beta$$

where α and β are adjustable parameters which can decrease with the step t. In this paper, we let

$$\alpha=\alpha-\frac{t}{epoch_{max}}\left(\alpha-0.1\right),\ \ \beta=\beta-\frac{t}{epoch_{max}}\left(\beta-0.05\right)$$

In the above equation, if an element is negative, then we set it to be 0.01

2.3.7. Normalization

For $i=1,2,\cdots N$, do normalization on each row of temporal matrix W_i so that the sum of row elements in temporal matrix is 1. Let the normalized matrix be B_{i+1}.

3. Numerical Experiments

In this section, we apply the genetic algorithm to both artificial bipartite networks and several well studied real-world bipartite networks, and analyze the results in terms of classification accuracy and ability of detecting meaningful communities. The algorithm is implemented by Matlab version 7.1.

3.1. Chain of Complete Bipartite Network

We test our algorithm on a type of exemplar networks, that is, chains of complete bipartite network. This network consists of many heterogeneous complete bipartite networks, connected through single nodes (**Figure 2**). Each complete bipartite network $C_i=\left(U_i,V_i,L_i\right)$ $\left(i=1,2,\cdots,K\right)$ is a bipartite network, where there is a link between any pair of nodes $\left(u,v\right),u\in U_i,v\in V_i$. Assume that C_i has s_i+t_i nodes and $L_i=s_i*t_i$ links, then the network has a total of $N=\sum_{i=1}^{K}\left(s_i+t_i\right)-K+1$ nodes and $M=\sum_{i=1}^{K}L_i$ links. The network has a clear link bipartite modular structure where each community corresponds to a single bipartite complete network, thus the optimal partition density is 1. Using the genetic algorithm above, we can easily detect the optimal partition and identify the overlapping nodes. In this paper, we use a network consists of two (3,4)- complete bipartite networks, one (4,5)- complete bipartite network, one (4,6)- complete bipartite network, and one (5,5) complete bipartite network, the optimal partition results are obtained and described in **Figure 2**.

3.2. Real-World Networks

In this subsection, we validate our algorithm on some real-world networks.

The Southern Women Network During the 1930s, ethnographers Davis, Stubbs Davis, St. Clair Drake, Gardner, and Gardner collected data on social stratification in the town of Natchez, Mississippi. One of their work is collecting data on women's attendance to social events in the town [24]. They constructed the famous women-event bipartite network and analyze it. Since then the women-event bipartite network has become a de facto standard for discussing bipartite networks in the social science [12] [15] [20] [21] [24]-[27].

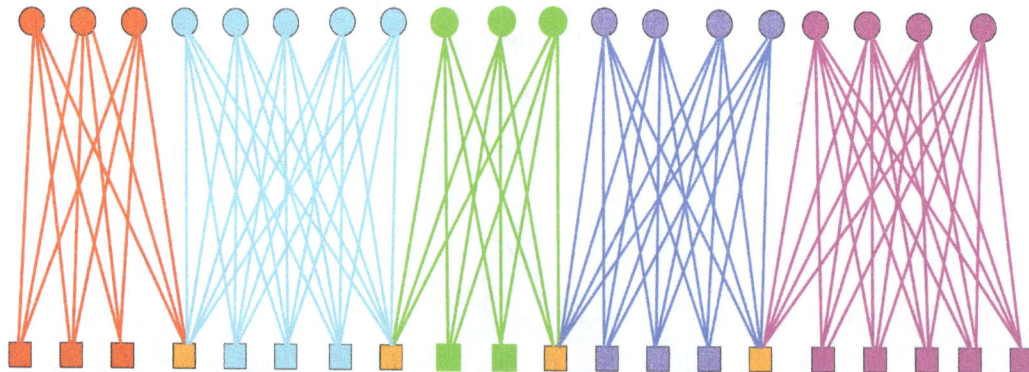

Figure 2. The chain of heterogeneous complete bipartite network. Each community is a complete bipartite network, and two adjacent communities are overlapped by one node.

Guimerà [15] has analyze the modules of both women and events by three methods: unweighted projection, weighted projection, and bipartite approach. The first method did not capture the true modular structure of the network. The second and third methods capture the two-module structure except one woman being partitioned wrong.

We applied the proposed method to the women-event network, using the parameters $K = 2$, $N = 200$, $p = 0.2$, $\theta = 0.3$, $\alpha = 0.7$, $\beta = 0.2$, $T = 800$. The result is illustrated in **Figure 3**. In this result, 18 women and 14 events are partitioned into two communities, where 4 events are overlapped. The average link density is 0.5610. In the women-event bipartite network, four event nodes B6, B7, B8, B9 colored by yellow are overlapped and belong to two communities. Comparing with the results obtained by Guimerà [15], the overlapped communities obtained by our method are more reasonable. When we partition the women-event network into four link communities using the parameters $K = 4$, $N = 200$, $p = 0.2$, $\theta = 0.3$, $\alpha = 0.7$, $\beta = 0.2$, $T = 800$, we can obtain the maximum average link density 0.683, The result is illustrated in **Figure 4**. Six yellow nodes are overlapped, where B_5, B_{11} belong to two communities, $B_6, B_7 B_8$ belong to three communities and B_9 belong to four communities.

3.3. The Scotland Corporate Interlock Network

The Scotland corporate interlock network describe the corporate interlocks in Scotland in the beginning of the twentieth century (1904-1905). The network consists of 244 nodes and 356 edges. The 244 nodes are divided into two parts, where 136 nodes indicate the board members who held multiple directorships, and 108 nodes indicate the firms). The edges exist between each firm and its board members. The largest component of the Scotland corporate interlock network contains 131 directors and 86 firms, forming many communities.

We applied the proposed method to the largest component of Scotland corporate interlock network, using the parameters $K = 20$, $N = 400$, $p = 0.2$, $\theta = 0.3$, $\alpha = 0.8$, $\beta = 0.2$, $T = 2000$. In the experiment, we divides the network into 20 communities, and the link community density is 0.24777. With the number of communities K increasing, the link community density obtained by our algorithm increase. When we use the para-

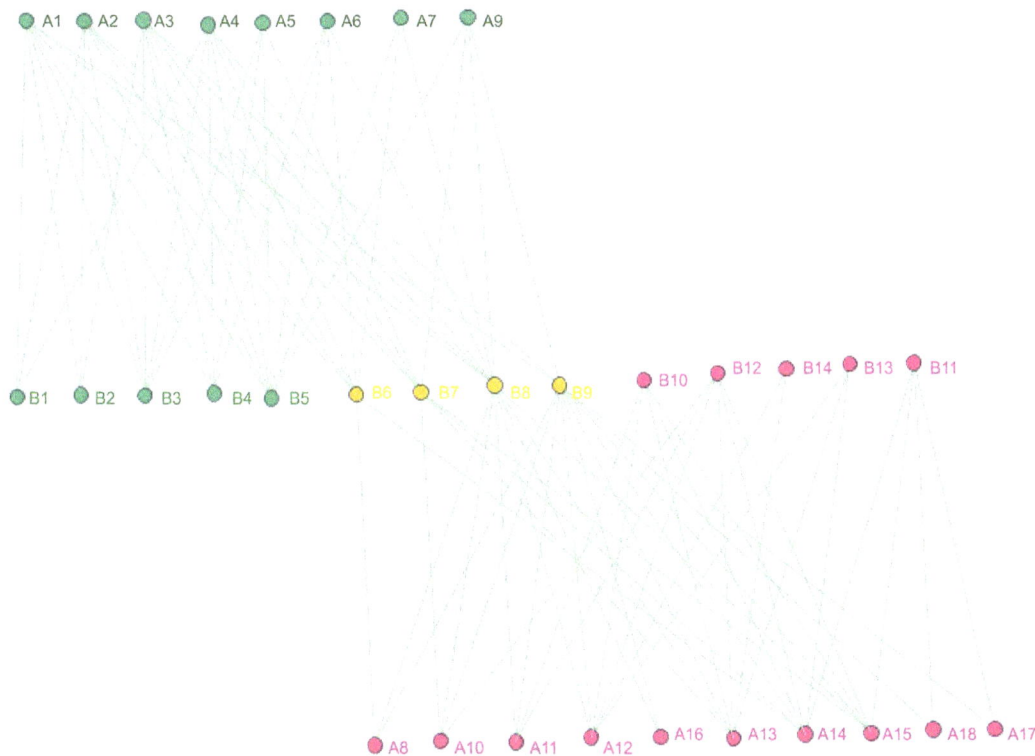

Figure 3. Result of the women-event networks partition into two link communities, where four yellow nodes B6; B7; B8; B9 belong to two communities.

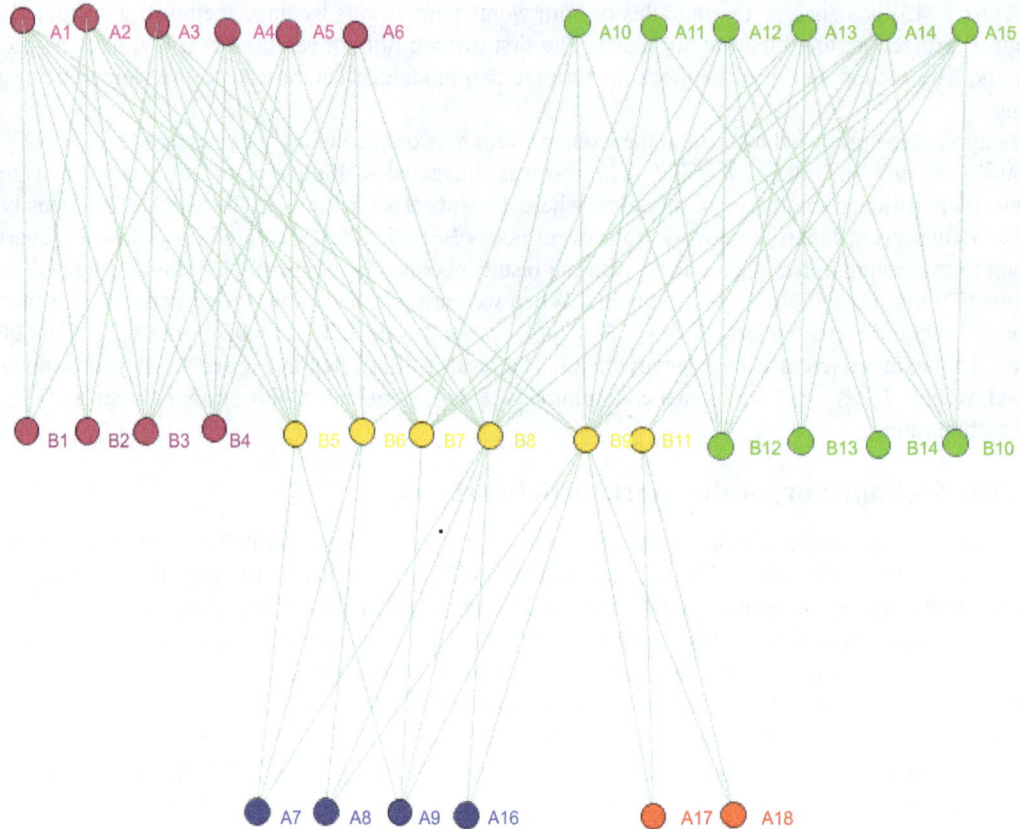

Figure 4. Result of the women-event networks partition into 4 link communities. Six yellow nodes are over-lapped, where B_5; B_{11} belong to two communities, B_6; B_7; B_8 belong to three communities and B_9 belong to four communities.

meters $K = 36$, $N = 100$, $p = 0.2$, $\theta = 0.3$, $\alpha = 0.8$, $\beta = 0.2$, $T = 2000$. We can obtained the maximum link community density 0.3553. If we increasing parameter K from 36 to 40, we can also partitioned the network into 36 link communities, the maximum link community density is also 0.3553. Since the real number of communities is 36 [25], Our results mean that we can find the optimal community solution by our algorithm.

4. Conclusion and Discussion

Bipartite network community structure is one of the main characteristics of bipartite networks and very helpful for understanding the functions of these networks. In this paper, we investigate the link community detection problem of bipartite network and propose a quantity function for link community detection of bipartite network. We formulate the link community identification problem of bipartite network into an integer programming model by maximizing the quantity function. Furthermore, we design a genetic algorithm for solving the link community detection problem and conduct validation experiments on some simulated and real-world networks. The extensive computational results demonstrate that our model and algorithm can detect overlapping communities. Using our model and algorithm, we can not only find the node overlapping communities but also the link overlapping communities in bipartite networks. Although we only investigate the unweighted bipartite networks, the model and algorithm can also be extended to deal with weighted bipartite networks.

Acknowledgements

This work is supported by National Natural Science Foundation of China under Grant No. 11131009. It is also supported by the Funding Project for Academic Human Resources Development in Institutions of Higher Learning Under the Jurisdiction of Beijing Municipality (CIT&TCD20130327).

References

[1] Albert, R. and Barabási, A.L. (2002) Statistical Mechanics of Complex Networks. *Reviews of Modern Physics*, **74**, 47-97. http://dx.doi.org/10.1103/RevModPhys.74.47

[2] Newman, M.E.J. (2003) The Structure and Function of Complex Networks. *SIAM Review*, **45**, 167-256. http://dx.doi.org/10.1137/S003614450342480

[3] Newman, M.E.J. and Girvan, M. (2004) Finding and Evaluating Community Structure in Networks. *Physical Review E*, **69**, Article ID: 026113. http://dx.doi.org/10.1103/physreve.69.026113

[4] Hu, Y., Chen, H., Zhang, P., Li, M., Di, Z. and Fan, Y. (2008) Comparative Definition of Community and Corresponding Identifying Algorithm. *Physical Review E*, **78**, Article ID: 026121. http://dx.doi.org/10.1103/PhysRevE.78.026121

[5] Fortunato, S. (2010) Community Detection in Graph. *Physics Reports*, **486**, 75-174. http://dx.doi.org/10.1016/j.physrep.2009.11.002

[6] Newman, M.E.J. (2012) Communities, Modules and Large-Scale Structure in Networks. *Nature Physics*, **8**, 25-31. http://dx.doi.org/10.1038/nphys2162

[7] Zhang, S., Jin, G., Zhang, X.S. and Chen, L. (2007) Discovering Functions and Revealing Mechanisms at Molecular Level from Biological Networks. *Proteomics*, **7**, 2856-2869. http://dx.doi.org/10.1002/pmic.200700095

[8] Ahn, Y.Y., Bagrow, J.P. and Lehmann, S. (2010) Link Communities Reveal Multi-Scale Complexity in Networks. *Nature*, **466**, 761-764. http://dx.doi.org/10.1038/nature09182

[9] Evans, T.S. and Lambiotte, R. (2009) Line Graphs, Link Partitions and Overlapping Communities. *Physical Review E*, **80**, Article ID: 016105. http://dx.doi.org/10.1103/PhysRevE.80.016105

[10] Evans, T.S. (2010) Clique Graphs and Overlapping Communities. *Journal of Statistical Mechanics: Theory and Experiment*, 12037. http://dx.doi.org/10.1088/1742-5468/2010/12/P12037

[11] Evans, T.S. and Lambiotte, R. (2010) Line Graphs of Weighted Networks for Overlapping Communities. *The European Physical Journal B*, **77**, 265-272. http://dx.doi.org/10.1140/epjb/e2010-00261-8

[12] Li, Z., Zhang, X.S., Wang, R.S., Liu, H. and Zhang, S. (2013) Discovering Link Communities in Complex Networks by an Integer Programming Model and a Genetic Algorithm. *PLoS ONE*, **8**, e83739. http://dx.doi.org/10.1371/journal.pone.0083739

[13] Zhang, S., Wang, R.S. and Zhang, X.S. (2007) Identification of Overlapping Community Structure in Complex Networks Using Fuzzy *c*-Means Clustering. *Physica A*, **374**, 483-490. http://dx.doi.org/10.1016/j.physa.2006.07.023

[14] He, D.X., Liu, D., Zhang, W., Jin, D. and Yang, B. (2012) Discovering Link Communities in Complex Networks by Exploiting Link Dynamics. *Journal of Statistical Mechanics*, **2012**, Article ID: P10015. http://dx.doi.org/10.1088/1742-5468/2012/10/P10015

[15] Guimmerà, R., Sale-Pardo, M. and Nunes Amaral, L.A. (2007) Module Identification in Bipartite and Directed Networks. *Physical Review E*, **76**, Article ID: 036102. http://dx.doi.org/10.1103/PhysRevE.76.036102

[16] Barber, M.J. (2007) Modularity and Community Detection in Bipartite Networks. *Physical Review E*, **76**, Article ID: 066102. http://dx.doi.org/10.1103/physreve.76.066102

[17] Zhan, W., Zhang, Z., Guan, J. and Zhou, S. (2011) Evolutionary Method for Finding Communities in Bipartite Networks. *Physical Review E*, **83**, Article ID: 066120. http://dx.doi.org/10.1103/PhysRevE.83.066120

[18] Murata, T. and Ikeya, T. (2010) A New Modularity for Detecting One-to-Many Correspondence of Communities in Bipartite Networks. *Advances in Complex Systems*, **13**, 19-31. http://dx.doi.org/10.1142/S0219525910002402

[19] Costa, A. and Hansen, P. (2014) A Locally Optimal Hierarchical Divisive Heuristic for Bipartite Modularity Maximization. *Optimization Letters*, **8**, 903-917. http://dx.doi.org/10.1007/s11590-013-0621-x

[20] Du, N., Wang, B., Wu, B. and Wang, Y. (2008) Overlapping Community Detection in Bipartite Networks. *Proceedings of the IEEE/WIC/ACM International Conference on Web Intelligence and Intelligent Agent Technology*, Sydney, 9-12 December 2008, 176-179. http://dx.doi.org/10.1109/WIIAT.2008.98

[21] Souam, F., Aïtelhadj, A. and Baba-Ali, R. (2013) Dual Modularity Optimization for Detecting Overlapping Communities in Bipartite Network. *Knowledge and Information Systems*, **40**, 455-488. http://dx.doi.org/10.1007/s10115-013-0644-8

[22] Zhang, Z.Y., Wang, Y. and Ahn, Y.Y. (2013) Overlapping Community Detection in Complex Network Using Symmetric Binary Matrix Factorization. *Physical Review E*, **87**, Article ID: 062803. http://dx.doi.org/10.1103/PhysRevE.87.062803

[23] Zhang, Z.Y. and Ahn, Y.Y. (2015) Community Detection in Bipartite Networks Using Weighted Symmetric Binary Matrix Factorization. *International Journal of Modern Physics C*, **26**, Article ID: 1550096.

[24] Freeman, L.C. (2003) Finding Social Groups: A Meta-Analysis of the Southern Women Data. In: Breiger, R., Carley, C. and Pattison, P., Eds., *Dynamic Social Network Modeling and Analysis*: *Workshop Summary and Papers*, National Research Council, The National Academies Press, Washington DC, 39-97.

[25] Chen, B.L., Chen, L., Zhou, S.R. and Xu, X.L. (2013) Detecting Community Structure in Bipartite Networks Based on Matrix Factorization. *International Journal of Wireless and Mobile Computing*, **6**, 599-607.
http://dx.doi.org/10.1504/IJWMC.2013.057576

[26] Liu, X. and Murata, T. (2010) An Efficient Algorithm for Optimizing Bipartite Modularity in Bipartite Networks. *Journal of Advanced Computational Intelligence and Intelligent Informatics*, **14**, 408-415.

[27] Murata, T. (2009) Detecting Communities from Bipartite Networks Based on Bipartite Modularities. *Proceedings of the 2009 International Conference on Computational Science and Engineering*, Vancouver, 29-31 August 2009, 50-57.

11

The Effect of Reputation to the Moral Hazard in C2C E-Market with Game Theory

Fangjun Mu

Management School, Jinan University, Guangzhou, China
Email: 784558439@qq.com

Abstract

It is well known that the reputation is the basis of a seller to survive and gain trust from customers in a competitive business environment. But as the existence of information asymmetry between buyer and seller, the moral hazard problem is the key obstacle that impedes the benefits of related shareholders and reduces the efficiency of total market. It is crucial to design a control mechanism to avoid the negative impact of moral hazard. This paper studies the principal and agent relationship between buyer and seller in C2C e-market; because of the influence of information asymmetry, many customers suffered from being cheated by sellers with defective products in practice. These frequent cases will deteriorate long term relationship between sellers and buyers. Here we focus on the analysis of the causes of moral risks and the effect of reputation on oral risk with repeated game theory. The purpose of this paper is to help both firms and customers effectively avoid morality risk and realize a win-win situation.

Keywords

C2C E-Commerce, Moral Hazard, Reputation Model, Game Theory

1. Introduction

C2C e-commerce plays an indispensable role in the Electronic Commerce market in China and the total transaction value is estimated to be 607 billion in 2014 [1]. But there is one serious problem impeding the development of C2C e-commerce for long time, that is the trustiness between strangers. There are many factors that influence the relationship between the seller and the buyer in C2C e-market. One of them is moral hazard, which means all egoistic behaviors of seller after making a deal with the buyer. This problem originally raised from insurance industry, where individual is easily to depend on the seller, as the buyer usually lacks the sense of taking some precautionary measures. In C2C e-business, the seller who will adulterate and do deliberate bribery is the reflection of moral hazard. The underlying reason of moral hazard is information asymmetry, which means the seller has more information about the quality, cost of product, while customer knows less. In the practice of C2C e-commerce, the seller usually will take an advantage of the quality of product, costs, and transportation costs and so on over customer. It is depicted that there are two types of information asymmetry depending on when and why it exists. One is adverse selection which happened before the coalition between buyer and seller, the other is moral hazard which happened after the deal.

In this paper, we will focus on the problem of moral hazard between the seller and buyer in C2C market and potentially intend to help solving the problem. One popular way is to introduce reputation to track the past behavior of seller [2]. A corporate reputation is an overall evaluation that reflects the extent to which people see the firm was substantially "good" or "bad" [3]. A good reputation is valuable because it can enhance trust and confidence so that the buyer feels that "it is safe to buy products and service from this seller". This outcome can also benefit the seller in its markets and various researches have also shown that sellers with good reputation are better able to attain and sustain superior profits outcomes over time.

Our primary research question is how the expected profits of the seller and the buyer depend on these two factors, one is the type of the seller, and the other is the reputation of the seller from the buyer. For example, does the seller always benefit from cheating or not. To answer this research question, we will examine a market where the reputation mechanism exists and check the influential mechanism. In this paper, we will set the reputation model of the seller in C2C E-Commerce market. We first characterize the situation that the type of seller is not common knowledge and then demonstrate that even though cheating has a direct benefit to the seller, it can sometimes hurt the seller, buyer, or both if the deal goes in the long run. Furthermore, we show the impact of reputation. In addition, we illustrate that the seller will always choose to be honest when the mechanism of reputation works. In a typical game-theoretic view of the relationship between buyer and seller, each player acts in order to maximize its own profit (rational player) without taking into account the overall optimal. Thus, incentive is offered to influence the behavior of the other one. Such an incentive is reputation.

This paper proceeds as follows: In Section 2 we review related literature. We then present the model in Section 3. We analyze the model to explain the existence of moral hazard and the coordination mechanism of reputation respectively in Section 4. Finally, we discuss extensions of our paper and provide our concluding remarks in Section 5.

2. Literature Review

C2C e-commerce specifies all business activities related to all buyer and buyer, like e-bay and TaoBao. There are many advantages like lower price, better service of C2C e-commerce, the main difference between C2C e-commerce is that the relationship between the seller and the buyer could be switched. The possibility of the occurring of lemon market means the reputation is also one aspectneed to be considered.

As a potential motivation, reputation could inspire seller to increase service quality. Since Adam Smith, reputation has been regarded as a very important mechanism to guarantee the implementation of contract in Economics [4], but it is widely used just since the combination with game theory. In management practice, the motivation of reputation is also very popular and has brought new management thinking to create and maintain a good reputation. For example, Kleps discussed that the company is a carrier of reputation, power origins from reputation, which determines endogeneity. But his argument is limited to B2B market. Here we are more focusing on the trade between customer and customer in e-commerce market, to discover the reputation mechanism is a bilateral problem.

The earlier research about reputation comes from economic, which emphasized the constraint of the agency market to the behavior of agency. As they are living in a competitive market, the potential value of the agency is determined by his past performance. The agency who cares about his reputation will be responsible for his behavior even there is no explicit motivational contract, they would work hard to increase the level of reputation, hoping they will gain more in the future.

Later researchers started to connect the reputation and the incentives of seller to build a complete model [5]. It is said the reputation of the market could be used as a replacement of explicit contract. When both parties in the game only cares about the immediate benefits, the optimal strategy is to not return debt, the optimal action for the bank is not to lend money, which is not beneficial for both parties. In the setting of repeated game, reputation provide implicit motivation for contracts, the player would like to compromise by giving up short term benefits to choose coordinate equilibrium.

The reputation model KMRW also proves that when the payoff of one player is not known by the other, this player has the incentive to build a good reputation to exchange for long run benefits [6]. In the setting of finite game, if not all players are rational, assuming the strategy of the irrational player is called tit for tat, then one cooperative equilibrium exists.

Among the few works that study the impact of reputation on the moral hazard behavior, KMRW model is the most relevant to this study. They consider a market with a seller and a buyer in a setting that both of them are customers, but without considering demand. Considering the demand uncertainties and information asymmetry [7], we can then extend the model and answer our research question, which cannot be answered using the prior model. Thus, we specifically develop a model to investigate how the impact of the reputation to the profit of the seller.

There are mainly three forms of moral hazard [8] in C2C e-commerce, one is returned goods, and the others are deliberate cheating and shoddy goods for quality goods. As the characteristic of C2C e-commerce is that the buyer could not touch the goods by himself, nor check them. For disposable shopping customers, who do not want to waste of time returning goods even though the product model or color does not meet their requirements, giving access to seller who would deregulate its product. In practice, many e-commerce websites have already regulated many items and described regulations trying to protect the benefits of buyers. But some buyers remain face the losses as they do not know full information of the model or products' attributes. In the meantime, the seller will also suffer from loss as they promise to accept returned goods unconditionally within 7 days. Furthermore, in the trade of tangible goods, the seller will present its product by photo or words, so the buyer could not get full information about the authenticity of the information and the quality of the product. For example, the biggest C2C e-commerce platform TaoBao, the seller all commit that their photos are absolutely objects photographed, but the buyer still do not have access to confirm it. It shows that information asymmetry is very common. Thirdly, the seller would sell shoddy goods for quality goods, as lack of standardized process to rating the quality of the products. Besides the preference of interest of customer influence the decision of customers. Then the seller with higher price and better products could hardly survive in the market, which is the cause of lemon market.

3. The Model

3.1. Assumption

In the setting of repeated game, we consider a market where both the sellers and the buyer are customers, which is quite popular in the real practice. There are two possible types of the seller, with the probability p that he has good reputation, and $1-p$ probability that his reputation is bad. The selling price of one good is P and the unit cost is C, the value of one good to the buyer is denoted as V, $V > P$, otherwise the buyer does not have the incentive to buy a good. Moreover, there are two actions which the seller could take no matter which type he is, to provide honest service or dishonest service.

Our model is on base of Ma (2005) that the integrity trade rate(the percentage of the honest trade among all trade), here the cost of seller with good reputation and bad reputation to take the action of being honest and dishonest is denoted as follows: C_{SG} and C_{FG}, C_{SB} and C_{FB}. "S" denotes the seller choose to be honest while "F" denotes the seller choose to be dishonest. "G" denotes the type of seller is good, while "B" denotes the type of seller is bad. As the seller of low reputation will have more management cost and more future risk, additionally, the seller with bad reputation is more familiar with cheating with the buyer, so here

Assumption 1: $0 < C_{SG} < C_{SB} < C_{FB} < C_{FG}$.

The information asymmetry in C2C e-business is reflected by that the seller knows his type while the buyer lack of knowledge about it. As is shown in **Figure 1**, if the seller with good reputation will choose to be honest, and the buyer thinks that the seller will not cheat her, so decide to make deal with him, the revenue of the seller is: $P - C - C_{SG}$, the revenue of the buyer is $V - P$; if the buyer thinks that the seller is cheating her then decide not to make a deal with the seller, then the seller with good reputation will suffer from loss: $-C_{SG}$. Similarly, we could conclude the payoff of buyer and seller when the type of seller is bad in **Figure 2**.

Assumption2: Suppose the unit value of the product provided by the seller within some periods values π, which is a function of seller's service levelα, seller's real strength θ and the uncertainty in C2C e-market, that

		Buyer	
		Deal	**No deal**
Seller of good reputation	Honest	$P-C-C_{SG}$, $V-P$	$-C_{SG}$, 0
	Dishonest	$P-C_{FG}$, $-P$	$-C_{FG}$, 0

Figure 1. The payoffs of the seller with good reputation G.

		Buyer	
		Deal	**No deal**
Seller of bad reputation	Honest	$P-C-C_{SB}$, $V-P$	$-C_{SB}$, 0
	Dishonest	$P-C_{FB}$, $-P$	$-C_{FB}$, 0

Figure 2. The payoffs of the seller with bad reputation B.

is $\pi = k\alpha + h\theta + \mu$, α is the private information of the seller, π is the common knowledge of both the seller and the buyer, besides θ and μ following nominal distribution, with means equal to 0 and variance equals σ_θ^2 and σ_μ^2 respectively.

Assumption3: The times that the buyer will purchase from the seller is stable to a constant m, then the benefits of the buyer is $w^b = m\pi$.

Assumption4: The sequence is as follows. Firstly the buyer will decide how many times to buy from this seller, then the seller will decide its service level.

The seller mainly benefits from the commission from purchasing times m, that is βm, the cost of the service provided by the seller is $c(\alpha)$, $c'(\alpha) > 0$, $c''(\alpha) > 0$. $c(\alpha) = (b\alpha^2)/2$, the revenue of the seller is $W_s(\alpha) = \beta m - (b\alpha^2)/2$.

4. Model Analysis

The purpose of introducing service level of seller is to reduce the risk of buyer, to maintain its benefits of buyer and guarantee the efficiency of market. Therefore the optimal service level to maximize the total benefit in C2C market should be:

$$\max{}_\alpha^\pi = m\pi + \beta m - (b\alpha^2)/2 \Rightarrow \alpha = mk/b \tag{1}$$

But as the first decision the buyer make is to choose the buying times from a specific seller, next the seller will decide its service level. Here βm is a constant, as a rational person, the seller will take following actions:

$$\max{}_a^\pi = \beta m - (b\alpha^2)/2 \Rightarrow \alpha' = 0 \tag{2}$$

$$\pi = \beta m - b\alpha \Rightarrow \alpha'' = 0 \tag{3}$$

In the case of single business, the seller will choose dishonest to maximize its own profits no matter what type it belongs to, and the buyer will not make a deal with seller after considering that, so this market does not exists. In the case of repeated business, the buyer will make decision on base of past shopping experience. As repeated game change the constraint mechanisms, the payoff of both parties will be different, so new equilibrium exists.

In the 1st business, the buyer thinks that the seller belong to good reputation, so the expected payoff of buyer is $(V-P)P_1 + (-P)(1-P_1) > 0$, here P_1 denotes the possibility that the seller was regarded to have good reputation at first time, only when $P_1 > P/V$, buyer will decide to make a deal with seller.

In this model, we suppose $P_1 > P/V$, the payment at the first stage of seller, here introduce a discount rate D, which will be counted in the second trade. Specifically, if the seller is the type of B and he will cheat at the first stage, then he his payoff is as high as $P - C_{FB}$, while this also induces the buyer to confirm they type of the seller. Then at the second stage the seller will choose to be honest after considering the behavior of buyer, then $-C_{FB} < -C_{SB}$.

The total payoff of the seller is:

$$U_1 = \left(P - C_{FB}\right)\left(1 + D\right) + \left(-C_{SB}\right) \tag{4}$$

Then considering the case when the seller of bad reputation try to hide its type firstly to gain the credibility of the buyer, in order to gain more benefit in the following deal, then the strategy of the buyer is (Deal, Deal), the total pay off of the seller is:

$$U_2 = \left(P - C - C_{SB}\right)\left(1 + D\right) + \left(P - C_{FB}\right) \tag{5}$$

When the seller choose to not cheat at the first deal, then $U_2 > U_1$, that is

$$U_2 - U_1 = \left(C_{FB} - C - C_{SB}\right)\left(1 + D\right) + \left(P + C_{SB} - C_{FB}\right) > 0, \text{ then } \delta = \frac{P - C}{C + C_{SB} - C_{FB}}, \text{ here } \delta_B = \frac{P - C}{C + C_{SB} - C_{FB}} \text{ is}$$

the threshold value of seller whose type is Bad when decide which strategy to take.

Similarly, we could calculate corresponding threshold value δ_G of seller whose type is Good when decide which strategy to take.

From the assumption that $C_{FG} - C_{SG} > C_{FB} - C_{SB}$, we could conclude $\delta_B < \delta_G$. As long as there exists one $\delta < \delta_B$, whatever the type, the seller will choose to be honest in order to gain long term profit.

After the game theory analysis, we can conclude, the seller in C2C market has the incentive to maintain its reputation in order to gain more profits in the future. That also accounts for the reasons why the seller will invest more on after sellers training, improving customers service level, caring about the quality of product and the comments of finished customer, is to keep a longer seller-buyer relationship.

5. Conclusions

As lack of well-designed evaluation system targeted at Chinese C2C e-market, the problem of moral hazard could not be avoided or solved thoroughly. The integrity between trade partners is the basis of E-business, so it is necessary to appeal to all partners participating in E-Commerce website, both buyers and sellers and the government to work together, to push the development of evaluation system on base of reputation, to connect the benefits of sellers with its reputation, to increase the cost of irregular actions in the E-market. Generally speaking, there are following actions that can be taken to realize the function of reputation as a credible signal. Firstly, from standpoint of the online platform, it is advisable to complete the systematic rating system in the market; Secondly, from the governing of the government, it is also helpful to regulate the market by taking actions to strengthen the implementation of identity recoding and reputation supervision; Thirdly, the success of completing the evaluating system to sellers heavily depends on the efforts imposed by customers who would like to join in the activity of evaluating the ratings of sellers.

Potential extensions of our work include the study of e-commerce with one seller and many buyers, or many buyers buying the same product from many sellers. Furthermore, advanced reputation constraints could be examined for e-commerce that exhibit additional complexities (e.g., multiple echelon inventories) [9] and information asymmetry [10] in both sides could be introduced.

References

[1]   Guan, X.J., Liu, J.-Z., et al. (2014) Chapter 10, Internal and External Trade, Table 10-10 Electronic Commerce Transactions. China Statistics Press, Beijing, 187.

[2]   Zhang, W.Y. (1996) Game Theory and Information Economics. Shanghai Sanlian People Press, Shanghai.

[3]   Dowling, G. (2004) Corporate Reputations: Should You Compete on Yours? California Management Review, 46, 19-36. http://dx.doi.org/10.2307/41166219

[4]   Fama, E. (1980) Agency Problems and the Theory of the Firm. Journal of Political Economy, 88, 288-307. http://dx.doi.org/10.1086/260866

[5]   Holmstrom, B. (1999) Managerial Incentive Problem: A Dynamic Perspective. The Review of Economic Studies, 66, 169-182.

[6]   Kreps, D., Milgrom, P., Roberts, J. and Wilson, R. (1982) Rational Cooperation in the Finitely Repeat Prisoners Dilemma. *Journal of Economic Theory*, **27**, 245-252. http://dx.doi.org/10.1016/0022-0531(82)90029-1

[7]   Raz, G. (2011) Supply Chain Sourcing Under Asymmetric Information. *Production and Operations Management*, **20**, 92-115.

[8]   Corbett, C.J., Decroix, G.A. and Ha, A.Y. (2005) Optimal Shared-Savings Contracts in Supply Chains : Linear Contracts and Double Moral Hazard. *European Journal of Operational Research*, **163**, 653-667.

[9]   Ozer, O. and Raz, G. (2011) Supply Chain Sourcing Under Asymmetric Information. *Production and Operations Management*, **20**, 92-115. http://dx.doi.org/10.1111/j.1937-5956.2010.01124.x

[10] Zhu, L. and You, J. (2011) Moral Hazard Strategy and Quality Contract Design in a Two-Echelon Supply Chain. *Journal of Systems Science and Systems Engineering*, **20**, 70-86. http://dx.doi.org/10.1007/s11518-011-5153-2

12

An Uncertain Programming Model for Competitive Logistics Distribution Center Location Problem

Bingyu Lan[1], Jin Peng[2], Lin Chen[3]

[1]College of Mathematics and Sciences, Shanghai Normal University, Shanghai, China
[2]Institute of Uncertain Systems, Huanggang Normal University, Hubei, China
[3]Institute of Systems Engineering, Tianjin University, Tianjin, China
Email: pengjin01@tsinghua.org.cn

Abstract

We employ uncertain programming to investigate the competitive logistics distribution center location problem in uncertain environment, in which the demands of customers and the setup costs of new distribution centers are uncertain variables. This research was studied with the assumption that customers patronize the nearest distribution center to satisfy their full demands. Within the framework of uncertainty theory, we construct the expected value model to maximize the expected profit of the new distribution center. In order to seek for the optimal solution, this model can be transformed into its deterministic form by taking advantage of the operational law of uncertain variables. Then we can use mathematical software to obtain the optimal location. In addition, a numerical example is presented to illustrate the effectiveness of the presented model.

Keywords

Competitive Location, Logistics Distribution Center, Uncertain Programming, Uncertainty Theory

1. Introduction

Research on distribution center location problem is a necessary component of the optimization of logistics distribution's system. The distribution center plays the role of a bridge that links customers with suppliers so as to transport goods from suppliers to customers. A lot of researches have been devoted to the distribution center location problem. For example, Lu and Bostel [1] investigated a facility location model for logistics system with reverse flows. Klose and Drexl [2] summarized different types of facility location models and their applications to distribution system design.

A large part of location problems have been studied in an ideal environment, in which only a unique facility offers services or products in the market. In practice, with the rapid development of economic, the environment is growing more complex. The facility has to compete with other players for more benefits. Thus the research on competitive location problem plays an important role in location theory. Competitive location problem differs from the general location problem because we must consider the competition between the existing facilities and new facilities. The briefly explain of the competitive location problem is that some facilities have been located in the market and the new facility will be located at the optimal place so as to compete with others for their market share.

Because of the importance of competitive location problem, many scholars devoted to the related research of competitive location problem in deterministic environment, that is, all parameters are known in advance and assumed to be fixed. Hotelling [3] initially introduced the competitive problem of two companies in the market, which laid the foundation for modern competitive location problem. Subsequently, a lot of scholars also studied this topic and made a progress in competitive location theory. In location space aspect, the competitive location problem was extended to the planar location problem by Drezner [4]. Moreover, it was also applied to the network location problem by Hakimi [5]. So Plastria [6] overviewed the papers of competitive location and clarified the location spaces into three traditional spatial settings: discrete space, continuous space and the network. In competition feature aspect, except for the simplest static competitive models, the dynamic models were proposed to describe the action cycles of competing players. Wong and Yang [7], and Yang and Wong [8] proposed a continuous equilibrium model, respectively. These models solved the competitive location problem with different assumptions of customer demands. With considering the future competition which known in the economic literature as Stackelberg equilibrium problem or leader-follower problem, Plastria and Vanhaverbeke [9] proposed three models to solve Stackelberg equilibrium problem in discrete space. The game theory was introduced into the competitive location problem by Küçükaydin et al. [10], in which they formulated a bilevel mixed-integer nonlinear programming model to solve the leader-follower problem. Sasaki et al. [11] employed a generic hub arc location model to locate arcs within the framework of Stackelberg. With respect to methodology, various approaches were proposed to obtain the estimate of the market share captured by each competitive facility, such as proximity approach [4], deterministic utility approach [12], cover-based approach [13], and gravity-based approach [14]. For a detailed view of the models of competitive location problem, see Drezner [15] and Plastria [6].

In practical life, there are many indeterminacy factors in competitive location problem. For example, the demand of customer for a kind of product is variable because it is influenced by other things liked weather. Hence, many scholars established models by stochastic method. Leonardi and Tadei [16] introduced the random utility model into location theory via assuming that the each customer described his utility from a random distribution of utility functions. Subsequently, Drezner and Drezner [17] used the random utility to calculate the expected market share of the competitive facility. In order to overcome the hard of complicated computation of the random utility model, Drezner et al. [18] proved that the random utility model could be approximated by the logit model and designed a procedure to find the best location. Except for the random utility function, some scholars also researched the competitive location problem with stochastic weights in network. Shiode and Drezner [19] analyzed a Stackelberg equilibrium problem on a tree network with stochastic weights and presented a procedure to find the solution. In addition, other papers studied this problem have been published, such as Drezner and Wesolowsky [20], Shiode and Ishii [21], and Wesolowsky [22].

In the above mentioned literatures, these researches cannot be proceeded smoothly without the assumption that there are enough history data to obtain probability distribution which is closed to the real frequency. However, sometimes the lack of history data posed difficulties for applying probability theory, especially when a new product was shipped to the customer by distribution center. In this case, we have to invite some experts to give the belief degree that each event will occur. In order to deal with belief degree, uncertainty theory was found by Liu [23] in 2007, which has become a new tool to describe the human uncertainty.

Within the framework of uncertainty theory, the research of uncertain facility location problem had made a great number of achievements. Gao [24] constructed two uncertain models to deal with single facility location problems on network. Wen et al. [25] proposed an uncertain facility location-allocation model by means of chance-constraints. Wang and Yang [26] investigated two types of uncertain programming models according to different decision criteria for modeling the hierarchical facility location problem in an uncertain environment.

Wu and Peng [27] presented an uncertain chance-constrained model to deal with logistics distribution center location problem under uncertain environment.

This paper addresses the problem that a logistics company enters a market by locating a new distribution center where there are many existing competitors in uncertain environment. The demands of customers and setup costs of the potential distribution centers are assumed to be uncertain variables. Then we construct the expected value model with the objective of maximizing the profit of the new distribution center. In order to obtain the optimal solution, the expected value model is transformed into its crisp equivalent model. At last we can use the mathematical software to find the optimal solution.

The innovations of this paper are as follows. We investigate the competitive logistics distribution center location problem under uncertain environment instead of logistics distribution center location problem in uncertain environment. It is different form the problem which dealt with by Wu and Peng [27]. This study is distinguished from the Revelle's [28] study by making three contributions. Firstly, we assume the customers' demands are uncertain variables rather than certain amount. Secondly, we study a specific facility rather than any facility, that is, the proposed model can be applied to select logistics distribution center. Thirdly, we describe the customers' patronizing behavior with a piecewise function instead of a 0 - 1 variable.

The remainder of this paper is organized as follows. We introduce some basic and necessary knowledge about uncertainty theory in Section 2. In Section 3, we state the competitive distribution center location problem and construct an expected value model. In Section 4, we transform the expected value model into its deterministic one. In Section 5, we give a numerical example to illustrate the modeling idea of this paper. At last, the Section 6 concludes the paper.

2. Preliminary

In order to understand the presented model of competitive location problem better, we introduce some necessary knowledge about uncertainty theory in this section.

Let us introduce the concept of uncertain measure first. Let $\mathcal{L}$ be a σ-algebra over a nonempty set Γ. $\Lambda \in \mathcal{L}$ is an event. A set function $\mathcal{M}$ from a σ-algebra $\mathcal{L}$ to an interval $[0,1]$ is an uncertain measure if it satisfies normality axiom, duality axiom, and subadditivity axiom.

1) (Normality Axiom) $\mathcal{M}\{\Gamma\} = 1$ for the universal set Γ;

2) (Duality Axiom) $\mathcal{M}\{\Lambda\} + \mathcal{M}\{\Lambda^c\} = 1$ for any $\Lambda \in \mathcal{L}$;

3) (Subadditivity Axiom) For every countable sequence of events $\{\Lambda_i\}$, $i = 1, 2, \cdots$, we have

$$\mathcal{M}\left\{\bigcup_{i=1}^{\infty} \Lambda_i\right\} \leq \sum_{i=1}^{\infty} \mathcal{M}\{\Lambda_i\}.$$

The triplet $(\Gamma, \mathcal{L}, \mathcal{M})$ is called an uncertainty space. The product axiom was defined to obtain an uncertain measure of compound event.

4) (Product Axiom) Let $(\Gamma_k, \mathcal{L}_k, \mathcal{M}_k)$ be uncertainty spaces for $k = 1, 2, \cdots$ The product uncertain measure $\mathcal{M}$ on product σ-algebra $\mathcal{L}$ is an uncertain measure satisfying

$$\mathcal{M}\left\{\prod_{k=1}^{\infty} \Lambda_k\right\} = \bigwedge_{k=1}^{\infty} \mathcal{M}_k\{\Lambda_k\}$$

where $\Lambda_k \in \mathcal{L}_k$ for $k = 1, 2, \cdots$, respectively.

In order to describe the quantities with uncertainty, Liu [23] defined the uncertain variable.

Definition 1. (*Liu* [23]) *An uncertain variable is a function* $\xi: (\Gamma, \mathcal{L}, \mathcal{M}) \to \Re$ *such that* $\{\xi \in B\}$ *is an event for any Borel set B.*

Liu [23] proposed uncertainty distribution to describe uncertain variable analytically.

Definition 2. (*Liu* [23]) *Let* ξ *be an uncertain variable. The uncertainty distribution of* ξ *is defined by*

$$\Phi(x) = \mathcal{M}\{\xi \leq x\}$$

for any number x in $\Re$.

Definition 3. (*Liu* [29]) *Let* $\Phi(x)$ *be an uncertainty distribution. If it is a continuous and strictly increasing function with respect to x at which* $0 < \Phi(x) < 1$, *and* $\lim_{x \to -\infty} \Phi(x) = 0$, $\lim_{x \to +\infty} \Phi(x) = 1$, *then uncertainty distribution* $\Phi(x)$ *is said to be regular.*

Definition 4. (*Liu* [29]) *Assume that uncertain variable ξ has a regular uncertainty distribution $\Phi(x)$. Then the inverse function $\Phi^{-1}(\alpha)$ is said to be the inverse uncertainty distribution of ξ.*

Example 1. *If ξ is a linear uncertain variable, then ξ has the uncertainty distribution*

$$\Phi(x) = \begin{cases} 0, & \text{if } x \leq a \\ \dfrac{x-a}{b-a}, & \text{if } a \leq x \leq b \\ 1, & \text{if } x \leq b, \end{cases}$$

denoted by $\mathcal{L}(a,b)$, where a and b are real numbers with $a < b$. The inverse uncertainty distribution of the linear uncertain variable $\mathcal{L}(a,b)$ is

$$\Phi^{-1}(\alpha) = (1-\alpha)a + \alpha b.$$

Definition 5. (*Liu* [30]) *The uncertain variables $\xi_1, \xi_1, \cdots, \xi_n$ are said to be independent if*

$$\mathcal{M}\left\{ \bigcap_{i=1}^{\infty} (\xi_i \in B_i) \right\} = \bigwedge_{i=1}^{\infty} \mathcal{M}\{\xi_i \in B_i\}$$

for any Borel sets $B_1, B_1, \cdots, B_n$.

A real-valued function $f(x_1, x_2, \cdots, x_n)$ is said to be strictly decreasing if

$$f(x_1, x_2, \cdots, x_n) \geq f(y_1, y_2, \cdots, y_n)$$

whenever $x_i \leq y_i$ for $i = 1, 2, \cdots, n$, and

$$f(x_1, x_2, \cdots, x_n) > f(y_1, y_2, \cdots, y_n)$$

whenever $x_i < y_i$ for $i = 1, 2, \cdots, n$.

Theorem 1. (*Liu* [29]) *Let $\xi_1, \xi_2, \cdots, \xi_n$ be independent uncertain variables with regular uncertainty distributions $\Phi_1, \Phi_2, \cdots, \Phi_n$, respectively. If $f(x_1, \cdots, x_n)$ is a strictly decreasing function, then the uncertain variable*

$$\xi = f(\xi_1, \xi_2, \cdots, \xi_n)$$

has an inverse uncertainty distribution

$$\Psi^{-1}(\alpha) = f\left(\Phi_1^{-1}(1-\alpha), \Phi_2^{-1}(1-\alpha), \cdots, \Phi_n^{-1}(1-\alpha) \right).$$

We review the important concept of the expected value, which represents the size of uncertain variable.

Definition 6. (*Liu* [23]) *Let ξ be an uncertain variable. Then the expected value of ξ is defined by*

$$E[\xi] = \int_0^{+\infty} \mathcal{M}\{\xi \geq r\} \, dr - \int_{-\infty}^{0} \mathcal{M}\{\xi \leq r\} \, dr$$

provided that at least one of the two integrals is finite.

Theorem 2. (*Liu* [29]) *Let ξ be an uncertain variable with regular uncertainty distribution Φ. Then*

$$E[\xi] = \int_0^1 \Phi^{-1}(\alpha) \, d\alpha.$$

Example 2. *Let $\xi \sim \mathcal{L}(a,b)$ be a linear uncertain variable. Then the expected value of ξ is*

$$E[\xi] = \frac{a+b}{2}.$$

3. The Model of Competitive Location Problem

In this section, we mainly propose the expected value model for competitive distribution center location problem within the framework of uncertain programming. Uncertain programming, proposed by Liu [31], is a type of mathematical programming which contains uncertain variables.

3.1. Problem Description

This paper investigates the competitive logistics distribution center location problem in uncertain environment. That is the problem in which a logistics company enters a market by locating a new distribution center where there are many existing distribution centers. The goal of the decision maker is to choose the location of the new distribution center so as to maximize its profit. The flow diagram of logistics distribution is shown in **Figure 1**. The potential distribution centers in **Figure 1** show that they can be selected to build a new distribution center. The setup costs of the potential distribution centers and the demands of customers are assumed to be uncertain variables with known uncertainty distributions. In addition, we assume that the customers patronize the nearest distribution center to meet their full demands.

3.2. Assumptions of Model

Before we begin to study competitive location problem with uncertain variables, we need to make some assumptions as follows (which are referred to Revelle [28] and Wu and Peng [27]):

1) There is one supplier and many existing distribution centers.

2) The supplier only supplies one kind of product.

3) There is no difference among the products provided by all distribution centers.

4) The location of the new distribution center can be selected from potential distribution centers.

5) The distances between the supplier and potential distribution centers, the distances between potential distribution centers and customers and the distances between existing distribution centers and customers are known in advance.

6) The allocation of customers demands is related to the distance. The full demands of customers will be assigned to the nearest distribution center.

In order to model the competitive location problem, we introduce the following indices and parameters:

i : the index of existing distribution centers, $i \in \{1, 2, \cdots, N\}$;

j : the index of potential distribution centers, $j \in \{1, 2, \cdots, M\}$;

k : the index of customers, $k \in \{1, 2, \cdots, K\}$;

c : the transportation cost of unit distance of the per product;

w : the profit of the unit product;

d_j : the distance between the supplier and the j-th potential distribution center;

d_{jk} : the distance between the j-th potential distribution center and the k-th customer;

D_{ik} : the distance between the i-th existing distribution center and the k-th customer;

h_j : the capacity of the j-th potential distribution center;

ξ_k : the demand of the k-th customer, which is assumed to be an uncertain variable;

Φ_k : the uncertainty distribution of ξ_k ;

η_j : the setup cost of the j-th potential distribution center, which is assumed to be an uncertain variable;

Ψ_j : the uncertainty distribution of η_j ;

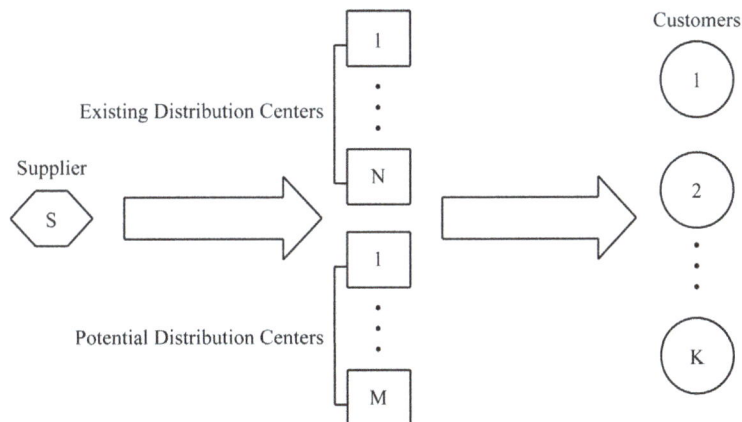

Figure 1. Logistics distribution process.

x_j: the quantity supplied to the j-th potential distribution center from the supplier, which is a decision variable;

y_{jk}: the quantity supplied to the k-th customer from the j-th potential distribution center, which is a decision variable;

v_j: 0 - 1 variable implies whether the j-th potential distribution center is chosen to build the new distribution center or not.

Remark 1: The meaning of variable v_j can be described as follows:

$$v_j = \begin{cases} 1, & \text{if the } j\text{-th potential distribution center is chosen} \\ 0, & \text{otherwise.} \end{cases}$$

In order to maximize the profit of the new distribution center, the decision maker must choose the appropriate site to build the new distribution center which attracts more customers. The majority of competitive location models assumed that customers will patronize the nearest distribution center. It is rationally for customers who want to sustain the less travel cost. In this paper, we consider that the customers choose the distribution center according to the distance between their sites and distribution centers rather than other conditions, such as price, service and attractiveness.

Thus we assume that the customers patronize the nearest distribution center, and this assumption which has been used by Revelle [28]. This patronizing behavior rule implies that the full demands of each customer are satisfied by the nearest distribution center. The meaning of patronizing behavior which was proposed by Revelle [28] is listed in the following. If the new distribution center is nearer than all existing distribution centers, then the customers choose the new distribution center to satisfy their demands. If they have the same distance, then the half demands of customers are satisfied by the new distribution center. Otherwise, the customers choose the existing distribution center. In Revelle's [28] model, the customers' patronizing behavior is embodied in objective function. It used two auxiliary variables to represent the facility which attracts full demands of customers and the facility which attracts half demands of customers, respectively. Our study is distinguished from the Revelle's [28] study by making the following important contribution: we use a function δ_{jk} to describe the customers' patronizing behavior. Specifically, the expression of the function is listed in the following.

$$\delta_{jk} = \begin{cases} 1, & \text{if } d_{jk} < \min_{1 \le i \le N}\{D_{ik}\} \\ \dfrac{1}{2}, & \text{if } d_{jk} = \min_{1 \le i \le N}\{D_{ik}\} \\ 0, & \text{otherwise.} \end{cases}$$

3.3. Expected Value Model

We note that the total profit of the new distribution center is made up of four parts. Thus the total profit is a function related to x, y and η, which can be written as

$$P(x,y,\eta) = w\sum_{k=1}^{K}\sum_{j=1}^{M} v_j \delta_{jk} y_{jk} - \sum_{j=1}^{M} d_j x_j c - \sum_{k=1}^{K}\sum_{j=1}^{M} d_{jk} y_{jk} c - \sum_{j=1}^{M} v_j \eta_j$$

where

$$x = (x_1, \cdots, x_M)$$

and

$$y = (y_{11}, \cdots, y_{1K}, \cdots, y_{M1}, \cdots, y_{MK})$$

are decision vectors, and

$$\eta = (\eta_1, \eta_2, \cdots, \eta_M)$$

is an uncertain vector.

Since we know the uncertain objective function $P(x, y, \eta)$ cannot be directly maximized, we can maximize its expected value, *i.e.*,

$$\max_{x,y} E\left[P(x, y, \eta)\right].$$

We employ the uncertain programming model to study the competitive logistics distribution center location problem. So we can build the expected value model as follow:

$$
\begin{cases}
\max\limits_{x,y} E\left[P(x, y, \eta)\right]. \\
\text{subject to}: \\
\quad E\left[v_j \delta_{jk} \xi_k\right] \geq y_{jk}, \quad j = 1, 2, \cdots, M; k = 1, 2, \cdots, K \\
\quad \sum\limits_{k=1}^{K} y_{jk} \leq x_j \leq v_j h_j, \quad j = 1, 2, \cdots, M \\
\quad \sum\limits_{j=1}^{M} v_j = 1 \\
\quad \delta_{jk} = \begin{cases} 1, & \text{if } d_{jk} < \min\limits_{1 \leq i \leq N}\{D_{ik}\} \\ \dfrac{1}{2}, & \text{if } d_{jk} = \min\limits_{1 \leq i \leq N}\{D_{ik}\}, \quad j = 1, 2, \cdots, M; k = 1, 2, \cdots, K \\ 0, & \text{otherwise} \end{cases} \\
\quad x_j, y_{jk} \geq 0, \quad j = 1, 2, \cdots, M; k = 1, 2, \cdots, K
\end{cases}
\tag{1}
$$

where $\delta_{jk} \neq 0$ and $v_j = 1$ show that $x_j > 0$ and $y_{jk} > 0$ for any $j = 1, 2, \cdots, M$ and $k = 1, 2, \cdots, K$. Otherwise, $x_j = 0$ and $y_{jk} = 0$.

In this expected value model, the first constraint means that the quantity supply is not exceed the demand of customer. The second constraint implies the volume of transport is less than the capacity of the new distribution center. The third one shows that we select single site to build the new distribution center. And the last one ensures the nonnegativity of decision variables x_j and y_{jk}.

4. The Crisp Equivalent Model

The key problem of the model is seeking for the optimal solution. Taking advantage of the operational law of uncertain variable, we can transform the expected value model (1) into its deterministic form. It is clearly that the total profit function is strictly decreasing with setup costs. According to Theorem 1 and Theorem 2, the objective function

$$E\left[P(x, y, \eta)\right]$$

can be converted into

$$w\sum_{k=1}^{K}\sum_{j=1}^{M} v_j \delta_{jk} y_{jk} - \sum_{j=1}^{M} d_j x_j c - \sum_{k=1}^{K}\sum_{j=1}^{M} d_{jk} y_{jk} c - \sum_{j=1}^{M} \delta_j \int_0^1 \Psi_j^{-1}(1-\alpha)\,d\alpha. \tag{2}$$

Similarly, the first constraint

$$E\left[v_j \delta_{jk} \xi_k\right] \geq y_{jk}, j = 1, 2, \cdots, M; k = 1, 2, \cdots, K$$

can be turned into

$$v_j \delta_{jk} \int_0^1 \Phi_k^{-1}(\alpha)\,d\alpha \geq y_{jk}, j = 1, 2, \cdots, M; k = 1, 2, \cdots, K. \tag{3}$$

It follows from the formulas (2) and (3) that expected value model (1) can be switched to the following equivalent model:

$$
\begin{cases}
\max_{x,y} \ w\sum_{k=1}^{K}\sum_{j=1}^{M} v_j \delta_{jk} y_{jk} - \sum_{j=1}^{M} d_j x_j c - \sum_{k=1}^{K}\sum_{j=1}^{M} d_{jk} y_{jk} c - \sum_{j=1}^{M} \delta_j \int_0^1 \Psi_j^{-1}(1-\alpha)\,d\alpha \\[2mm]
\text{subject to:} \\[2mm]
v_j \delta_{jk} \int_0^1 \Phi_k^{-1}(\alpha)\,d\alpha \ge y_{jk},\ j=1,2,\cdots,M; k=1,2,\cdots,K \\[2mm]
\sum_{k=1}^{K} y_{jk} \le x_j \le v_j h_j,\ \ j=1,2,\cdots,M \\[2mm]
\sum_{j=1}^{M} v_j = 1 \\[2mm]
\delta_{jk} = \begin{cases} 1, & \text{if } d_{jk} < \min_{1\le i\le N}\{D_{ik}\} \\[1mm] \dfrac{1}{2}, & \text{if } d_{jk} = \min_{1\le i\le N}\{D_{ik}\},\ j=1,2,\cdots,M; k=1,2,\cdots,K \\[1mm] 0, & \text{otherwise} \end{cases} \\[2mm]
x_j, y_{jk} \ge 0,\ \ j=1,2,\cdots,M; k=1,2,\cdots,K.
\end{cases}
\tag{4}
$$

Clearly, we note that the equivalent model (4) is a deterministic programming model. As all know, the software Lingo cannot show the integral function. Therefore, we can find the optimal solution by using the mathematical software Matlab. In addition, in uncertainty theory, the inverse distribution function is easy to calculate for us. Thus we can calculate the inverse distribution function before we use the software Lingo to find the optimal solution. We can choose one of them to solve this programming. For convenience, the following example is seeking for solution by Lingo.

5. Numerical Example

In order to illustrate the modeling idea and the effectiveness of this model, we give a numerical example in this section. Suppose that there is a supplier in a city. And there are 3 distribution centers to distribute the new model of the televisions to 7 customers. A logistics company want to select an optimal site from 4 potential distribution centers after survey. The distances d_j, d_{jk} and D_{ik} are listed in **Table 1**, **Table 2** and **Table 3**, respectively. The unit transportation cost $c = 0.01$ and the unit product profit $w = 5$ are given via survey. ξ_k, η_j are assumed to be independent linear uncertain variables with known uncertainty distributions which are presented in **Table 4** and **Table 5**, respectively. In addition, the capacities of potential distribution centers are listed in **Table 4**.

Table 1. The distance d_j between supplier and potential distribution centers.

j	1	2	3	4
d_j	150	200	210	170

Table 2. The distance D_{ik} between existing distribution centers and customers.

$i\backslash k$	1	2	3	4	5	6	7
1	10	20	30	28	15	15	25
2	14	25	25	32	30	12	20
3	18	28	8	13	23	25	20

Table 3. The distance d_{jk} between potential distribution centers and customers.

$j\backslash k$	1	2	3	4	5	6	7
1	15	25	18	12	20	8	18
2	22	15	16	10	15	21	6
3	21	8	24	20	16	25	21
4	18	12	20	14	12	20	12

Table 4. The capacities h_j and setup costs η_j of potential distribution centers.

j	1	2	3	4
h_j	200	240	180	210
η_j	$\mathcal{L}(20,24)$	$\mathcal{L}(24,32)$	$\mathcal{L}(16,20)$	$\mathcal{L}(22,28)$

Table 5. The demands ξ_k of customers.

k	1	2	3	4	5	6	7
ξ_k	$\mathcal{L}(50,56)$	$\mathcal{L}(70,76)$	$\mathcal{L}(80,90)$	$\mathcal{L}(60,76)$	$\mathcal{L}(68,72)$	$\mathcal{L}(68,72)$	$\mathcal{L}(54,60)$

According to the model (4), the expected value model of this numerical example is listed in the following formula

$$
\begin{cases}
\max_{x,y} 5\sum_{k=1}^{7}\sum_{j=1}^{4} v_j \delta_{jk} y_{jk} - 0.01\sum_{j=1}^{4} d_j x_j - 0.01\sum_{k=1}^{7}\sum_{j=1}^{4} d_{jk} y_{jk} - \sum_{j=1}^{4}\delta_j \int_0^1 \Psi_j^{-1}(1-\alpha)\,\mathrm{d}\alpha \\[2mm]
\text{subject to:} \\[2mm]
v_j \delta_{jk} \int_0^1 \Phi_k^{-1}(\alpha)\,\mathrm{d}\alpha \ge y_{jk},\, j=1,2,3,4;k=1,2,\cdots,7 \\[2mm]
\sum_{k=1}^{7} y_{jk} \le x_j \le v_j h_j,\; j=1,2,3,4 \\[2mm]
\sum_{j=1}^{4}\delta_j = 1 \\[2mm]
\delta_{jk} = \begin{cases} 1, & \text{if } d_{jk} < \min_{1\le i\le 3}\{D_{ik}\} \\[1mm] \dfrac{1}{2}, & \text{if } d_{jk} = \min_{1\le i\le 3}\{D_{ik}\},\; j=1,2,3,4;k=1,2,\cdots,7 \\[1mm] 0, & \text{otherwise} \end{cases} \\[2mm]
x_j, y_{jk} \ge 0,\; j=1,2,3,4;k=1,2,\cdots,7.
\end{cases}
\tag{5}
$$

Then we use Lingo to get the optimal solution which is listed in the following.

$$
\delta_{22} = \delta_{32} = \delta_{42} = 1, \delta_{14} = \delta_{24} = 1, \delta_{25} = \frac{1}{2}, \delta_{16} = 1, \delta_{17} = \delta_{27} = \delta_{47} = 1;
$$

$$
v_1 = v_3 = v_4 = 0, v_2 = 1;
$$

$$
x_1 = x_3 = x_4 = 0, x_2 = 233;
$$

$$
y_{22} = 73, y_{24} = 68, y_{25} = 35, y_{27} = 57.
$$

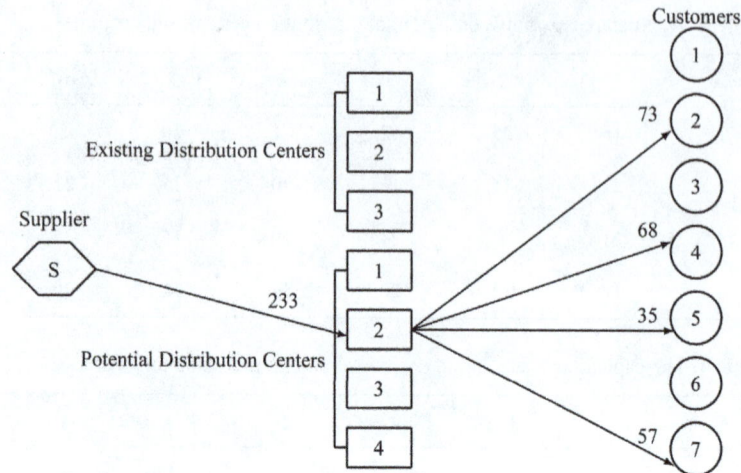

Figure 2. Location and distribution plan.

The solution $v_2 = 1$ shows that the second one is selected to build new distribution center, $x_2 = 233$ means that the supplier must supply 233 products to the new distribution center, and $y_{22} = 73$, $y_{24} = 68$, $y_{25} = 35$, $y_{27} = 57$, state that the quantity of product are supplied form new distribution center to customers. The above solutions show that we can choose the second potential distribution center to built the new distribution center and the total profit is 644.58. Meanwhile we can know how to distribute goods for customers according to the plan which is shown in **Figure 2**.

6. The Crisp Equivalent Model

In the process of practical logistics network optimization, uncertain factors often appear in competitive logistics distribution center location problem because of lacking of or even without historical data. This paper investigated a useful model to handle competitive logistics distribution center location problem with uncertain customers demands and uncertain setup costs. The mathematical model of this problem was established by uncertain programming based on the expected value criterion. In order to solve this model, we took advantage of the properties of uncertain variable. Then the expected value model was transformed into its crisp equivalent model, and we used mathematical software Lingo to find its optimal solution. At last, a numerical example was presented to illustrate the effectiveness of the proposed model.

This paper only considers the demands of customers and setup costs of new distribution center are uncertain variables. Indeed, other uncertain factors in competitive logistics distribution center location problem are worthy of studying. We can further focus on the uncertain utility which can be used to describe the uncertainty of customers' patronizing behavior. Furthermore, we can seek for the expression of uncertain utility. In this paper, we only center on the static competition. It is necessary for further research to consider dynamic competition problem. Thus we can establish dynamic uncertain programming model for uncertain dynamic competitive facility location problem.

Acknowledgements

This work was supported by the Projects of the Humanity and Social Science Foundation of Ministry of Education of China (No. 13YJA630065), and the Key Project of Hubei Provincial Natural Science Foundation (No. 2012FFA065).

References

[1]   Lu, A. and Bostel, N. (2007) A Facility Location Model for Logistics Systems including Reverse Flows: The Case of Remanufacturing Activities. *Computers & Operations Research*, **34**, 299-323. http://dx.doi.org/10.1016/j.cor.2005.03.002

[2]   Klose, Z. and Drexl, A. (2005) Facility Location Models for Distribution System Design. *European Journal of Operational Research*, **162**, 4-29. http://dx.doi.org/10.1016/j.ejor.2003.10.031

[3]   Hotelling, H. (1929) Stability in Competition. *Economic Journal*, **39**, 41-57. http://dx.doi.org/10.2307/2224214

[4]   Drezner, Z. (1982) Competitive Location Strategies for Two Facilities. *Regional Science and Urban Economics*, **12**, 485-493. http://dx.doi.org/10.1016/0166-0462(82)90003-5

[5]   Hakimi, S. (1983) On Locating New Facilities in a Competitive Environment. *European Journal of Operational Research*, **12**, 29-35. http://dx.doi.org/10.1016/0377-2217(83)90180-7

[6]   Plastria, F. (2001) Static Competitive Facility Location: An Overview of Optimisation Approaches. *European Journal of Operational Research*, **129**, 461-470. http://dx.doi.org/10.1016/S0377-2217(00)00169-7

[7]   Wong, S. and Yang, H. (1999) Determining Market Areas Captured by Competitive Facilities: A Continuous Equilibrium Modeling Approach. *Journal of Regional Science*, **39**, 51-72. http://dx.doi.org/10.1111/1467-9787.00123

[8]   Yang, H. and Wong, S. (2000) A Continuous Equilibrium Model for Estimating Market Areas of Competitive Facilities with Elastic Demand and Market Externality. *Transportation Science*, **34**, 216-227. http://dx.doi.org/10.1287/trsc.34.2.216.12307

[9]   Plastria, F. and Vanhaverbeke, L. (2008) Discrete Models for Competitive Location with Foresight. *Computers & Operations Research*, **35**, 683-700. http://dx.doi.org/10.1016/j.cor.2006.05.006

[10]   Küçükaydin, H., Aras, N. and Altnel, I. (2012) A Leader-Follower Game in Competitive Facility Location. *Computers & Operations Research*, **39**, 437-448. http://dx.doi.org/10.1016/j.cor.2011.05.007

[11]   Sasaki, M., Campbell, J., Krishnamoorthy, M. and Ernst, A. (2015) A Stackelberg Hub Arc Location Model for a Competitive Environment. *Computers & Operations Research*, **47**, 27-41. http://dx.doi.org/10.1016/j.cor.2014.01.009

[12]   Drezner, T. (1994) Locating a Single New Facility among Existing Unequally Attractive Facilities. *Journal of Regional Science*, **34**, 237-252. http://dx.doi.org/10.1111/j.1467-9787.1994.tb00865.x

[13]   Drezner, T., Drezner, Z. and Kalczynski, P. (2011) A Cover-Based Competitive Location Model. *Journal of the Operational Research Society*, **62**, 100-113. http://dx.doi.org/10.1057/jors.2009.153

[14]   Huff, D. (1966) A Programmed Solution for Approximating an Optimum Retail Location. *Land Economics*, **42**, 293-303. http://dx.doi.org/10.2307/3145346

[15]   Drezner, T. (2014) A Review of Competitive Facility Location in the Plane. *Logistics Research*, **7**, 114-126. http://dx.doi.org/10.1007/s12159-014-0114-z

[16]   Leonardi, G. and Tadei, R. (1984) Random Utility Demand Models and Service Location. *Regional Science and Urban Economics*, **14**, 399-431. http://dx.doi.org/10.1016/0166-0462(84)90009-7

[17]   Drezner, T. and Drezner, Z. (1996) Competitive Facilities: Market Share and Location with Random Utility. *Journal of Regional Science*, **36**, 1-15. http://dx.doi.org/10.1111/j.1467-9787.1996.tb01098.x

[18]   Drezner, T., Drezner, Z. and Wesolowsky, G. (1998) On the Logit Approach to Competitive Facility Location. *Journal of Regional Science*, **38**, 313-327. http://dx.doi.org/10.1111/1467-9787.00094

[19]   Shiode, S. and Drezner, Z. (2003) A Competitive Facility Location Problem on a Tree Network with Stochastic Weights. *European Journal of Operational Research*, **149**, 47-52. http://dx.doi.org/10.1016/S0377-2217(02)00459-9

[20]   Drezner, Z. and Wesolowsky, G. (1981) Optimal Location Probabilities in the ℓ_p Distance Weber Problem. *Transportation Science*, **15**, 85-97. http://dx.doi.org/10.1287/trsc.15.2.85

[21]   Shiode, S. and Ishii, H. (1991) A Single Facility Stochastic Location Problem under a Distance. *Annals of Operations Research*, **31**, 469-478. http://dx.doi.org/10.1007/BF02204864

[22]   Wesolowsky, G. (1977) Probabilistic Weights in the One Dimensional Facility Location Problem. *Management Science*, **24**, 224-229. http://dx.doi.org/10.1287/mnsc.24.2.224

[23]   Liu, B. (2007) Uncertainty Theory. Spinger-Verlag, Berlin. http://dx.doi.org/10.1007/978-3-540-73165-8_5

[24]   Gao, Y. (2012) Uncertain Models for Single Facility Location Problems on Networks. *Applied Mathematical Modelling*, **36**, 2592-2599. http://dx.doi.org/10.1016/j.apm.2011.09.042

[25]   Wen, M., Qin, Z. and Kang, R. (2014) The α-Cost Minimization Model for Capacitated Facility Location-Allocation Problem with Uncertain Demands. *Fuzzy Optimization and Decision Making*, **13**, 345-356. http://dx.doi.org/10.1007/s10700-014-9179-z

[26]   Wang, K. and Yang, Q. (2014) Hierarchical Facility Location for the Reverse Logistics Network Design under Uncertainty. *Journal of Uncertain Systems*, **8**, 255-270.

[27]   Wu, Z. and Peng, J. (2014) A Chance-Constrained Model of Logistics Distribution Center Location under Uncertain Environment. *Advances in Information Sciences and Service Sciences*, **6**, 33-42.

[28]   Revelle, C. (1986) The Maximum Capture or Sphere of Influence Location Problem: Hotelling Revisited on a Network.

Journal of Regional Science, **26**, 343-358. http://dx.doi.org/10.1111/j.1467-9787.1986.tb00824.x

[29] Liu, B. (2010) Uncertainty Theory: A Branch of Mathematics for Modeling Human Uncertainty. Spinger-Verlag, Berlin. http://dx.doi.org/10.1007/978-3-642-13959-8

[30] Liu, B. (2009) Some Research Problems in Uncertainty Theory. *Journal of Uncertain Systems*, **3**, 3-10.

[31] Liu, B. (2009) Theory and Practice of Uncertain Programming. Spinger-Verlag, Berlin.

13

Hybrid Formulation of the Multi-Item Capacitated Dynamic Lot Sizing Problem

Mayank Verma*, Renduchintala Raghavendra Kumar Sharma

Department of Industrial and Management Engineering, Indian Institute of Technology Kanpur, Kanpur, India
Email: *mayankverma.p@gmail.com, rrks@iitk.ac.in

Abstract

It is shown that when backorders, setup times and dynamic demand are included in capacitated lot sizing problem, the resulting classical formulation and one of the transportation formulations of the problem (referred to as CLSP_BS) are equivalent. And it is shown that both the formulations are "weak" formulations (as opposed to "strong" formulation). The other transportation version is a strong formulation of CLSP_BS. Extensive computational studies are presented for medium and large sized problems. In case of medium-sized problems, strong formulation produces better LP bounds, and takes lesser number of branch-and-bound (B&B) nodes and less CPU time to solve the problem optimally. However for large-sized problems strong formulation takes more time to solve the problem optimally, defeating the benefit of strength of bounds. This essentially is because of excessive increase in the number of constraints for the large sized problems. Hybrid formulations are proposed where only few most promising strong constraints are added to the weak formulation. Hybrid formulation emerges as the best performer against the strong and weak formulations. This concept of hybrid formulation can efficiently solve a variety of complex real life large-sized problems.

Keywords

Capacitated Lot Sizing, Linear Relaxation, Strong Formulation, Hybrid Formulation, Branch-and-Bound

1. Introduction and Literature Survey

Researchers have extensively studied the lot sizing problem in last five decades, but finding an effective and practicable solution to this problem in real time remains to be a challenge faced by the production planners in a

---

*Corresponding author. Current Affiliation: Atomic Energy Regulatory Board, Mumbai, India.

manufacturing setup. This was evident from our recent visit to a leading Indian automobile manufacturer, located in the southern part of the country. Nearly all versions of capacitated lot sizing problem are NP-Hard; and hence better heuristics are required for solution.

Even the basic model of single item capacitated lot sizing problem (CLSP) is well known to be NP Hard [1]. With the addition of setup times, the problem becomes so hard that even finding a feasible solution is NP complete [2]. Chen & Thizy [3] showed that even without considering setup times, just the inclusion of setup cost makes the lot sizing problem strongly NP Hard. Hence it becomes actually intricate to solve a CLSP with considerations of realistic situations like capacity, backorders, setup times, setup costs, setup carryovers, etc.

The focus of this work is on the capacitated version of the lot sizing problems; the readers are hence referred to [4] that extensively reviews various research done on CLSP. Karimi *et al.* [4] concluded that although lot sizing had been one of the favorite research areas in last few decades but still the realistic and practical variants of the CLSP, specially which considers backorders, setup times and setup carry-overs, had not received much attention. Also Quadt & Kuhn [5] gave an extensive review of the literature on the different extensions of the capacitated lot sizing problem. It is again evident from the review work of Quadt & Kuhn [5] that models with inclusion of backorders have received least attention in the past. Hence despite a common outlook that lot sizing is an over-researched problem, it still remains to be one of the preferred research areas; this again becomes apparent from the most recent works of [6]-[8], to name only a few.

Apart from the classical (denoted by PC) formulation of CLSP given by [9], CLSP is also modelled as the transportation problem (denoted by PT) type of formulation or plant location type of formulation [10] and the shortest path (denoted by PS) type of formulation [11]. Alfieri *et al.* [10] took up the relative analysis of the PC, PT and PS models; however their analysis did not consider setup times and backorders. Later Denizel *et al.* [12] added setup times to their model and verified the proposed relationship between strength of different types of formulations.

Multi-item multi-period capacitated lot sizing problem with dynamic demands, backorders and setup times (CLSP_BS) is considered in this work. Apart from the formulation PC, two PT formulations (PT^a and PT^b) of CLSP_BS are considered in this work; and the relative strength of these three formulations is investigated. This is important because it has an impact on the choice of model formulation and the corresponding solution procedure. Also as it is observed by Alfieri *et al.* [10] that although PT provides better lower bounds, it takes more computational time as compared to PC. Hence choosing an appropriate formulation to solve the lot sizing problem is critical. Chen & Thizy [3] and Barany *et al.* [13] did mathematical comparison of the Lagrangian and linear relaxations for the classical version of the multi-item CLSP. They however did not consider the variables of backorders and setup times in their model.

Limiting the size of the problem by selective inclusion of variables and constraints in different forms has been attempted and discussed by some researchers in the recent past. Tightening of the MIP using extended formulations drastically amplifies the size of the problem. Hence Van Vyve and Wolsey [14] developed approximate extended formulations, where in order to get a good quality lower bound, partial reformulations were applied. Through a control parameter the trade-off between strength and size of formulation is manipulated. They basically used a coefficient modification based heuristic algorithm within the branch-and-cut enumeration framework. Pochet and Wolsey [15] gave that adding a limited number of (l, S) inequalities to the regular formulation and to the backlogging extensions gave improved upper bounds or same bounds as the transportation formulation in the presence of Wagner-Whitin costs. Akartunali and Miller [16] dealt with the multi level single resource big bucket problem with extension to backlogging and used the cutting planes to tighten their formulation. Their flexible and easy heuristics generate multiple solutions and competitive lower bounds. A complete mathematical survey of the compact extended formulations applied into the combinatorial optimization problems is done very recently by Conforti *et al.* [17]. They provide some mathematical tools for studying the extended formulations. The minimum size up to which an extended formulation could be compacted was also devised.

An interesting concept of hybrid formulations is introduced in this work, where a small percentage of the most promising strong constraints are added to the weak PT formulation to note that the hybrid formulations swiftly produce better bounds than the weak formulation and gives optimal solution in least time as compared to weak or the strong formulations. Hence for the large sized problem instances, hybrid formulations actually are the best performers.

Here we provide the breakup of this paper. In Section 2, we provide the classical and two types of transportation formulations. The relationship between cost structures of the two formulations is also shown which makes the classical and transportation formulations equivalent. In Section 3, linear programming (LP) relaxations of the three types of formulations are proposed. Section 4 provides empirical investigations on medium sized problems to note the behavior of strong and weak formulations. In Section 5 we attempt the large sized problems and note a reverse behavior of strong and weak formulations. Hybrid formulations are then proposed, which is shown to be the best performer as against strong and weak formulations. We conclude in Section 6 by highlighting the important contributions of this work.

2. Formulations

Here we discuss the classical and two transportation formulations of CLSP_BS. Conversion of classical formulation to the transportation formulation is done using a relationship between the cost parameters and the variables. Note the difference in the names of variables and parameters used in this paper compared to that existing in the literature. An attempt has been made to simplify the notations; viz. starting all variable names with X and Y, parameters cost by C, time by T, etc.

2.1. Classical Formulation of CLSP_BS: "PC"

The standard formulation of CLSP available in the literature is extended to include the proposed variables and situations. Note suffix "PC" added to the parameters—inventory and backorder costs, indicating their association only with the classical formulation of the problem.

Index:

i	: Item type, $i = 1, \cdots, I$
t	: Planning period, $t = 1, \cdots, T$

Variables:

XP_{it}	: Number of items "i" to be produced during the period "t"
$XINV_{it}$	: Number of items "i" carried as inventory at the end of period "t"
XBO_{it}	: Number of items "i" that will be backordered from period "t"
YS_{it}	: Binary variable for setup of the resource for item "i" during the period "t"
	= 0 (if there is no setup required), 1 otherwise

Parameters:

CP_{it}	: Unit cost of producing item "i" in period "t"
CS_{it}	: Unit cost of setup, for item "i" in period "t"
$CINV\_PC_{it}$	: Unit cost of holding inventory of item "i" in period "t".
CBO_PC_{it}	: Unit cost of backordering item "i", that was demanded during the period "t"
CAP_{it}	: Capacity available to produce item "i" during the period "t"
$CAPT_t$	: Capacity available in time units, in a period "t"
D_{it}	: Demand of item "i" during the period "t"
TP_i	: Time required to process the item "i"
TS_i	: Time required to setup the production for item "i"

PC : Minimize $Z(PC)$

$$Z(PC) = \sum_{i=1}^{I} \sum_{t=1}^{T} \left[CP_{it} XP_{it} + CS_{it} YS_{it} + CINV\_PC_{it} XINV_{it} + CBO\_PC_{it} XBO_{it} \right] \tag{1}$$

Subject to:

$$XP_{it} + XINV_{i,t-1} + XBO_{it} = D_{it} + XINV_{it} + XBO_{i,t-1} \qquad \forall i,t \tag{2}$$

$$\sum_{i=1}^{I} \left(TP_i XP_{it} + TS_i YS_{it} \right) \leq CAPT_t \qquad \forall t \tag{3}$$

$$XP_{it} \leq CAP_{it} YS_{it} \qquad \forall i,t \tag{4}$$

$$XINV_{it}, XP_{it}, XBO_{it} \geq 0 \qquad \forall i,t \tag{5}$$

$$XINV_{i0}, XINV_{iT} = 0 \qquad \forall i \tag{6}$$

$$XBO_{i0}, XBO_{i,T} = 0 \qquad \forall i \tag{7}$$

$$YS_{it} \in [0,1] \qquad \forall i,t \tag{8}$$

(1) the objective of the problem, is to minimize the cost of production, setup, inventory and backlogging, summed over all items and time periods. (2) is the inventory balance constraint for each item and period. (3) is the time capacity limiting constraint, which ensures that the total time utilized in doing production and setups can atmost be equal to the maximum time available in any period. (4) is the production capacity constraint, ensuring the production quantity to be always less than or equal to the maximum production capacity available for all items and periods. (5) restricts non negativity over the production quantity and the quantity of items carried as inventory and backorders. (6) and (7) forces the initial and final inventory and backorders to be zero. (8) forces the setup variable to be binary for all items and for all the periods.

2.2. Transportation Formulation of CLSP_BS: "PT"

CLSP_BS can be reshaped using variable redefinition technique, in the form of a transportation problem. Note the suffix "PT" with the cost parameters in this formulation. Two versions of CLSP_BS (PT^a and PT^b) are presented in the following sub-sections.

Index:

i	: Item type, $i = 1, \cdots, I$
t, r	: Planning period; $t = 1, \cdots, T$; $i = 1, \cdots, I$

Variables:

X_{itr}	: Number of items "i" produced in period "t" to satisfy demand of period "r"
YS_{it}	: Binary variable for setup of the resource for item "i" during the period "t"
	= 0 (if there is no setup required), 1 otherwise

Parameters:

$CINV\_PT_{itr}$	: Unit cost of holding inventory of item "i" from period "t" to period "r"; $(r \geq t)$.
CBO_PT_{itr}	: Unit cost of backordering item "i", which is produced in period "t", but was required at period "r"; $(r \leq t)$.
D_{ir}	: Demand of item "i" during the period "r"

2.2.1. Transportation Formulation 1 [PTᵃ]

PT^a: Minimize $Z\left(PT^a\right)$

$$Z\left(PT^a\right) = \sum_{i=1}^{I}\sum_{r=1}^{T}\sum_{t=1}^{T}\left(CP_{it} + CINV\_PT_{itr} + CBO\_PT_{itr}\right)X_{itr} + \sum_{i=1}^{I}\sum_{t=1}^{T}CS_{it}YS_{it} \tag{1a'}$$

Subject to:

$$\sum_{t=1}^{T}X_{itr} = D_{ir} \qquad\qquad \forall i, r \tag{2'}$$

$$\sum_{i=1}^{I}\left[TP_i\left(\sum_{r=1}^{T}X_{itr}\right) + TS_i YS_{it}\right] \leq CAPT_t \quad \forall t \tag{3'}$$

$$\sum_{r=1}^{T}X_{itr} \leq CAP_{it}YS_{it} \qquad\qquad \forall i, t \tag{4'}$$

$$X_{itr} \geq 0 \qquad\qquad \forall i, t, r \tag{5'}$$

and (8).

(1') defines the objective seeking to minimize the cost of production, setup, inventory and backorders summed over all items and time periods. (2') is the demand constraint ensuring production during the planning horizon to satisfythe demand of period "r". (3') limits the sum of production and the setup times to at most be equal to the maximum time available during the planning horizon. (4') ensures that the maximum production is equal to the capacity available for all the items and periods. (5') imposes a non-negativity on the variable.

2.2.2. Transportation Formulation 2 [PTᵇ]

A new constraint is defined here, considering the fact that the production of any item "i" in any period "t" to satisfy its demand of period "r" can at most be equal to the demand. This can be written mathematically as:

$$X_{itr} \leq D_{ir}YS_{it} \qquad\qquad \forall i, t, r \tag{6'}$$

This constraint also ensures a setup (and hence a probable production) in "t" to satisfy the demand of "r". In formulation PT^b, we simply add (6′) to the formulation PT^a as follows:

PT^b: Minimize $Z\left(PT^b\right)$

$$Z\left(PT^b\right) = \sum_{i=1}^{I}\sum_{r=1}^{T}\sum_{t=1}^{T}\left(CP_{it} + CINV\_PT_{itr} + CBO\_PT_{itr}\right)X_{itr} + \sum_{i=1}^{I}\sum_{t=1}^{T}CS_{it}YS_{it} \qquad (1b')$$

Subject to: (2′)-(6′) and (8).

It is observed that constraint (6′) is mostly used in the literature where capacitated lot sizing problem is modeled as a transportation problem. But in presence of (2′) and (4′), which takes care of the demand and setup respectively, actually (6′) is not required. Soas brought in the later part of this paper (Section 4, **Table 3**), same objective value is obtained by solving PT^a and PT^b. (6′) being a strong constraint, its inclusion in the problem should be beneficial in terms of bounds. That is, relaxation of PT^b is expected to produce better bounds than that of PT^a. This behavior will be analyzed in the next sections.

2.3. Equivalence of Costs

Note that the costs defined in this work are general to also accommodate different setup cost for the same item in different periods; or different production cost for the same item in different periods, etc. Though such costs may not actually occur, but any such possibility is accounted for in the model. To generate inventory and the backorder costs for the transportation formulation, following relations are used:

$$CINV\_PT_{itr} = \sum_{p=t}^{r-1}CINV\_PC_{ip} \quad \text{for } r > t; \text{ and } CINV\_PT_{itr} = 0 \text{ for } r \leq t. \qquad (9)$$

$$CBO\_PT_{itr} = \sum_{p=r}^{t-1}CBO\_PC_{ip} \quad \text{for } r < t; \text{ and } CBO\_PT_{itr} = 0 \text{ for } r \geq t. \qquad (10)$$

3. Relaxations

3.1. Relaxations of Classical Formulation PC

In PC, when we relax the binary constraint YS_{it} to vary continuously between zero and one, we have its linear programming (LP) relaxation PC_r. Mathematically:

$$0 \leq YS_{it} \leq 1 \qquad\qquad \forall i, t \qquad (11)$$

PC_r: Minimize (1); subject to: (2)-(6), (7) and (11).

3.2. Relaxations of Transportation Formulation

Again in the transportation formulation of the CLSP_BS when we relax the binary constraint YS_{it} to vary continuously between zero and one, it is referred to as a linear programming (LP) relaxation. Mathematically this is denoted by (11).

The LP relaxation PT_r^a may be referred to as: Minimize (1); subject to: (2′)-(5′) and (11). Also, the LP relaxation PT_r^b is referred to as: Minimize (1); subject to: (2′)-(6′) and (11).

4. Computational Experiences and Analysis

4.1. Experimental Setup

Here we perform empirical investigations on the randomized data sets, for the three formulations—PC, PT^a and PT^b. The experimental set up and the procedure, along with results and analysis is detailed in this section. A variety of small and large problem sizes—10 × 10 [(number of items) × (number of time periods)], 20 × 10, 10 × 20, 20 × 20 is considered. Number of binary variables for these problems is of the order O (I × T), I and T being the number of items and number of time periods in the planning horizon respectively. For each of these sizes, we solved 50 problem sets, each set containing a binary formulation and an LP relaxation of the classical and two types of transportation formulations. Experiments are performed using CPLEX 10.0 solver of GAMS 22.3 on a standalone 2.79 GHz core 2 Duo CPU with 1.96 GB RAM.

4.2. Creating Test Instances

The cost parameters used in the two formulations (PC and PT) are generated randomly, taking general guidance on generating random problems from the standard literatures viz. [18]-[20] and [5] the details of which is provided in this sub-section. We choose uniform distribution to generate parameters as this is traditionally used in the literature indicated above. **Table 1** states the range of values of different parameters that are taken to form the random problems.

Inventory and backorder costs for the transportation formulation are calculated using Equations (9) and (10) respectively. As evident from the table that backorder costs is assumed to be higher than inventory carrying costs, as backorder cost also account for the loss of goodwill for the customer who could not be instantly served. It also implies that our model is comparatively more open to carry inventory, compared to incur backorders, as prompt satisfaction of the demand is primarily important in today's competitive environment. The relation between demand and production capacity can also be stated in terms of tightness factor, which can be defined as the ratio of average periodic demand and production capacity. For an uncapacitated problem the value of tightness factor will be 0; while for the case when required capacity (demand) is exactly equal to the available capacity, the tightness factor is 1. For the data considered here, the tightness factor is about 0.8, which is rigid enough to produce some infeasible problems; such infeasible problems are eliminated from the problem sets.

4.3. Order of Problem

In four sizes of problems considered the number of binary variables is of the order (IT), the order of continuous variables in classical formulation PC is (3IT) and that of continuous variables in the transportation formulations is (IT^2). Similarly, while the number of constraints in classical and PT^a transportation formulation is (2IT + T), the number of constraints in the PT^b swells to $(IT^2 + 2IT + T)$. Hence as the problem size grows, number of variables and constraints increase accordingly.

4.4. Analysis of Results

Table 2 shows the average time taken by the different formulations and their LP relaxations. One may note that as the problem size increases computational time increases for all formulations and their relaxations. This table is provided here just to give a general idea to the reader about the computational time taken to solve a particular formulation or its relaxation. In order to actually compare the formulations and their relaxations on different

Table 1. Range of parameters.

Parameter	Range	Parameter	Range
Inventory cost $CINV\_CL_{it}$	U (10, 20)	Processing time TP_i	U (50, 80)
Backorder cost CBO_CL_{it}	U (20, 30)	Setup time TS_i	U (250, 300)
Setup cost CS_{it}	U (500, 600)	Demand D_{it}	U (20, 50)
Production cost CP_{it}	U (80, 100)	Production capacity CAP_{it}	U (30, 60)

Table 2. Average time (in seconds) to solve formulations PC and PT.

Problem size	PC		PT^a		PT^b	
	PC	PC_8	PT^a	PT_r^a	PT^b	PT_r^a
10 × 10	0.09	0.07	0.13	0.08	0.15	0.09
20 × 10	0.15	0.07	0.36	0.11	0.32	0.13
10 × 20	0.17	0.09	1.60	0.15	0.87	0.19
20 × 20	0.60	0.11	145.4	0.27	21.3	0.35

aspects, we perform a t-test the result of which is shown in the next tables.

In **Table 3**, for all the problem sizes we compare (using t test) different characteristics of PC with PT^a, and also those of PT^a with PT^b formulation. Parameters on which formulations are compared are the optimal values obtained by each of these formulations, the CPU time taken to obtain these optimal values, number of nodes traversed in the branch-and-bound (B&B) tree to reach this optimal value, the objective values of LP relaxation of the formulations, and CPU time taken to calculate the LP relaxation. Note that in **Table 3**, against the row comparing PT^a-PC, "Optimal" or "LP bounds" is the "t" calculated for difference between objective values of (PT^a/PC) and 1; and "CPU time" is the "t" calculated for difference between (CPU time for PT^a/CPU time for PC) and 1. Similarly for the parameter "B&B nodes".

Note from **Table 3** that for all the problem sizes considered, the optimal values obtained from the three formulations PC, PT^a and PT^b are all the same. Hence these formulations are actually the three ways to formulate the same problem. For all problem sizes, PT^a take more computational time, compared to PC.

Before moving on to compare the CPU times taken to calculate optimal values for PT^a and PT^b, note the significance of their LP bounds. Note that the objective value obtained by the LP relaxation of PC and PT^a are exactly the same, but there is a significant difference between LP bounds of PT^a and PT^b. For all problem sizes, PT^b invariably produces better (larger) LP bounds as compared to the PT^a. This is the reason of earmarking PT^b as the strong formulation and PT^a as the weak formulation of CLSP_BS.

As PT^b is a strong formulation rendering better LP bounds, it is able to calculate the optimal values in significantly less number of B&B nodes, invariably for all problem sizes. The significance level for the better LP bounds and lesser B&B nodes, both increases as the problem size increases, implying that the strength of a formulation is realized more for the larger problem sizes, hence useful for the practical lot sizing problems. t values indicate that LP relaxation of PC is solved faster than that of PT^a. This is the reason why computation of optimal solution takes significantly large time for PT^a and lesser time for PC, because quality of LP bounds is the same in both cases. However LP relaxation of PT^b takes more time to solve as compared to PT^a, but still PT^b is solvable optimally in significantly better CPU times than PT^a, because of the better quality bounds of PT^b than PT^a.

5. Hybrid Formulations

5.1. Large Sized Problems

We observed in the previous section that strong formulation PT^b produce better bounds than the weak formulation PT^a and hence CLSP_BS is optimally solvable in lesser number of B&B nodes and lesser CPU time by the formulation PT^b, compared to PT^a. In previous section, medium sized problems where the largest one of size 20 × 20 (with 400 binary variables) were solved. In order to perform a test on yet large sized problems, 50 sets for each of the four sizes—50 × 10, 50 × 20, 100 × 10 and 100 × 20 are solved. These problem instances are again solved using CPLEX 10.0 solver of GAMS 22.3 on a standalone 2.79 GHz Core 2 Duo CPU with 1.96 GB RAM.

Table 3. t values comparing optimal, LP bounds and CPU times of 3 different formulations.

Problem size	Compare	Optimal	CPU time (optimal)	B&B nodes	LP bounds	CPU time (LP bound)
10 × 10	PT^a-PC	1	16.788	6.584	1	6.781
	PT^b-PT^a	1	10.875	−3.984	54.398	8.750
10 × 20	PT^a-PC	1	7.398	4.875	1	17.985
	PT^b-PT^a	1	−11.147	−59.371	79.708	14.709
20 × 10	PT^a-PC	1	18.121	7.203	1	25.177
	PT^b-PT^a	1	−7.090	−28.848	79.448	16.771
20 × 20	PT^a-PC	1	5.007	5.277	1	7.674
	PT^b-PT^a	1	−23.687	−81.024	88.889	7.071

5.2. Initial Analysis and Motivation for Hybrid Formulation

In **Table 4** we note the average time taken by the different formulations and their LP relaxations. Note that the problem size 50 × 10 was solvable in a few seconds. However for the other 3 sizes, the number of variables increases 2 and 4 times respectively and hence CLSP_BS being an NP-hard problem the solver got out of memory after running for a few hours, while attempting to reach optimality. A small duality gap of 0.1% as the stopping criteria is introduced for the three large sizes—100 × 10, 50 × 20, and 100 × 20. These sizes were then solvable in a few seconds after the introduction of duality gap. One may note that as the problem size increases computational time increases for all formulations and their relaxations. This table is provided here just to give a general idea to the reader about the computational time taken to solve a particular formulation or its relaxation. In order to actually compare the formulations and their relaxations on different aspects, we perform a *t*-test the result of which is shown in the next tables.

In **Table 5**, for all problem sizes we compare different characteristics of PC with PT^a and also PT^a with PT^b. PT^H is the hybrid formulation, which will be discussed in Section 5.4. For problem size 50 × 10, the quality of LP bounds of PT^b are so good that despite its LP relaxation taking longer time to get solved, PT^b is optimally solved in lesser number of nodes and lesser CPU time as compared to the formulation PT^a.

Table 4. Average time (in seconds) to solve formulations PC and PT.

Problem size	PC		PT^a		PT^b		Duality gap (%)
	PC	PC_8	PT^a	PT_r^a	PT^b	PT_r^b	
50 × 10	0.5	0.13	9.06	0.18	4.1	0.24	0
100 × 10	0.5	0.15	1.79	0.31	2.3	0.44	0.1
50 × 20	0.44	0.15	4.07	0.5	5.05	0.76	0.1
100 × 20	1.05	0.26	11.4	1	18.8	1.6	0.1

Table 5. *t* values comparing optimal, LP bounds and CPU times of 3 different formulations.

Problem size	Compare	CPU time (optimal)	B&B nodes	LP bounds	CPU time (LP bound)
50 × 10	PT^a-PC	7.531	7.675	1	27.648
	PT^b-PT^a	−14.830	−35.520	109.499	32.325
50 × 20	PT^a-PC	30.504	6.657	1	156.346
	PT^b-PT^a	11.371	−50.763	155.975	121.182
	PT^a-PT^H	9.009	9.816	−68.922	−9.692
	PT^b-PT^H	13.335	−7.027	158.385	126.453
	PT^H-PC	26.835	6.406	68.772	136.843
100 × 10	PT^a-PC	16.066	5.028	1	57.267
	PT^b-PT^a	12.282	−15.482	197.802	59.158
	PT^a-PT^H	9.773	6.944	−93.715	−3.765
	PT^b-PT^H	14.607	−2.072	178.070	40.876
	PT^H-PC	15.963	5.037	93.493	45.822
100 × 20	PT^a-PC	21.269	6.111	1	60.954
	PT^b-PT^a	12.841	−30.449	256.225	49.772
	PT^a-PT^H	6.859	5.403	−122.891	−3.834
	PT^b-PT^H	19.776	−27.953	241.0896	56.672
	PT^H-PC	30.568	8.602	122.604	77.767

However the same does not remain true for the larger sizes 50×20, 100×10 and 100×20. Now the CPU time taken to solve LP relaxation of PT^b takes much more time as compared to that of PT^a (apperent from the increased t-values). The quality of bounds remain to be good for PT^b as against PT^a. Due to this, while PT^b is optimally solvable in lesser number of nodes, it takes larger CPU time as compared to the problem PT^a. So the advantage of using a strong formulation, which was apparent for the medium sized problems, seems to have lost.

5.3. Most Promising Strong Constraints (MPS_Constr)

We develop a new set of strong demand constraints to tackle the problem of huge computational times taken by the LP relaxations of the strong constraints. New sets of modified demand constraint were built using the following procedure. To make a hybrid formulation, we add some percentage of most promising strong constraints to the weak formulation. But to decide the appropriate percentage of the total strong constraints, we varied it from 20% to 10% to 5% etc. In each such step, the bounds obtained by LP relaxation were observed to deteriorate; but there was a significant improvement in computational time for solving the LP relaxations of the hybrid formulation. After some iterations of trial and error, 5% of the strongest constraints were selected to be added to the weak formulation of CLSP_BS for making hybrid formulations. At this level, the bounds were only a little inferior, but LP relaxations took substantially less computational time, which significantly improves the overall computational times for solving the problem.

Building MPS_Constr

Sort the strong constraints in increasing order of demand $\left(D_{ir} \right)$. Set cut off number $N = INT\,(0.05*K)$, where K is number of demand points ($=ixr$). Let the demand associated with this number N be $cut\_off\_D_k\_(5\%)$. Prepare a set of most promising demands (MPD) $= \left\{ k : D_k \le cut\_off\_D_k\_(5\%) \right\}$. Now, MPS_constr $= \left\{ YS_{it} D_k \ge X_{itr} \forall i, t, r; \text{ where } k \in MPD \right\}$.

5.4. Hybrid Formulation and Relaxation

Hybrid formulation of CLSP_BS, referred here as PT^H is described as follows:
   PT^H: Minimize (1a'); Subject to (2')-(5'), (8) and MPS_constr.
   The constraint (8) can be replaced with (11) in the above formulation to get the LP relaxation as follows:
   PT^H_r : Minimize (1a'); Subject to (2')-(5'), (11) and MPS_constr.

5.5. Empirical Investigation for Hybrid Formulations

Here we provide results to verify the efficacy of hybrid formulation against strong and weak formulations. On the same experimental setup and problem-sets that were attempted in Section 5.2, for the sizes—50×20, 100×10 and 100×20, we solve hybrid formulation and its LP relaxation. Comparison of bounds, nodes and computational time given by hybrid formulation is compared with that of the strong and weak formulations using t-test.

Table 5 provides the comparison of hybrid formulation with PT^a and PT^b for the three larger sizes 50×20, 100×10 and 100×20. While strong formulation PT^b is optimally solvable in lesser number of B&B nodes than PT^a, it takes longer time to get solved because LP relaxation of PT^b takes longer CPU time than that of PT^a. is shown in **Table 5**. Note from the significant t values that hybrid formulation is the best performer in terms of computational time, when compared with strong and weak formulations.

As compared to PT^a, hybrid PT^H gives better LP bounds and optimally solves the problem in lesser B&B nodes and lesser CPU times. When compared to PT^b, PT^H generates inferior LP bounds but solves the problem optimally in lesser CPU time. The reason of this behavior is that although LP bounds of hybrid formulation are inferior (due to which PT^H has to traverses more number of B&B nodes), but its bounds at each node are obtained significantly quicker than the bounds of the strong formulation. This all is evident from the t values given below.

The primary reason because of which LP relaxation of strong formulation takes large computational time is heavy increase in the number of constraints, as its size grows beyond a level. While number of constraints in PT^a remains 2010, 2020 and 4020, it hugely swells to 12,010, 22,020 and 44,020 respectively in case of PT^b. More are the number of constraints, larger the time taken to solve the problem. Through the concept of hybrid formulations we try to limit the increase in number of constraints, while at the same time take the benefit of strongest constraints. Hence there is a modest increase in number of constraints for PT^H compared to PT^a, but sweeping decrease in number of constraints as against PT^b.

6. Conclusions

This work has attempted to cater to a variant of multi-item multi-period CLSP, which considers dynamic demand, backorders and setup times (CLSP_BS). Apart from considering a practically important variation of CLSP, the generality of the cost structure increases the applicability of this model into a variety of real life situations. A classical formulation and two transportation (strong and weak) formulations of this problem are compared.

The main emphasis of this work is the introduction of hybrid formulation of CLSP_BS, which is proved to be the best performer. It was observed that for large sized instances of CLSP_BS, strong formulation took lesser number of branch-and-bound nodes but more CPU time to optimally solve the problem, compared to the weak formulations. The reason of this behavior is an extreme increase in number of constraints for the strong formulation, due to which solving LP relaxation at each B&B node becomes time intensive. Hence despite obtaining better bounds and the problem getting solved in lesser number of B&B nodes, the problem actually takes longer time to get solved optimally. In hybrid formulation, we add few strongest constraints to the weak formulation, so that the quality of bounds of the strong formulation is retained while maintaining lesser number of constraints to obtain optimal solution faster. In this way, better characteristics of both strong and weak formulations are retained in the hybrid formulation. This concept of hybrid formulations can prove to be very efficient in solving the real life large sized problem instances in a variety of applications.

References

[1]   Florian, M., Lenstra, J.K. and Rinnooy Kan, A.G. (1980) Deterministic Production Planning Algorithms and Complexity. *Management Science*, **26**, 669-679. http://dx.doi.org/10.1287/mnsc.26.7.669

[2]   Maes, J., McClain, J.O. and Van Wassenhove, L.N. (1991) Multilevel Capacitated Lotsizing Complexity and LP-Based Heuristics. *European Journal of Operation Research*, **53**, 131-148. http://dx.doi.org/10.1016/0377-2217(91)90130-N

[3]   Chen, W.H. and Thizy, J.M. (1990) Analysis of Relaxations for the Multi-Item Capacitated Lot-Sizing Problem. *Annals of Operations Research*, **26**, 29-72. http://dx.doi.org/10.1007/BF02248584

[4]   Karimi, B., Fatemi Ghomi, S.M.T. and Wilson, J.M. (2003) The Capacitated Lot Sizing Problem: A Review of Models and Algorithms. *Omega*, **31**, 365-378. http://dx.doi.org/10.1016/S0305-0483(03)00059-8

[5]   Quadt, D. and Kuhn, H. (2008) Capacitated Lot-Sizing with Extensions: A Review. *4OR*, **6**, 61-83. http://dx.doi.org/10.1007/s10288-007-0057-1

[6]   Kim, S.I., Han, J., Lee, Y. and Park, E. (2010) Decomposition Based Heuristic Algorithm for Lot-Sizing and Scheduling Problem Treating Time Horizon as a Continuum. *Computers & Operations Research*, **37**, 302-314. http://dx.doi.org/10.1016/j.cor.2009.05.007

[7]   Melo, R.A. and Wolsey, L.A. (2010) Uncapacitated Two-Level Lot-Sizing. *Operations Research Letters*, **38**, 241-245. http://dx.doi.org/10.1016/j.orl.2010.04.001

[8]   Akbalik, A. and Penz, B. (2009) Exact Methods for Single-Item Capacitated Lotsizing Problem with Alternative Machines and Piece-Wise Linear Production Costs. *International Journal of Production Economics*, **119**, 367-379. http://dx.doi.org/10.1016/j.ijpe.2009.03.010

[9]   Manne, A.S. (1958) Programming of Economic Lot Sizes. *Management Science*, **4**, 115-135. http://dx.doi.org/10.1287/mnsc.4.2.115

[10]   Alfieri, A., Brandimarte, P. and D'Orazio, S. (2002) LP-Based Heuristics for the Capacitated Lot-Sizing Problem: The Interaction of Model Formulation and Solution Algorithm. *International Journal of Production Research*, **40**, 441-458. http://dx.doi.org/10.1080/00207540110081461

[11]   Pochet, Y. and Wolsey, L.A. (1988) Lot-Size Models with Backlogging: Strong Reformulations and Cutting Planes. *Mathematical Programming*, **40**, 317-335. http://dx.doi.org/10.1007/BF01580738

[12]   Denizel, M., Altekin, F.T., Sural, H. and Stadtler, H. (2008) Equivalence of the LP Relaxations of Two Strong Formulations for the Capacitated Lot-Sizing Problem with Setup Times. *OR Spectrum*, **30**, 773-785. http://dx.doi.org/10.1007/s00291-007-0094-3

[13]   Barany, I., Van Roy, T.J. and Wolsey, L.A. (1984) Strong Formulations for Multi-Item Capacitated Lot Sizing. *Management Science*, **30**, 1255-1261. http://dx.doi.org/10.1287/mnsc.30.10.1255

[14] Van Vyve, M. and Wolsey, L.A. (2006) Approximate Extended Formulations. *Mathematical Programming*, **105**, 501-522. http://dx.doi.org/10.1007/s10107-005-0663-7

[15] Pochet, Y. and Wolsey, L.A. (2006) Production Planning by Mixed Integer Programming. Springer.

[16] Akartunalı, K. and Miller, A.J. (2009) A Heuristic Approach for Big Bucket Multi-Level Production Planning Problems. *European Journal of Operational Research*, **193**, 396-411. http://dx.doi.org/10.1016/j.ejor.2007.11.033

[17] Conforti, M., Cornuéjols, G. and Zambelli, G. (2010) Extended Formulations in Combinatorial Optimization. *4OR*, **8**, 1-48. http://dx.doi.org/10.1007/s10288-010-0122-z

[18] Trigeiro, W.W., Thomas, L.J. and McClain, J.O. (1989) Capacitated Lot Sizing with Setup Times. *Management Science*, **35**, 353-366. http://dx.doi.org/10.1287/mnsc.35.3.353

[19] Ozdamar, L. and Barbarosoglu, G. (1999) Hybrid Heuristics for the Multi-Stage Capacitated Lot Sizing and Loading Problem. *Journal of the Operational Research Society*, **50**, 810-825. http://dx.doi.org/10.1057/palgrave.jors.2600773

[20] Küçükyavuz, S. and Pochet, Y. (2009) Uncapacitated Lot Sizing with Backlogging: The Convex Hull. *Mathematical Programming*, **118**, 151-175. http://dx.doi.org/10.1007/s10107-007-0186-5

14

Mathematical Model and Algorithm for the Task Allocation Problem of Robots in the Smart Warehouse

Zhenping Li, Wenyu Li

School of Information, Beijing Wuzi University, Beijing, China
Email: lizhenping66@163.com, leeyu0513@163.com

Abstract

In the smart warehousing system adopting cargo-to-person mode, all the items are stored in the movable shelves. There are some warehouse robots transporting the shelves to the working platforms for completing order picking or items replenishment tasks. When the number of robots is insufficient, the task allocation problem of robots is an important issue in designing the warehousing system. In this paper, the task allocation problem of insufficient warehouse robots (TAPIR) is investigated. Firstly, the TAPIR problem is decomposed into three sub-problems: task grouping problem, task scheduling problem and task balanced allocation problem. Then three sub-problems are respectively formulated into integer programming models, and the corresponding heuristic algorithms for solving three sub-problems are designed. Finally, the simulation and analysis are done on the real data of online bookstore. Simulation results show that the mathematical models and algorithms of this paper can provide a theoretical basis for solving the TAPIR problem.

Keywords

Smart Warehouse, Warehouse Robot, Task Allocation Problem, Mathematical Model, Heuristic Algorithm

1. Introduction

In the smart warehousing system adopting cargo-to-person mode, all the items are stored in the movable shelves neatly arranged in the warehouse, and picking workers stand in front of stationary picking platforms. There are some isomorphic warehouse robots running along the marked line to complete order picking or items replenishment tasks [1]. When starting order picking or items replenishment tasks, the computer system controls one or more warehouse robots to the sites of assigned movable shelves; the warehouse robots lift up their assigned shelves and transport them to the corresponding picking platforms. After the picking workers complete the order picking or items replenishment tasks, the warehouse robots transport the movable shelves back to their own original locations. In the warehousing management system, complex transportation works are done by the warehouse robots instead of human; the working efficiency is improved in some ways [2].

Since different items of one order might be laid in different shelves. When picking an order, the robots usually need to transport more than one shelf to the picking platforms. In another words, the warehouse robots need to complete multiple tasks for picking one order. In order to improve the picking efficiency, it needs the multi-robot cooperation to complete multiple tasks. The task allocation problem of warehouse robots greatly affects the working efficiency of smart warehousing system.

Some researchers investigated the tasks allocation problem of abundant robots (TAPAR) where the number of robots was larger than the number of tasks to be completed in picking one order. Guo adopts the market auction method to solve the task allocation problem of abundant mobile robots (TAPAR) in a smart logistic centre [3]. Li uses the genetic algorithm and ant colony algorithm to study the TAPAR problem [4]. Christopher *et al.* propose the resource allocation problem of multi-vehicle system in the smart warehouse by using Alphabet Soup method [5]. And Li designs the heuristic algorithm for solving the TAPAR problem [6].

In practical, the number of robots is often smaller than the number of tasks to be completed in picking one order, which means the robots are insufficient. By now, little researches focused on the task allocation problem of insufficient mobile robots (TAPIR).

In this paper, the TAPIR problem in smart warehousing system based on cargo-to-person mode is investigated. We partition the task allocation problem of insufficient robots into three sub-problems: task grouping problem, task scheduling problem and task balanced allocation problem. The mathematical models of three sub-problems are built, and the corresponding heuristic algorithms are designed to solve three sub-problems. Finally, the simulation and analysis are done on the real data of online bookstore S.

2. Problem Description and Mathematical Models

2.1. Problem Description

In a smart warehousing system based on cargo-to-person mode, there are m available warehouse robots, k movable shelves, and n picking platforms. Each shelf and picking platform is arranged in a fixed site. Each shelf stores several types of items. The unit-distance cost of a robot walking loaded with a shelf is c_1, the unit-distance cost of a robot walking unloaded is c_2, and the fixed cost of every warehouse robot running one time is r. Let d_{ij} represents the distance of a robot running from shelf s_i to shelf s_j, and u_j represents the distance of a robot transporting shelf s_j to its corresponding picking platform.

The operation cost of a warehouse robot completing a given task is composed of three parts: the related cost, the self-cost and the fixed cost. Related cost refers to the cost that the warehouse robot needs to move unloaded from current position to the position of assigned task. Self-cost refers to the cost that the warehouse robot completing a task. The fix cost refers to the cost running a warehouse robot one time. Suppose the current position of robot F_l is in the position of shelf s_i, if we assign robot F_l to finish task s_j, then the operation cost of robot F_l to finish task s_j can be expressed as follows:

$$c_{lj} = d_{ij}c_2 + u_j c_1 + r$$

where d_{ij} is the distance of robot F_l running from current position s_i to the position of assigned task s_j, u_j is the double distance from position of task s_j to picking platform, r is the fixed cost for controlling robot F_l completing one task.

Given an order to be picked in the smart warehouse, there are p tasks to be completed for picking this order ($m < p$). Assume that the set of tasks for picking this order is $S = \{s_1, s_2, \cdots, s_p\}$.

Suppose there are e orders to be picked in a period of time which are listed in a set: $O = \{O_1, O_2, \cdots, O_e\}$, and the orders are picked one by one, in order to minimize the total operation cost, how to assign all tasks for each warehouse robot and give its working schedule?

The task allocation problem of insufficient warehouse robots (TAPIR) can be decomposed into three sub-problems: task grouping problem, task scheduling problem, and task balanced allocation problem. Firstly, all tasks of picking one order are partitioned into several groups, where the number of groups equals to the number of robots. Secondly, the scheduling scheme of each group tasks is given according to every robot. Finally, assign each warehouse robot to complete one group of tasks.

In the following subsections, we will construct the mathematical models for these three sub-problems.

2.2. Mathematical Model for the Task Grouping Problem

For the task allocation problem of insufficient warehouse robots (TAPIR), since the number of robots is smaller than the number of tasks for picking an order, so some warehouse robots need to complete more than one tasks when picking one order. Before assign the tasks to robots, we firstly partition all tasks into several groups, where the number of group equals to the number of robots.

In order to reduce the total operation cost and ensure the task allocation of warehouse robots balanced, we define the similarity r_{ij} between task s_i and task s_j. The similarity r_{ij} can be calculated by formula (1), where d_{ij} is the distance for a robot to move from the site of task s_i to the site of task s_j, d_{max} is the maximum distance between any pair of tasks ($d_{max} = \max\left\{d_{ij}\right\}$).

$$r_{ij} = \frac{d_{max} - d_{ij}}{d_{max}} \tag{1}$$

It is easy to verify that $0 \le r_{ij} \le 1$. If we assign two tasks with larger similarity value to one robot, the related cost will be lower. On the contrary, if we assign two tasks with smaller similarity value to one robot, the related cost will be higher. Based on this idea, we can partition the tasks into several groups according to the similarity value. To ensure the task allocation balanced among all robots, we restrict the number of tasks in each group is no more than a given value D (for example $D = \left\lfloor \frac{p}{m} \right\rfloor + 1$).

Define the decision variables:

$$x_{iv} = \begin{cases} 1, & \text{task } s_i \text{ is assigned in the } v\text{th group;} \\ 0, & \text{otherwise } i = 1, 2, \cdots, p; v = 1, 2, \cdots, m. \end{cases}$$

The mathematical model of task grouping problem for picking one order can be formulated as an integer programming model.

$$\max z = \sum_{v=1}^{m} \left(\frac{\sum_{i=1}^{p}\sum_{j=1}^{p} r_{ij} x_{iv} x_{jv}}{\left(\sum_{i=1}^{p} x_{iv}\right)^2} \right) \tag{2}$$

$$s.t. \begin{cases} \sum_{v=1}^{m} x_{iv} = 1, & i = 12, \cdots, p & (3) \\ \sum_{i=1}^{p} x_{iv} \ge 1, & v = 1, 2, \cdots, m & (4) \\ \sum_{i=1}^{p} x_{iv} \le D, & v = 1, 2, \cdots, m & (5) \\ x_{iv} = 0, 1, & i = 1, 2, \cdots, p; v = 1, 2, \cdots, m & (6) \end{cases}$$

Objective function (2) maximizes sum of the average similarity of tasks assigned to the same group; Constraints (3) ensure that each task is just assigned into one group; Constraints (4) and Constraints (5) ensure that each group contains at least one task and no more than D tasks; Constraints (6) indicate that the variables are binary.

2.3. Mathematical Model for the Task Scheduling Problem

Because the self-cost of transporting a task to its corresponding picking platform is a constant no matter which warehouse robot completes this task, the fix cost of each robot is also a constant no matter which task it completes, so the total operation cost of warehouse robots depends mainly on the total related cost that the warehouse robots completing all of the tasks.

If a warehouse robot located in s_0 is assigned to complete a group of tasks $\{s_1, s_2, \cdots, s_q\}$, the total related cost depends on the scheduling order of the tasks. For example, if the scheduling order of the tasks is $s_1 \rightarrow s_2 \rightarrow \cdots \rightarrow s_q$, then the total related cost is $(d_{0,1} + d_{1,2} + \cdots + d_{q-1,q}) \cdot c_2$. When the scheduling order of tasks changed, the total related cost will change.

Given a warehouse robot and a group of tasks, the task scheduling problem is to find the optimal scheduling order of the tasks, so that the total related cost for the robot completing all tasks is minimal.

Since the related cost depends on the distance between two successive tasks, so the task scheduling problem is similar to Hamiltonian path problem, where the robot should start its routing from initial position s_0, and visit every site of the task in $\{s_1, s_2, \cdots, s_q\}$ one time. After completing all tasks, the robot will stay in the site of the last task, which means that it needs not return to its initial position s_0.

Suppose that warehouse robot F_l (locating in site s_0) is assigned to complete the tasks $\{s_1, s_2, \cdots, s_q\}$ of group k. Suppose that s_{q+1} is a dummy end location, which corresponding to the last location of tasks $\{s_1, s_2, \cdots, s_q\}$. Define $d_{i,q+1} = 0 \ (i = 0, 1, \cdots, q)$.

To find the optimal scheduling order of tasks, we defined the variables as follows:

f_{ij} : the flow that run from task s_i to task s_j, $i = 0, 1, \cdots, q; j = 1, 2, \cdots, q+1$

$$x_{ij} = \begin{cases} 1, & \text{after warehouse robot } F_l \text{ finishing task } s_i, \text{ it directly begin task } s_j; \ i = 0, 1, 2, \cdots, q; j = 1, 2, \cdots, q, q+1 \\ 0, & \text{Otherwise.} \end{cases}$$

We define that $x_{ij} = \begin{cases} 1, & f_{ij} > 0, \\ 0, & f_{ij} = 0. \end{cases}$

The task scheduling problem can be formulated into the following integer programming model.

$$\min w_{lk} = c_2 \sum_{i=0}^{q} \sum_{j=1}^{q+1} d_{ij} x_{ij} \tag{7}$$

$$s.t \begin{cases} \sum_{j=1}^{q+1} x_{ij} = 1, & i = 0, 1, 2, \cdots, q & (8) \\[2mm] \sum_{i=0}^{q} x_{ij} = 1, & j = 1, 2, \cdots, q+1 & (9) \\[2mm] \sum_{k=0}^{q} f_{ki} - \sum_{j=0}^{q+1} f_{ij} = 1, & i = 1, 2, \cdots, q & (10) \\[2mm] \sum_{j=1}^{q} f_{0j} = q, & & (11) \\[2mm] x_{ij} \leq f_{ij} \leq q \cdot x_{ij} & i = 0, 1, \cdots, q; j = 1, 2, \cdots, q+1 & (12) \\[2mm] x_{ij} \in \{0, 1\} & i = 0, 1, \cdots, q; j = 1, 2, \cdots, q+1 & (13) \end{cases}$$

Objective function (7) minimizes the total related cost that the warehouse robot F_l completes all tasks in group k; Constraints (8) (9) ensure that each task must be completed by robot F_l; Constraint (10) denote that when a task is completed, the total flow run into it will be absorbed 1 by the task. Constraint (11) ensures the sum of flows come out from location s_0 equals q. Constraint (12) represents the relationship between x_{ij} and f_{ij}; Constraint (13) indicates that the variables x_{ij} are binary.

2.4. Mathematical Model for the Task Balanced Allocation Problem

After calculating the related cost w_{lk} of warehouse robot $F_l (l = 1, 2, \cdots, m)$ to complete the tasks in group $k (k = 1, 2, \cdots, m)$, we obtained a related cost matrix $W = (w_{ij})_{m \times m}$. Since we have partitioned the tasks into m groups, each robot needs to complete one and only one group of tasks. The tasks balanced allocation problem is to assign each robot completing only one group of tasks so that the total related costs is minimal.

Define the decision variables:

$$y_{lk} = \begin{cases} 1, & \text{assign warehouse robot } F_l \text{ to complete the } k\text{th group of tasks;} \\ 0, & \text{otherwise } l = 1, 2, \cdots, m;\ k = 1, 2, \cdots, m \end{cases}$$

The mathematical model of tasks balanced allocation problem can be formulated into an integer programming model as follows:

$$\min z = \sum_{l=1}^{m} \sum_{k=1}^{m} w_{lk} y_{lk} \tag{14}$$

$$s.t. \begin{cases} \sum_{l=1}^{m} y_{lk} = 1, & k = 1, 2, \cdots, m \tag{15} \\ \sum_{k=1}^{m} y_{lk} = 1, & l = 1, 2, \cdots, m \tag{16} \\ y_{lk} \in (0,1), & l = 1, 2, \cdots, m,\ k = 1, 2, \cdots, m \tag{17} \end{cases}$$

Objective function (14) minimizes the total related cost that warehouse robots complete all groups of tasks; Constraints (15) ensure that each group of tasks can only be assigned to one warehouse robot; Constraints (16) represent that each warehouse robot can only complete one group of tasks; Constraints (17) indicate that the variables are binary.

3. Algorithms for Solving Three Sub-Problems

In Section 2, we have decomposed the TAPIR problem into three sub-problems and formulate them as integer programming models. For small size of sub-problems, we can solve the integer programming model by Lingo software. Since some sub-problem (such as the second sub-problem) is NP-hard, for larger size of the problem, it is not practical to solve them by Lingo software. In this section, we will respectively design heuristic algorithms for solving each sub-problem of Section 2.

3.1. Algorithm for Solving the Task Grouping Problem

Given the number of robots m and the position of each robot, the set of tasks $\{s_1, s_2, \cdots, s_p\}$ need to be completed in picking one order. The heuristic algorithm for grouping all tasks into m groups can be described as follows:

Initialization: $T = \{s_1, s_2, \cdots, s_p\}$, $group = 0$, where T is the set of ungrouped tasks, $group$ denotes the number of non-empty groups.

Step 1: Calculate the similarity matrix $R = (r_{ij})_{p \times p}$, where r_{ij} is the similarity between task s_i and task s_j, which can be calculated by Equation (1).

Step 2: Find the smallest value of matrix R, $r_{i_0, j_0} = \min_{i,j} \{r_{ij}\}$, let $G_1 = \{s_{i_0}\}$, $G_2 = \{s_{j_0}\}$; $group = 2$;

$T = T \setminus \{s_{i_0}, s_{j_0}\}$.

Step 3: If $group = m$, go to step 4, else.

For each $s_i \in T$, and $k = 1, \cdots, group$, calculate the average similarity between the ungrouped task s_i and all tasks in group k as follows

$$AR(i,k) = \frac{\sum_{s_j \in G_k} r_{ij}}{|G_k|}.$$

The sum of average similarity between task s_i and all non-empty groups

$$SAR(i) = \sum_{k=1}^{group} AR(i,k).$$

Find the task with minimum sum of average similarity

$$i_{min} = \arg\min_{i} SAR(i).$$

Update the number of non-empty group, add the tasks in the new non-empty group, and update the set of un-grouped tasks.

$$group := group + 1;$$
$$G_{group} := \{s_{i_{min}}\};$$
$$T := T \setminus \{s_{i_{min}}\};$$

Go to step 3.

Step 4: If $T = \Phi$, go to step 5. Else add the ungrouped tasks to the group, with which it has the maximum average similarity.

For each $s_i \in T$, and $k = 1, \cdots, m$, where $|G_k| < D$.

Calculate the average similarity between s_i and each task in group k.

$$AR(i,k) = \frac{\sum\limits_{s_j \in G_k} r_{ij}}{|G_k|}.$$

Find the group with maximum average similarity value.

$$AR(i^*, k^*) = \arg\max_{i,k} AR(i,k).$$

Update the set T and G_{k*}

$$T := T \setminus \{s_{i^*}\}, \quad G_{k^*} := G_{k^*} \cup \{s_{i^*}\}.$$

Go to Step 4.

Step 5: Output the tasks in each group: $G_1, G_2, \cdots, G_m$.

3.2. Algorithm for Solving the Task Scheduling Problem

To solve the task scheduling problem, we design a heuristic algorithm:

Define the following variables:

$route(l,k)$ represents the scheduling route, whose initial value is empty;

$location$ represents the location of robot F_l, whose intimal position is s_{l0};

w_{lk} represents the total related cost for robot F_l completing tasks in $G_k = \{s_{k_1}, s_{k_2}, \cdots, s_{k_{g_k}}\}$;

Input: The locations of m robots and the sites of each task in m groups.

Step 1: For each robot $F_l (1 \leq l \leq m)$ and each task group $G_k (1 \leq l \leq m)$, suppose the location of robot F_l is s_{l0}, the number of tasks in group k is g_k, $G_k = \{s_{k_1}, s_{k_2}, \cdots, s_{k_{g_k}}\}$.

Let $TN = \{k_1, k_2, \cdots, k_{g_k}\}$ be the index set of tasks.

$$route(l,k) = \Phi; \quad location = 0; \quad w_{lk} = 0.$$

Calculate the distance matrix $D(l,k) = (d_{ij})_{(g_k+1) \times (g_k+1)}$, where d_{0j} is the distance from s_{l0} to s_{k_j}, d_{ij} is the distance from s_{k_i} to s_{k_j}, $(i, j = 1, 2, \cdots, g_k)$.

Step 2: If $TN = \Phi$, go to step 3.

Else, find the minimal element of matrix $D(l,k)$ in the $location$ row and columns of TN,

$j^* = \arg\min_{j \in T} \{d_{location, j}\}$, $route(l,k) := route(l,k) \cup \{j^*\}$, $TN := TN \setminus \{j^*\}$, $location := j^*$, $w_{lk} := w_{lk} + c_2 d_{location, j^*}$,

go to step 2.

Step 3: Output the total related cost w_{lk}, the optimal scheduling order of tasks $route(l,k)$, and the final location of robot F_l.

3.3. Algorithm for Solving the Task Balanced Allocation Problem

After scheduling the tasks of each group, we obtain the related cost matrix $W = (w_{lk})$, where w_{lk} represents the related cost of robot F_l completing the tasks of groups k. Then the balance allocation problem is converted into an assignment problem, which can be solved by Hungarian Algorithm [7]. We can also obtain the optimal solution by solving the integer linear programming model directly.

4. Simulations Analysis

In this section, we do simulation on a smart warehouse of online bookstore S. At present, online bookstore S uses the smart ware housing system based on cargo-to-person mode. In the warehouse, 96 classes of books are stored in 24 movable shelves. Each shelf stores 4 classes of books. There are 3 picking platforms in the smart warehouse. Currently, the orders are picked one by one. **Figure 1** depicts the layout of the warehouse, the square boxes represent 24 shelves and their positions, three rectangular boxes represent three picking platforms; three cycles with arrow represent the warehouse robots. The coordinates of shelves and picking platforms are listed in **Table 1**. The initial positions of three robots are the same as their nearest shelves.

There are 10 orders to be picked. The tasks sets of 10 orders are listed in **Table 2**. The unit cost of a robot walking loaded with a shelf is 3 dollars, the unit cost of a robot walking unloaded is 1.9 dollars, and the fixed cost of controlling each warehouse robot completing a task is 10 dollars. In order to minimize the total cost in the process of picking 10 orders, how to allocate and schedule the tasks to each warehouse robot?

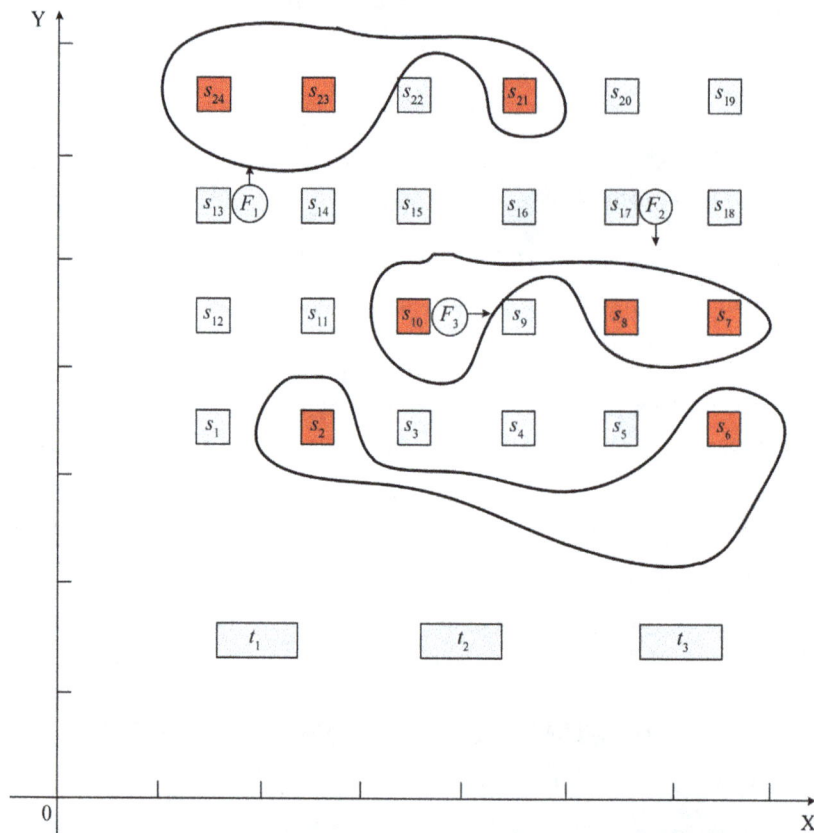

Figure 1. The layout of the warehouse, the square boxes represent shelves, three rectangular boxes represent three picking platforms three cycles with arrow represent the warehouse robots. 8 tasks of the first order depicted in red squares are partitioned into three groups. The tasks in the same group are surrounded by a closed curve (Coordinate Unit: 2 m).

Table 1. Coordinates of each shelf and each picking platform.

Shelf	s_1	s_2	s_3	s_4
Coordinate	(3, 7)	(5, 7)	(7, 7)	(9, 7)
Shelf	s_5	s_6	s_7	s_8
Coordinate	(11, 7)	(13, 7)	(13, 9)	(11, 9)
Shelf	s_9	s_{10}	s_{11}	s_{12}
Coordinate	(9, 9)	(7, 9)	(5, 9)	(3, 9)
Shelf	s_{13}	s_{14}	s_{15}	s_{16}
Coordinate	(3, 11)	(5, 11)	(7, 11)	(9, 11)
Shelf	s_{17}	s_{18}	s_{19}	s_{20}
Coordinate	(11, 11)	(13, 11)	(13, 13)	(11, 13)
Shelf	s_{21}	s_{22}	s_{23}	s_{24}
Coordinate	(9, 13)	(7, 13)	(5, 13)	(3, 13)
Picking platform	t_1	t_2	t_3	
Coordinate	(4, 3)	(8, 3)	(12, 3)	

Table 2. The tasks of picking each order.

Order	Tasks of picking the order
1	$\{s_2, s_6, s_7, s_8, s_{10}, s_{21}, s_{23}, s_{24}\}$
2	$\{s_2, s_5, s_{10}, s_{16}\}$
3	$\{s_4, s_8, s_{12}, s_{18}, s_{20}, s_{22}, s_{24}\}$
4	$\{s_3, s_9, s_{14}, s_{18}, s_{19}, s_{22}, s_{24}\}$
5	$\{s_1, s_3, s_5, s_{11}, s_{14}, s_{15}, s_{19}, s_{23}\}$
6	$\{s_2, s_5, s_9, s_{11}, s_{16}, s_{19}, s_{23}\}$
7	$\{s_4, s_7, s_{10}, s_{13}, s_{17}, s_{18}\}$
8	$\{s_2, s_8, s_{14}, s_{22}, s_{23}\}$
9	$\{s_2, s_6, s_{11}, s_{12}, s_{14}, s_{19}, s_{22}, s_{23}\}$
10	$\{s_3, s_6, s_7, s_{12}, s_{17}, s_{21}\}$

Since the orders are picked one by one, for each order to be picked, we can calculate the distance between any pair of tasks, and obtain the similarity matrix by Equation (1).

For the uth $(u = 1, 2, \cdots, 10)$ order, we first partition the tasks into three groups by the heuristic algorithm in section 3.1.

Then we obtain the scheduling order for each group tasks and every robot by the heuristic algorithm in section 3.2. Using the Hungarian Algorithm, we get the tasks assignment and scheduling results, and the total operation cost.

After assigning the tasks of the uth order to the robots, we update the initial position of each robot, and begin to assign the tasks of the $(u+1)$ th order. Repeat the steps until all tasks of 10 orders are assigned to robots.

The simulation results of the first order is depicted in **Figure 1**, where 8 tasks are partitioned into three groups, the tasks in the same group are surrounded by a closed curve. The optimal tasks assignment and the scheduling routes are $F_1 \rightarrow s_{24} \rightarrow s_{23} \rightarrow s_{21}$, $F_2 \rightarrow s_6 \rightarrow s_2$, $F_3 \rightarrow s_{10} \rightarrow s_8 \rightarrow s_7$. Which mean that robot F_1 need to complete tasks s_{24}, s_{23}, s_{21} in sequence, then stop at position s_{21}. Robot F_2 need to complete tasks s_6, s_2 in sequence, then stop at position s_2. Robot F_3 need to complete tasks s_{10}, s_8, s_7 in sequence, then stop at position s_7.

The task grouping results of all orders are listed in **Table 3**. The task allocation and optimal scheduling routes of all orders are shown in **Table 4**, the total operation costs of each order are listed in **Table 4**.

According to the task assignment and scheduling routes, and the operation cost in **Table 4**, the total operation cost that the warehouse robots complete all tasks of 10 orders is 4330.2 dollars.

If the number of warehouse robots is larger than the maximum number of tasks for picking one order, the problem becomes TAPAR. The tasks need not be grouped, and the total operation cost for picking these 10 orders will be 4010.4 dollars [2].

From the task allocation results of 10 orders, we can see that the total operation cost in TAPIR is 319.8 dollars more than the total operation cost in TAPAR [2]. However, the number of warehouse robots in the TAPIR is 7 less than that in TAPAR. Since the cost of increasing a warehouse robot is very large, in practice, the number of warehouse robots is often insufficient. If we consider holding cost of robots in the TAPIR, we can find that the warehouse had better hold fewer robots.

Table 3. Task grouping results of 10 orders.

Order	Group 1	Group 2	Group 3	Sum of average similarity
1	$\{s_{21}, s_{23}, s_{24}\}$	$\{s_7, s_8, s_{10}\}$	$\{s_2, s_6\}$	2.416667
2	$\{s_2\}$	$\{s_5, s_{16}\}$	$\{s_{10}\}$	2.625000
3	$\{s_{20}, s_{22}, s_{24}\}$	$\{s_{18}\}$	$\{s_4, s_8, s_{12}\}$	2.333333
4	$\{s_{19}, s_{22}, s_{24}\}$	$\{s_{18}\}$	$\{s_3, s_9, s_{14}\}$	2.333334
5	$\{s_{15}, s_{19}, s_{23}\}$	$\{s_5, s_{11}, s_{14}\}$	$\{s_1, s_3\}$	2.319444
6	$\{s_{16}, s_{19}, s_{23}\}$	$\{s_{11}\}$	$\{s_2, s_5, s_9\}$	2.428572
7	$\{s_{13}, s_{17}\}$	$\{s_4, s_{10}\}$	$\{s_7, s_{18}\}$	2.416666
8	$\{s_{22}, s_{23}\}$	$\{s_{14}\}$	$\{s_2, s_8\}$	2.500000
9	$\{s_{19}, s_{22}, s_{23}\}$	$\{s_{11}, s_{12}, s_{14}\}$	$\{s_2, s_6\}$	2.333333
10	$\{s_3, s_{12}, s_{17}\}$	$\{s_6, s_7\}$	$\{s_{21}\}$	2.472222

Table 4. The task assignment and optimal scheduling routes, and the operation cost of each order.

Order	Task assignment and optimal scheduling routes	Operation cost
1	$F_1 \to s_{24} \to s_{23} \to s_{21}$; $F_2 \to s_6 \to s_2$; $F_3 \to s_{10} \to s_8 \to s_7$	517.2
2	$F_1 \to s_{16} \to s_5$; $F_2 \to s_2$; $F_3 \to s_{10}$	222.6
3	$F_1 \to s_{20} \to s_{22} \to s_{24}$; $F_2 \to s_4 \to s_8 \to s_{12}$; $F_3 \to s_{18}$	508.2
4	$F_1 \to s_{24} \to s_{22} \to s_{19}$; $F_2 \to s_{14} \to s_9 \to s_3$; $F_3 \to s_{18}$	493.6
5	$F_1 \to s_{19} \to s_{23} \to s_{15}$; $F_2 \to s_3 \to s_1$; $F_3 \to s_5 \to s_{11} \to s_{14}$	512.8
6	$F_1 \to s_{16} \to s_{19} \to s_{23}$; $F_2 \to s_2 \to s_5 \to s_9$; $F_3 \to s_{11}$	457
7	$F_1 \to s_{13} \to s_{17}$; $F_2 \to s_7 \to s_{18}$; $F_3 \to s_{10} \to s_4$	381.6
8	$F_1 \to s_{22} \to s_{23}$; $F_2 \to s_8 \to s_2$; $F_3 \to s_{14}$	361.2
9	$F_1 \to s_{23} \to s_{22} \to s_{19}$; $F_2 \to s_2 \to s_6$; $F_3 \to s_{14} \to s_{11} \to s_{12}$	514
10	$F_1 \to s_{21}$; $F_2 \to s_6 \to s_7$; $F_3 \to s_{12} \to s_3 \to s_{17}$	362

5. Conclusions

The task allocation problem of insufficient warehouse robots (TAPIR) in the smart warehouse is an important issue in the smart warehousing system designing. By now, few researches focus on the TAPIR problem. In this paper, we investigate the TAPIR problem, and divide it into three sub-problems: task grouping problem, task scheduling problem and task balanced allocation problem. We formulate each sub-problem into an integer programming model and design the heuristic algorithm for solving each sub-problem. The mathematical models and algorithms can be used in designing the smart warehousing system.

In this paper, we assume that the orders are picked one by one. When the number of orders is large, the orders can be batched firstly [8], and then consider the orders in each batch as one combined order. Using the mathematical models and heuristic algorithms of this paper, we can find the optimal task allocation and scheduling routes for picking each combined order.

In this paper, we have not considered the cost of holding a robot. If we consider this cost, we can analyze the relation between the total cost and the number of warehouse robots and find the optimal number of robots a warehouse should hold. This is the problem we are researching on.

Acknowledgements

This work was supported by the National Natural Science Foundation of China (11131009, F012408), the Funding Project for Academic Human Resources Development in Institutions of Higher Learning Under the Jurisdiction of Beijing Municipality (CIT&TCD20130327), and Major Research Project of Beijing Wuzi University. Funding Project for Technology Key Project of Municipal Education Commission of Beijing (ID: TSJHG201310037036); Funding Project for Beijing key laboratory of intelligent logistics system; Funding Project of Construction of Innovative Teams and Teacher Career Development for Universities and Colleges Under Beijing Municipality (ID: IDHT20130517); Funding Project for Beijing philosophy and social science research base specially commissioned project planning (ID: 13JDJGD013).

References

[1]   Li, Z. and Li, W. (2014) Study on Optimization of Storage Bays of Smart Warehouses of Online Bookstores. *Logistics Technology*, **12**, 340-342. (In Chinese)

[2]   Zou, S. (2013) The Present and Future of Warehouse Robot. *Logistics Engineering and Management*, **6**, 171-172. (In Chinese)

[3]   Guo, Y. (2010) Auction-Based Multi-Agent Task Allocation in Smart Logistic Center. Ph.D. Thesis, Harbin Institute of Technology, Shenzhen. (In Chinese)

[4]   Li, G. (2012) Task Allocation of Warehouse Robots Based on Intelligence Optimization Algorithm. Ph.D. Thesis, Harbin Institute of Technology, Shenzhen. (In Chinese)

[5]   Hazard, C.J., Wurman, P.R. and D'Andrea, R. (2006) Alphabet Soup: A Tested for Studying Resource Allocation in Multi-Vehicle Systems. *American Association for Artificial Intelligence*.

[6]   Li, Z. and Li, W. (2015) Research on the Task Allocation Problem of Warehouse Robots in the Smart Warehouse. *Proceedings of the* 12*th International Symposium on Operations Research and Its Applications*, Luoyang, 21-24 August 2015, 29-33.

[7]   Kuhn, H.W. (2005) The Hungarian Method for the Assignment Problem. *Naval Research Logistics*, **1**, 7-21. http://dx.doi.org/10.1002/nav.20053

[8]   Li, Z. and Li, W. (2014) Study on Supplier Warehouse Location Combination Problem of Chain Supermarkets. *Logistics Technology*, **9**, 137-139. (In Chinese)

15

When to Sell an Asset Where Its Drift Drops from a High Value to a Smaller One

Pham Van Khanh

Institute of Economics and Corporate Group, Hanoi, Vietnam
Email: van_khanh1178@yahoo.com

Abstract

To solve the selling problem which is resembled to the buying problem in [1], in this paper we solve the problem of determining the optimal time to sell a property in a location the drift of the asset drops from a high value to a smaller one at some random change-point. This change-point is not directly observable for the investor, but it is partially observable in the sense that it coincides with one of the jump times of some exogenous Poisson process representing external shocks, and these jump times are assumed to be observable. The asset price is modeled as a geometric Brownian motion with a drift that initially exceeds the discount rate, but with the opposite relation after an unobservable and exponentially distributed time and thus, we model the drift as a two-state Markov chain. Using filtering and martingale techniques, stochastic analysis transform measurement, we reduce the problem to a one-dimensional optimal stopping problem. We also establish the optimal boundary at which the investor should liquidate the asset when the price process hit the boundary at first time.

Keywords

Optimal Stopping Time, Posterior Probability, Threshold, Markov Chain, Jump Times, Martingale, Brownian Motion

1. Introduction

In this paper we consider the following problem: How to find the optimal stopping time to sell a stock (or an asset) when the expected return of a stock is assumed to be a constant larger than the discount rate up until some random, and unobservable, time τ, at which it drops to a constant smaller than the discount rate.

An investor wants to hold the position as long as the inertia is present by taking advantage of the drift which is exceeding the discounted rate (or interest rate). On the other hand, when the inertia disappears the investor would like to exit the position by selling the asset.

The under study problem in this paper was also addressed in [1] where the buying problem with the same assumption was solved. The results of [1] showed that the optimal buying time was the first passage time over some unknown level for the a posteriori probability process π_t defined below and by simulating it was found that the optimal time to buy an asset was the time which the asset price process had just passed the trough.

The author of [2] studied a problem of finding an optimal stopping strategy to liquidate an asset with unknown drift; more exactly he wanted to find the best time to sell a stock when its drift was a discrete random variable which took the given values. The first time the posterior mean of the drift passes below a non-decreasing boundary that is the unique solution of a particular integral equation is shown to be optimal.

Some classical optimal stopping time problem has been considered in [3]. These are applied in mathematical finance but these are basic problem, and it is difficult to apply in real world.

For related studies of stock selling problems, see [4] [5] and for studies of basic optimal stopping problems see [3]. The method we use to study in this paper is the martingale theory, the transformation theory of measuring and the optimal stopping time is referenced in the literature [3] [6] [7].

In this paper, the asset price is modeled as a linear Brownian motion with a drift that drops from one constant to a smaller constant at some unobservable time. This drift is modeled as a Markov chain with two states which are denoted by 0 and 1 where 0 is denoted for price decrease and 1 is denoted for price increase.

We define the asset price model in Section 2, and the optimal selling problem is set up. In Section 3, we study the simulation to examine our studies and finally, Section 4 is conclusion.

2. The Model

We take as given a complete probability space (Ω, F, P). On this probability space, let the change-point τ be a random variable with distribution

$$P(\tau = 0) = \pi, P(\tau > t \mid t > 0) = e^{-\lambda t}$$

where λ is the intensity of the transition from state 1 to state 0 and assume that λ is positive and that belongs to [0; 1). Denote the drift of the price process a_t, $t \geq 0$, can be modeled as a Markov chain with two states a_l denoted by state 0 and a_h denoted by state 1 such that $P(a = a_h) = \pi_0$; $P(a = a_l) = 1 - \pi_0$ at time 0, $a_l < r < a_h$ where r is discounted rate which is a given constant and process a_t, $t \geq 0$ can only transit from state 1 to state 0 with transition density matrix as follows $Q = \begin{bmatrix} 0 & 0 \\ \lambda & -\lambda \end{bmatrix}$. Next, let W be a Brownian motion which is independent of τ. The asset price process X is modeled by a geometric Brownian motion with a drift that drops from a_h to a_l at time τ. More precisely,

$$dX_t = a_t X_t dt + \sigma X_t dW_t$$

and $X_0 = x > 0$, where $a_t = a_h - (a_h - a_l)I(t \geq \tau)$ i.e. $a_t = a_h$ if $t < \tau$ and $a_t = a_l$ if $t \geq \tau$ the volatility $\sigma > 0$ is a constant.

At the time of $t > 0$ we define the a posteriori probability process π_t by

$$\pi_t = P\{a = a_l \mid \mathcal{F}_t^X\}$$

where $\{\mathcal{F}_t^X\}_{0 \leq t \leq T}$ is a filter generated by X and τ. The process π_t indicate the probability of event that the price process decreases. We consider the optimal stopping problem:

Find $\mathcal{F}^X$-stopping time $\tau, 0 \leq \tau \leq T$ such that:

$$V = \sup_{0 \leq \tau \leq T} E\left[e^{-r\tau} X_\tau\right]. \tag{2.1}$$

Similar the buying problem, posterior probability process π_t satisfying the following stochastic differential (see theorem 9.1, [6]):

$$d\pi_t = \lambda(1 - \pi_t)dt - \pi_t \frac{a_h X_t - [a_h \pi_t + a_l(1 - \pi_t)]X_t}{\sigma X_t} d\bar{W}_t$$

or

$$d\pi_t = \lambda\left(1-\pi_t\right)dt - \pi_t\left(1-\pi_t\right)\left(\frac{a_h - a_l}{\sigma}\right)d\overline{W}_t$$

where $d\overline{W}_t$ is a P-Brownian motion with respect to $\mathcal{F}_t^X$ given by

$$d\overline{W}_t = \frac{dX_t - \left[\left(1-\pi_t\right)a_l + \pi_t a_h\right]X_t dt}{\sigma X_t} = \frac{\left\{a_t - \left[\left(1-\pi_t\right)a_l + \pi_t a_h\right]\right\}X_t dt + \sigma X_t dW_t}{\sigma X_t}$$

$$= \frac{\left\{a_t - \left[\left(1-\pi_t\right)a_l + \pi_t a_h\right]\right\}}{\sigma}dt + dW_t.$$

Moreover, in terms of $\overline{W}_t$ we have

$$\frac{dX_t}{X_t} = E\left[a_t\middle|\mathcal{F}_t^X\right]dt + \sigma d\overline{W}_t = \left[\left(1-\pi_t\right)a_l + \pi_t a_h\right]dt + \sigma d\overline{W}_t.$$

Processes X_t and π_t satisfy the following system equations:

$$\begin{cases} \dfrac{dX_t}{X_t} = \left(a_l + \pi_t\left(a_h - a_l\right)\right)dt + \sigma d\overline{W}_t \\[2mm] d\pi_t = \lambda\left(1-\pi_t\right)dt + \pi_t\left(1-\pi_t\right)\left(\dfrac{a_h - a_l}{\sigma}\right)d\overline{W}_t. \end{cases}$$

Put $\Phi_t = \dfrac{\pi_t}{1-\pi_t}; \omega = \dfrac{a_h - a_l}{\sigma}$ and Ito's formula gives that:

$$d\Phi_t = \left(\frac{\lambda_2}{1-\pi_t} + \omega^2 \pi_t \Phi_t\right)dt - \omega\Phi_t d\overline{W}_t.$$

We define new process $\left\{\tilde{W}_t\right\}$ as follow:

$$d\tilde{W}_t = -\left(\omega\pi_t + \sigma\right)dt + d\overline{W}_t$$

and a new measure P^* satisfying:

$$\frac{dP^*}{dP} = \exp\left\{\frac{1}{2}\int_0^T\left(\sigma + \omega\pi_t\right)^2 dt - \int_0^T\left(\sigma + \omega\pi_t\right)d\overline{W}_t\right\}$$

$$= \exp\left\{-\frac{1}{2}\int_0^T\left(\sigma + \omega\pi_t\right)^2 dt - \int_0^T\left(\sigma + \omega\pi_t\right)d\tilde{W}_t\right\}.$$

By Girsanov theorem, $\tilde{W}_t$ is a P^*-Brownian motion. Furthermore,

$$d\Phi_t = \left(\lambda + \left(\lambda - \sigma\omega\right)\Phi_t\right)dt - \omega\Phi_t d\tilde{W}_t \Rightarrow \Phi_t = Z_t\left[\Phi_0 + \lambda\int_0^t Z_s^{-1}ds\right];$$

where

$$Z_t = \exp\left\{\left(\lambda_2 + \sigma\omega - \frac{\omega^2}{2}\right)t - \omega\tilde{W}_t\right\}.$$

The price process X_t satisfying the following stochastic differential

$$\frac{dX_t}{X_t} = \left[\left(1-\pi_t\right)a_h + \pi_t a_l\right]dt + \sigma\left(d\tilde{W}_t + \left(\omega\pi_t + \sigma\right)dt\right)$$

or in term of $\tilde{W}_t$

$$\frac{dX_t}{X_t} = \left(a_h + \sigma^2\right)dt + \sigma d\tilde{W}_t.$$

The solution of this stochastic equation is

$$X_t = X_0 \exp\left\{\left(a_h + \frac{\sigma^2}{2}\right)t + \sigma\tilde{W}_t\right\}.$$

Now we consider the process:

$$\eta_t = \exp\left\{-\frac{1}{2}\int_0^t \left(\sigma - \omega\pi_t\right)^2 dt - \int_0^t \left(\sigma - \omega\pi_t\right)d\tilde{W}_t\right\}$$

then η_t is a $\mathcal{F}^X$-martingale and $\dfrac{d\eta_t}{\eta_t} = -\left(\omega\pi_t + \sigma\right)d\tilde{W}_t$ where $\eta_0 = 1$.

Let U_t is a process which defined by $dU_t = -\lambda U_t dt - \sigma U_t d\tilde{W}_t$, we have $U_t = \dfrac{e^{(a_h-\lambda)t}}{X_t}$, $U_0 = \dfrac{1}{X_0}$.

Put $Y_t = \dfrac{(1+\Phi_t)U_t}{(1+\Phi_0)U_0}$, $Y_0 = 1$ and according to Itô's formula:

$$\frac{dY_t}{Y_t} = -\left(\omega\pi_t + \sigma\right)d\tilde{W}_t.$$

From this we have $Y_t = \eta_t \forall t$ (a.s.), thus:

$$e^{-rt}\eta_\tau X_\tau = e^{-rt}Y_\tau X_\tau = \frac{X_0}{1+\Phi_0}e^{(a_h-\lambda-r)\tau}\left(1+\Phi_\tau\right).$$

Denote

$$X_0 = x, \Phi_0 = \phi$$

then

$$E_P e^{-rt}X_\tau = E_{p^*}\eta_T e^{-rt}X_\tau = E_{p^*}\eta_\tau e^{-rt}X_\tau = \frac{x}{1+\phi}E_{p^*}e^{(a_h-\lambda-r)\tau}\left(1+\Phi_\tau\right).$$

To solve the problem (2.1) we solve the following auxiliary problem:

$$G(\phi) = \sup_{\tau\in\mathfrak{M}} E_{p^*} e^{(a_h-\lambda-r)\tau}\left(1+\Phi_t\right). \tag{2.2}$$

Put

$$Z_t = e^{(a_h-\lambda-r)\tau}\left(1+\Phi_t\right).$$

The optimal stopping time is the first hitting time of the process Φ_t to the area $[B,\infty)$ with some B. Moreover pairs (B,F) satisfying the flowing free boundary problem:

$$\begin{cases} \mathcal{L}G + \left(a_h - \lambda - r\right)G = 0 & 0 < z < B \\ G(z) = 1 + z & z \geq B \\ G'(B) = 1 \\ G(0+) < \infty \end{cases} \tag{2.3}$$

where $\mathcal{L}$ is infinitesimal generated operator.

Differential equation in (2.3) has the general solution as follows: $G(z) = C_1 G_1(z) + C_2 G_2(z)$ where

$$G_1(x) = xe^{\frac{\lambda - \lambda x \ln x + \omega\sigma x \ln x}{\omega^2 x}} \text{Whittaker } W\left(\frac{\omega^2 + \omega\sigma - \lambda}{\omega^2}, -\frac{\sqrt{4\omega^2\lambda - 4\omega^2(a_h + a_l - 2r) + \omega^4 + 4(a_h - a_l - \lambda)^2}}{2\omega^2}, \frac{2\lambda}{\omega^2 x}\right)$$

$$G_2(x) = xe^{\frac{\lambda - \lambda x \ln x + \omega\sigma x \ln x}{\omega^2 x}} \text{Whittaker } M\left(\frac{\omega^2 + \omega\sigma - \lambda}{\omega^2}, -\frac{\sqrt{4\omega^2\lambda - 4\omega^2(a_h + a_l - 2r) + \omega^4 + 4(a_h - a_l - \lambda)^2}}{2\omega^2}, \frac{2\lambda}{\omega^2 x}\right).$$

Changing variables and using some analytic transformations we obtain:

$$\alpha = \frac{2\lambda}{\omega^2} > 0, \ \beta = \alpha - \frac{2\sigma}{\omega} > 0, \ \gamma = \sqrt{(\beta-1)^2 + \frac{8(\lambda + r - a_h)}{\omega^2}} > |\beta - 1|$$

then

$$G_1(z) = \int_0^\infty e^{-\alpha u} u^{(\beta+\gamma-3)/2}(1+zu)^{(\gamma-\beta+1)/2} \, du \quad \text{and} \quad G_2(z) = \int_0^{\frac{1}{z}} e^{\alpha u} u^{(\beta+\gamma-3)/2}(1-zu)^{(\gamma-\beta+1)/2} \, du.$$

We also have

$$G_2(z) > (1-\sqrt{z})^{(\gamma-\beta+1)/2} \int_0^{\frac{1}{\sqrt{z}}} u^{(\beta+\gamma-3)/2} du = \frac{2}{\beta+\gamma-1}(1-\sqrt{z})^{(\gamma-\beta+1)/2} z^{-\frac{(\beta+\gamma-1)}{4}} > \varepsilon z^{-\frac{(\beta+\gamma-1)}{4}} \to +\infty$$

as $z \to 0+$ since $0 < z < 1$ and $-\frac{\beta+\gamma-1}{4} < 0$.

We have

$$G_1'(z) = \frac{\gamma-\beta+1}{2} \int_0^\infty e^{-\alpha t} t^{(\beta+\gamma-1)/2}(1+zt)^{(\gamma-\beta-1)/2} \, dt > 0$$

since $r + \lambda > a_h$ therefore $\gamma = \sqrt{(\beta-1)^2 + \frac{8(\lambda + r - a_h)}{\omega^2}} > \beta - 1$ and $G_1(z)$ is an increasing function.

Moreover

$$G_1''(z) = \frac{\gamma-\beta+1}{2} \frac{\gamma-\beta-1}{2} \int_0^\infty e^{-\alpha t} t^{(\beta+\gamma+1)/2}(1+zt)^{(\gamma-\beta-3)/2} \, dt < 0$$

since

$$\gamma = \sqrt{(\beta-1)^2 + \frac{8(\lambda + r - a_h)}{\omega^2}} > \sqrt{(\beta-1)^2 + \frac{8(\lambda + a_l - a_h)}{\omega^2}}$$

$$= \sqrt{(\beta-1)^2 + \frac{4 \times 2(\lambda - \sigma\omega)}{\omega^2}} = \sqrt{(\beta-1)^2 + 4\beta} = \beta + 1.$$

These mean that the function $G_1(z)$ is increasing and convex on $(0, \infty)$.

Figure 1 shows the graph of function $G_1(z)$, we can check the increase and convex properties of it. The graph of $G_2(z)$ is shown in **Figure 2**, we can see that it tends to infinite when z as $0+$.

But $G(0+) < \infty$ therefore $C_2 = 0$ and $G(z) = C_1 \int_0^\infty e^{-\alpha u} u^{(\beta+\gamma-3)/2}(1+zu)^{(\gamma-\beta+1)/2} \, du.$

According to (2.3) we have

$$\begin{cases} C_1 \int_0^\infty e^{-\alpha u} u^{(\beta+\gamma-3)/2}(1+Bu)^{(\gamma-\beta+1)/2} \, du = 1 + B \\ C_1 \dfrac{\gamma-\beta+1}{2} \int_0^\infty e^{-\alpha t} u^{(\beta+\gamma-1)/2}(1+Bu)^{(\gamma-\beta-1)/2} \, du = 1 \end{cases}$$

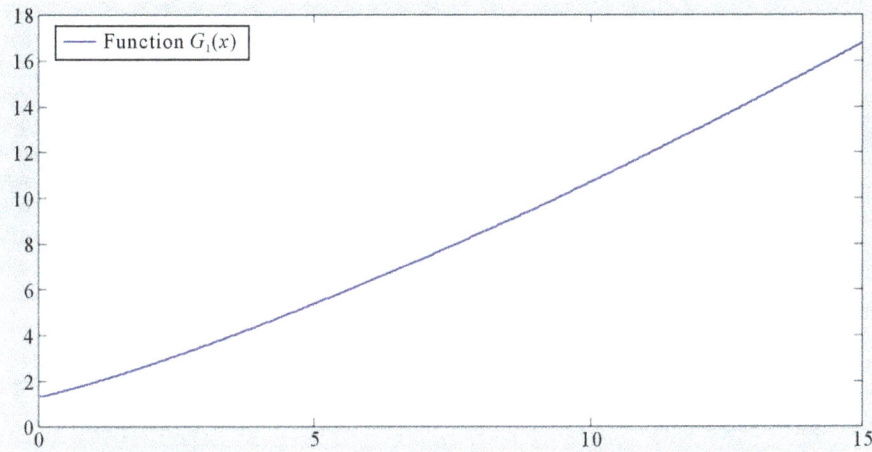

Figure 1. Graph of the function $G_1(x)$.

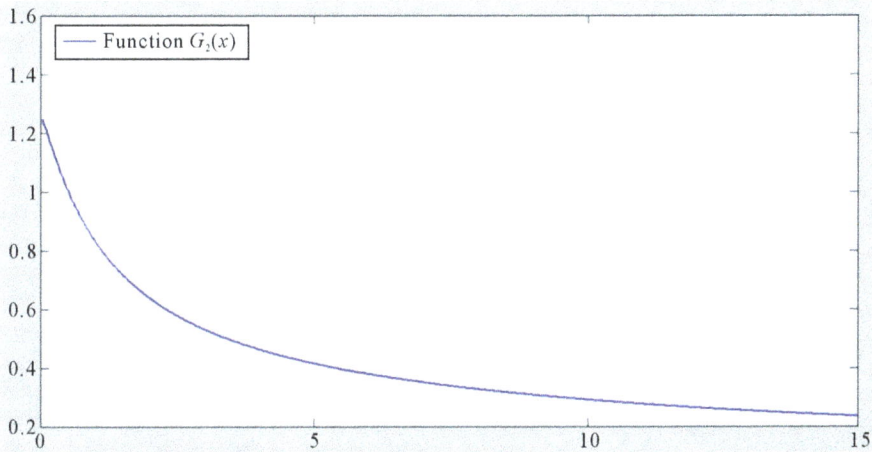

Figure 2. Graph of the function $G_2(x)$.

So B is the solution of the following equation:

$$2\int_0^\infty e^{-\alpha u} u^{(\beta+\gamma-3)/2} \left(1+Bu\right)^{(\gamma-\beta+1)/2} du = (1+B)(\gamma-\beta+1)\int_0^\infty e^{-\alpha u} u^{(\beta+\gamma-1)/2} \left(1+Bu\right)^{(\gamma-\beta-1)/2} du. \qquad (2.4)$$

Lemma 2.1. *The free boundary Equation* **(2.4)** *has unique positive solution B.*

Proof: The Equation (2.4) is equivalent to

$$2\int_0^\infty e^{-\alpha u} u^{(\beta+\gamma-3)/2} \left(1+xu\right)^{(\gamma-\beta+1)/2} du - (1+x)(\gamma-\beta+1)\int_0^\infty e^{-\alpha u} u^{(\beta+\gamma-1)/2} \left(1+xu\right)^{(\gamma-\beta-1)/2} du = 0.$$

Denote:

$$h(x) = 2\int_0^\infty e^{-\alpha u} u^{(\beta+\gamma-3)/2} \left(1+xu\right)^{(\gamma-\beta+1)/2} du - (1+x)(\gamma-\beta+1)\int_0^\infty e^{-\alpha u} u^{(\beta+\gamma-1)/2} \left(1+xu\right)^{(\gamma-\beta-1)/2} du.$$

The graph of $h(x)$ is shown in **Figure 3**. We shall prove that the function $h(x)$ satisfying $\lim_{x\to 0+} h(x) > 0$, $\lim_{x\to +\infty} h(x) = -\infty$ and $h(x)$ is decreasing and therefore the equation $h(x)=0$ has unique solution on $(0,+\infty)$.

We have

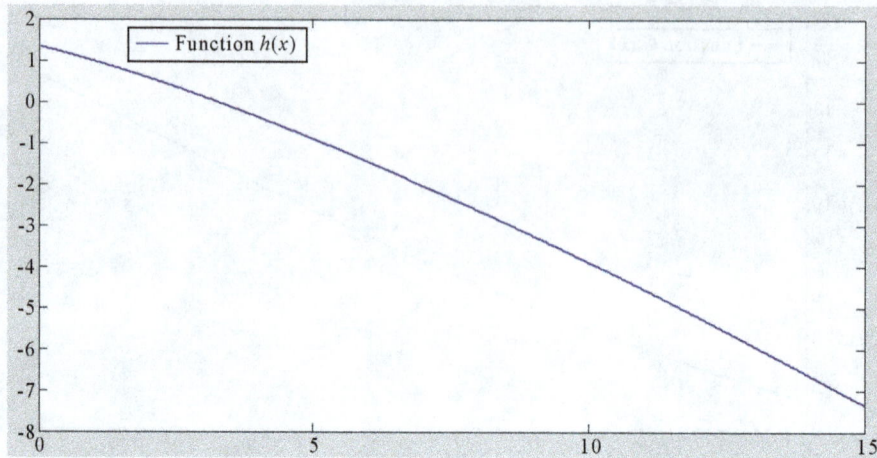

Figure 3. Graph of the function $h(x)$.

$$h(0+) = 2\int_0^\infty e^{-\alpha u} u^{(\beta+\gamma-3)/2} du - (\gamma-\beta+1)\int_0^\infty e^{-\alpha u} u^{(\beta+\gamma-1)/2} du$$

$$= 2\int_0^\infty e^{-\alpha u} u^{(\beta+\gamma-3)/2} du - (\gamma-\beta+1)\int_0^\infty e^{-\alpha u} u^{(\beta+\gamma-1)/2} du$$

$$= \frac{4}{\beta+\gamma-1}\int_0^\infty e^{-\alpha u} du^{(\beta+\gamma-3)/2} - (\gamma-\beta+1)\int_0^\infty e^{-\alpha u} u^{(\beta+\gamma-1)/2} du$$

$$= \frac{4}{\beta+\gamma-1}\left[e^{-\alpha u} u^{(\beta+\gamma-1)/2}\Big|_0^\infty + \alpha\int_0^\infty e^{-\alpha u} u^{(\beta+\gamma-1)/2} du \right] - (\gamma-\beta+1)\int_0^\infty e^{-\alpha u} u^{(\beta+\gamma-1)/2} du$$

$$= \left[\frac{4\alpha}{\beta+\gamma-1} - (\gamma-\beta+1) \right]\int_0^\infty e^{-\alpha u} u^{(\beta+\gamma-1)/2} du.$$

It follows that

$$h(0+) = \left[\frac{4\alpha}{\beta-1+\gamma} + (\beta-1-\gamma) \right]\int_0^\infty e^{-\alpha u} u^{(\beta+\gamma-1)/2} du = \left[\frac{4\alpha + (\beta-1)^2 - \gamma^2}{\beta-1+\gamma} \right]\int_0^\infty e^{-\alpha u} u^{(\beta+\gamma-1)/2} du$$

$$= \left[\frac{\dfrac{8\lambda}{\omega^2} - \dfrac{8(\lambda+r-a_h)}{\omega^2}}{\beta-1+\gamma} \right]\int_0^\infty e^{-\alpha u} u^{(\beta+\gamma-1)/2} du = \frac{8(a_h-r)}{\omega^2(\beta-1+\gamma)}\int_0^\infty e^{-\alpha u} u^{(\beta+\gamma-1)/2} du > 0$$

$$h'(x) = (\gamma-\beta+1)\int_0^\infty e^{-\alpha u} u^{(\beta+\gamma-1)/2} (1+xu)^{(\gamma-\beta-1)/2} du - (\gamma-\beta+1)\int_0^\infty e^{-\alpha u} u^{(\beta+\gamma+1)/2} (1+xu)^{(\gamma-\beta-1)/2} du$$

$$- (1+x)\frac{\gamma-\beta-1}{2}(\gamma-\beta+1)\int_0^\infty e^{-\alpha u} u^{(\beta+\gamma+1)/2} (1+xu)^{(\gamma-\beta-3)/2} du$$

$$= -(1+x)\frac{\gamma-\beta-1}{2}(\gamma-\beta+1)\int_0^\infty e^{-\alpha u} u^{(\beta+\gamma+1)/2} (1+xu)^{(\gamma-\beta-3)/2} du.$$

Because

$$\alpha = \frac{2\lambda}{\omega^2} > 0, \ \beta = \alpha - \frac{2\sigma}{\omega} = \frac{2(\lambda+a_l-a_h)}{\omega^2} > 0, \ \gamma = \sqrt{(\beta-1)^2 + \frac{8(\lambda+r-a_h)}{\omega^2}} > |\beta-1|$$

and

$$\gamma = \sqrt{(\beta-1)^2 + \frac{8(\lambda+r-a_h)}{\omega^2}} > \sqrt{(\beta-1)^2 + \frac{8(\lambda+a_l-a_h)}{\omega^2}} = \sqrt{(\beta-1)^2 + 4\beta} = \beta+1$$

we obtain

$$h'(x) = -(1+x)\frac{\gamma-\beta-1}{2}(\gamma-\beta+1)\int_0^\infty e^{-\alpha u} u^{(\beta+\gamma+1)/2}(1+xu)^{(\gamma-\beta-3)/2}\, du < 0 \quad \forall x > 0.$$

We will prove that $\lim_{x\to+\infty} h(x) = -\infty$. Indeed, with the large enough x we have $1+xu \sim xu$ so

$$h(x) = 2\int_0^\infty e^{-\alpha u} u^{(\beta+\gamma-3)/2}(1+xu)^{(\gamma-\beta+1)/2}\, du - (1+x)(\gamma-\beta+1)\int_0^\infty e^{-\alpha u} u^{(\beta+\gamma-1)/2}(1+xu)^{(\gamma-\beta-1)/2}\, du$$

$$\sim 2x^{(\gamma-\beta+1)/2}\int_0^\infty e^{-\alpha u} u^{(\beta+\gamma-3)/2} u^{(\gamma-\beta+1)/2}\, du - (1+x)(\gamma-\beta+1)x^{(\gamma-\beta-1)/2}\int_0^\infty e^{-\alpha u} u^{(\beta+\gamma-1)/2} u^{(\gamma-\beta-1)/2}\, du$$

$$= 2x^{(\gamma-\beta+1)/2}\int_0^\infty e^{-\alpha u} u^{(\beta+\gamma-3)/2} u^{(\gamma-\beta+1)/2}\, du - (\gamma-\beta+1)x^{(\gamma-\beta+1)/2}\int_0^\infty e^{-\alpha u} u^{\gamma-1}\, du$$

$$= (1+\beta-\gamma)x^{(\gamma-\beta+1)/2}\int_0^\infty e^{-\alpha u} u^{\gamma-1}\, du \to -\infty$$

since $1+\beta-\gamma < 0$, $\gamma-\beta+1 > 0$, $\int_0^\infty e^{-\alpha u} u^{\gamma-1}\, du = const > 0$.

Consequently, $\lim_{x\to 0+} h(x) > 0$, $\lim_{x\to+\infty} h(x) = -\infty$ and $h(x)$ is decreasing so the equation $h(x) = 0$ has unique solution on $(0,+\infty)$. The theorem is proved. $\square$

Theorem 2.2. *Stopping time* $\tau_B = \inf\{t \geq 0 : \Phi_t \geq B\}$ *is the optimal stopping time for* (2.1).

Proof: Let

$$L(x) = \begin{cases} \dfrac{1+B}{G_1(B)} G_1(x) & x < B \\ 1+x & x \geq B \end{cases}$$

and we will prove that $L(x) > 1+x \quad \forall x < B$, indeed

$$\frac{1+B}{G_1(B)} G_1(x) > 1+x \Leftrightarrow \frac{1+B}{G_1(B)} > \frac{1+x}{G_1(x)}.$$

Now, we examine the function $g(x) = \dfrac{1+x}{G_1(x)}$.

Take the derivative we obtain

$$g'(x) = \frac{1}{F_1^2(x)}\left(\int_0^\infty e^{-\alpha u} u^{(\beta+\gamma-3)/2}(1+xu)^{(\gamma-\beta+1)/2}\, du - (1+x)\frac{\gamma-\beta+1}{2}\int_0^\infty e^{-\alpha u} u^{(\beta+\gamma-1)/2}(1+xu)^{(\gamma-\beta-1)/2}\, du\right)$$

$$= \frac{h(x)}{F_1^2(x)} > \frac{h(B)}{F_1^2(B)} = 0 \Rightarrow g(B) > g(x) \Rightarrow \frac{1+B}{G_1(B)} > \frac{1+x}{G_1(x)}.$$

This follows

$$L(x) \geq 1+x \quad \forall x.$$

Using the Dynkin's formula to the process $Y_t = e^{(a_h-\lambda-r)t} L(\Phi_t)$ we have:

$$dY_t = e^{(a_h-\lambda-r)t}\left(a_h - r + (a_l-r)\Phi_t\right)1_{\{\Phi_t > B\}}\, dt - e^{(a_h-\lambda-r)t}\omega\Phi_t L'(\Phi_t)\, d\tilde{W}_t.$$

Because B satisfying $B \geq \dfrac{a_h - r}{r - a_l} := d$ so the drift of Y is positive and therefore Y is super martingale and

$Y_{t \wedge \tau_B}$ is martingale. By optional theorem we have:

$$\mathbb{E}_\phi e^{(a_h - \lambda - r)\tau}\left(1 + \Phi_\tau\right) \leq \mathbb{E}_\phi Y_\tau \leq \mathbb{E}_\phi Y_0 = L(\phi) \Rightarrow G(\phi) \leq L(\phi).$$

By $\tau = \tau_B$ we have $\mathbb{E}_\phi Y_{\tau_B} = \mathbb{E}_\phi Y_0 = L(\phi)$ moreover

$$G(\phi) = \sup_\tau \mathbb{E}_y^Q Y_\tau \geq \mathbb{E}_y^Q Y_{\tau_B} = L(\phi). \text{ So } G(\phi) = L(\phi).$$

We will show that B satisfy the condition: $B \geq \dfrac{a_h - r}{r - a_l} := d$. Indeed, by general optimal stopping theory all

points satisfy the form

$$\left(\mathcal{L} + a_h - \lambda - r\right)\left(1 + \phi\right) = a_h - r + \left(a_l - r\right)\phi$$

with positive value will be in continuation area

$$C = \left\{\phi : G(\phi) > 1 + \phi\right\}.$$

The optimal stopping time is the first hitting time of $\left\{\Phi_t\right\}_{t \geq 0}$ to the area:

$$D = \left\{\phi : G(\phi) = 1 + \phi\right\} \text{ or } \tau_B = \inf\left\{t \geq 0 : \Phi_t \geq B\right\}.$$

Thus function G satisfy the following condition:

$$G(\phi) = \sup_{\tau : \Phi_t \geq d} \mathbb{E}_\phi e^{(a_h - \lambda - r)}\left(1 + \Phi_\tau\right).$$

We define the function:

$$G^d(\phi) = \begin{cases} \dfrac{1 + d}{G_1(\phi)} G_1(\phi) & \phi < d \\[2mm] 1 + \phi & \phi \geq d \end{cases}$$

Now, we assume $B < d$, because $h(x)$ is a decreasing function so $h(d) < 0$. Then the left derivative of

$\left(G^d(d-)\right)' > \left(G^d(d+)\right)'$. Specially, $G^d(\phi) < 1 + \phi$ with some $\phi < d$. This follows $e^{(a_h - \lambda_2 - r)t}G^d(\Phi_t)$ is su-

per martingale and $e^{(a_h - \lambda - r)t \wedge \tau_d}G^d(\Phi_{t \wedge \tau_d})$ martingale. For the stopping time τ satisfying $\Phi_\tau \geq d$ we have:

$$\mathbb{E}_\phi e^{(a_h - \lambda - r)}\left(1 + \Phi_\tau\right) = \mathbb{E}_\phi e^{(a_h - \lambda - r)}G^d(\Phi_\tau) \leq G^d(\phi) \Rightarrow G(\phi) \leq G^d(\phi).$$

This contradicts to the existence of ϕ such that $G^d(\phi) < 1 + \phi$ when $G(\phi) \geq 1 + \phi \quad \forall \phi$. Finally, we achieve $B \geq d$.

The optimal stopping time τ_B is the first hitting time of $\left\{\Phi_t\right\}_{t \geq 0}$ to the area

$$D = \left\{\phi : G(\phi) = 1 + \phi\right\}.$$

But $G(\phi) \geq 1 + \phi \quad \forall \phi$ therefore

$$\tau_B = \inf\left\{t \geq 0 : \Phi_t \geq B\right\},$$

by this, we have finished the provement.

3. Simulation Study

To make visual for the above theory we simulate the asset price process, the posterior probability process π_t

process $\Phi(t) = \dfrac{\pi_t}{1 - \pi_t}$ (notice that $\Phi(t)$ is an increasing function of π_t) and the selling threshold B. Some parameters is used in our simulating are $a_l = -0.05$; $r = 0.06$; $a_h = 0.2$; $\sigma = 0.04$; $\lambda = 0.5$ and the time interval is [0, 1].

As can be seen in the figures from 4 to 8 if the price is increasing then the $\Phi(t)$ and π_t are decreasing and conversely.

Figure 4 shows the price process has increased since the time 0.2 so the $\Phi(t)$ decreased from the respective time and it can not hit the red line denoted the threshold, therefore the optimal selling time in this case is 1.

Figure 5 simulate a price process which is fluctuated from time 0 to 0.14 and decrease dramatically at the time 0.14 so the process $\Phi(t)$ increase sharply from this and crossover the threshold, it follows that the optimal time to liquidate the asset is about 0.17. At this time the price of the asset is lower than the origin but if we hold it we will sell it at a much more loss in the future.

Another simulation is shown in **Figure 6**. Clearly, whenever the price process is increasing, the $\Phi(t)$ and the posterior probability process are decreasing and the liquidated time is 0.77. At this time the price is not the highest but it is significantly higher than the original value which is 1.5.

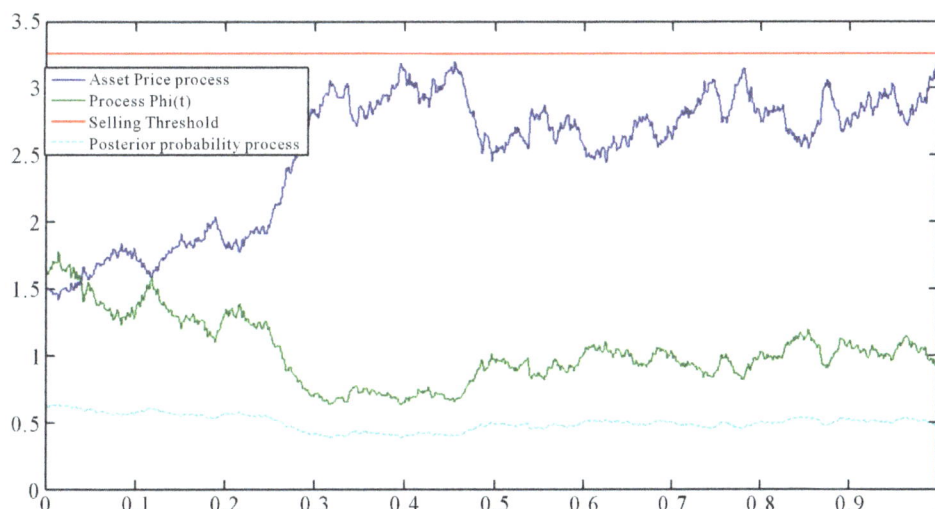

Figure 4. A simulation of asset price process, the posterior probability process, process $\Phi(t)$, the threshold probability and the optimal stopping time. In this case, the process $\Phi(t)$ always under the threshold probability so the optimal stopping time is the final time 1.

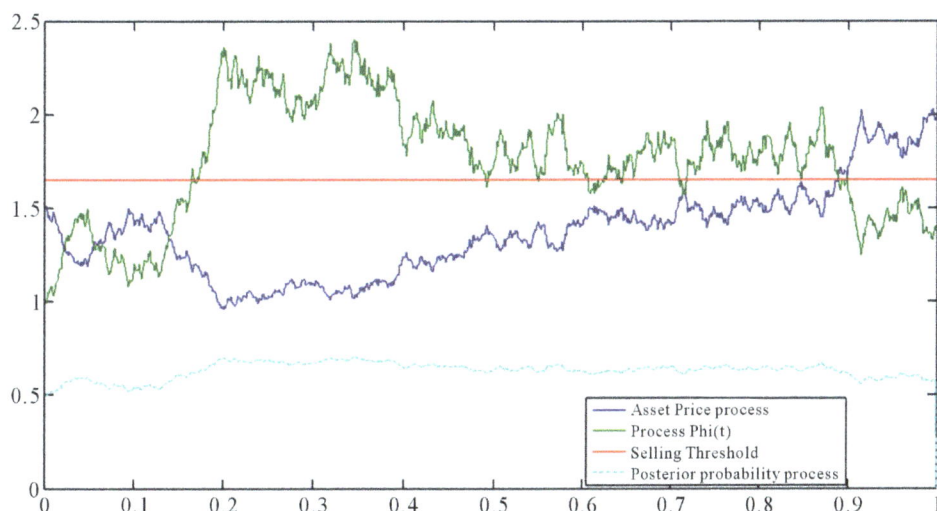

Figure 5. A simulation of asset price process, the posterior probability process, process $\Phi(t)$, the threshold probability and the optimal stopping time. In this case, the first time that the process $\Phi(t)$ over passes the threshold probability at the time 0.17 so the optimal stopping time is 0.17.

In **Figure 7**, we can see the same scenario with the simulation in **Figure 6**. The time to liquidate in this case is 0.795, the price is about 1.75 whereas the started price was 1.5. We can see

$$S_t e^{-rt} = 1.75 * e^{-0.06*0.795} = 1.668 > 1.5 = S_0$$ it means that we benefit by this trade affair.

The same scenario with the simulation in **Figure 1**, the simulation results in **Figure 8** show the price illustrates an uptrend from time 0 to the end that the process $\Phi(t)$ can not pass over the selling threshold B, consequently, the optimal time to sell in this situation is 1.

4. Conclusion

This research considers the problem of how to find the optimal time to liquidate an asset when the asset price is modeled by the geometric Brownian motion which has a change point. In particular, the drift of the process drops from a high value to a smaller one and this drift process can be modeled as two-state Markov process. The results of this research indicate that a optimal selling decision is made when the probability of downtrend surpassed some certain threshold. We also simulate the price process with a number of parameters and conduct

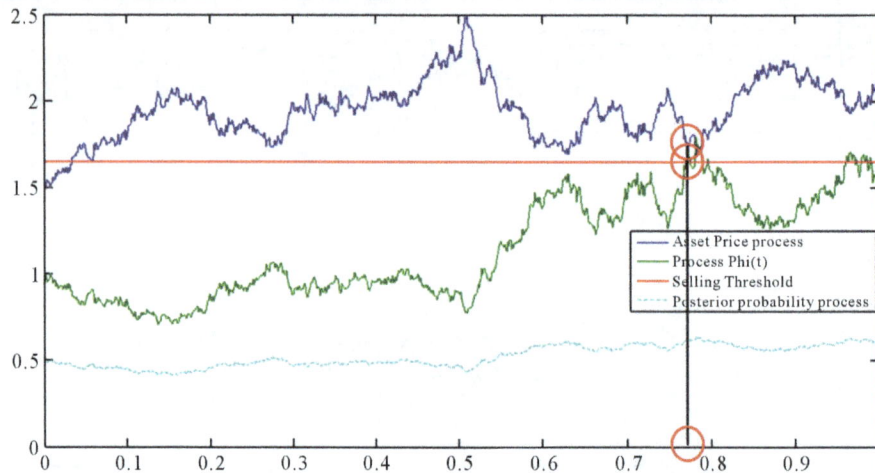

Figure 6. A simulation of asset price process, the posterior probability process, process $\Phi(t)$, the threshold probability and the optimal stopping time. In this case, the first time that the process $\Phi(t)$ over passes the threshold probability at the time 0.77 so the optimal stopping time is 0.77.

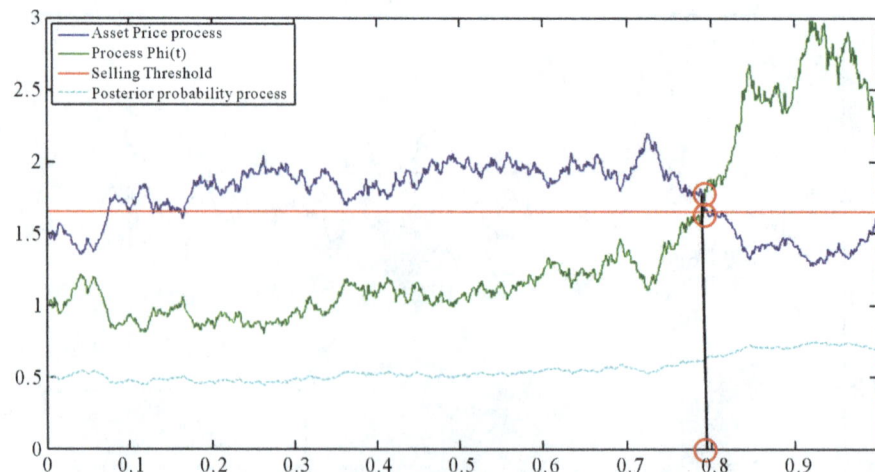

Figure 7. A simulation of asset price process, the posterior probability process, process $\Phi(t)$, the threshold probability and the optimal stopping time. In this case, the first time that the process $\Phi(t)$ over passes the threshold probability at the time 0.795 so the optimal stopping time is 0.795.

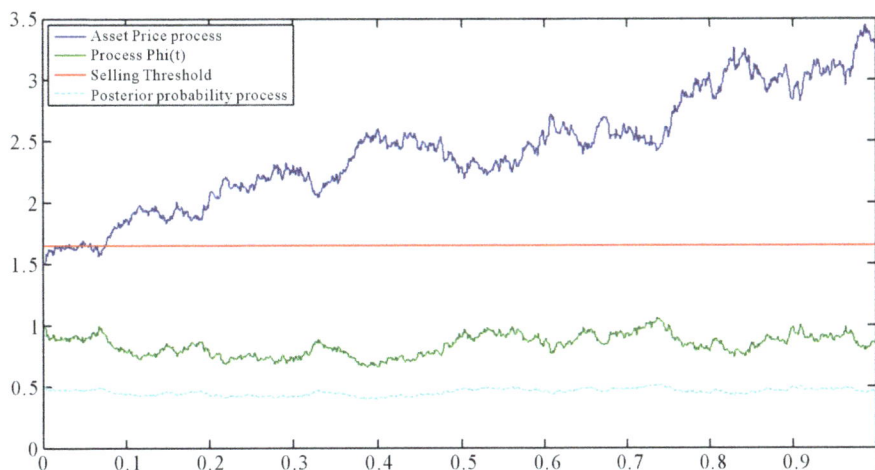

Figure 8. A simulation of asset price process, the posterior probability process, process $\Phi(t)$, the threshold probability and the optimal stopping time. In this case, the process $\Phi(t)$ always under the threshold probability so the optimal stopping time is the final time 1, the same with the case in **Figure 4**.

numerical solution to the experimental selling threshold. In next studies, we will consider problems in which the price growth rate is a Markov process which has more than 2 states and establish some properties as well as distribution of stopping time.

Acknowledgements

This research is funded by Vietnam National Foundation for Science and Technology Development (NAFOSTED) under grant number 10103-2012.17.

References

[1]   Khanh, P. (2014) Optimal Stopping Time to Buy an Asset When Growth Rate Is a Two-State Markov Chain. *American Journal of Operations Research*, **4**, 132-141. http://dx.doi.org/10.4236/ajor.2014.43013

[2]   Khanh, P. (2012) Optimal Stopping Time for Holding an Asset. *American Journal of Operations Research*, **4**, 527-535. http://dx.doi.org/10.4236/ajor.2012.24062

[3]   Peskir, G. and Shiryaev, A.N. (2006) Optimal Stopping and Free-Boundary Problems (Lectures in Mathematics ETH Lectures in Mathematics. ETH Zürich (Closed)). Birkhäuser, Basel.

[4]   Shiryaev, A.N., Xu, Z. and Zhou, X.Y. (2008) Thou Shalt Buy and Hold. *Quantitative Finance*, **8**, 765-776. http://dx.doi.org/10.1080/14697680802563732

[5]   Guo, X. and Zhang, Q. (2005) Optimal Selling Rules in a Regime Switching Model. *IEEE Transactions on Automatic Control*, **9**, 1450-1455. http://dx.doi.org/10.1109/TAC.2005.854657

[6]   Lipster, R.S. and Shiryaev, A.N. (2001) Statistics of Random Process: I. General Theory. Springer-Verlag, Berlin, Heidelberg.

[7]   Shiryaev, A.N. (1978) Optimal Stopping Rules. Springer Verlag, Berlin, Heidelberg.

Permissions

All chapters in this book were first published in AJOR, by Scientific Research Publishing; hereby published with permission under the Creative Commons Attribution License or equivalent. Every chapter published in this book has been scrutinized by our experts. Their significance has been extensively debated. The topics covered herein carry significant findings which will fuel the growth of the discipline. They may even be implemented as practical applications or may be referred to as a beginning point for another development.

The contributors of this book come from diverse backgrounds, making this book a truly international effort. This book will bring forth new frontiers with its revolutionizing research information and detailed analysis of the nascent developments around the world.

We would like to thank all the contributing authors for lending their expertise to make the book truly unique. They have played a crucial role in the development of this book. Without their invaluable contributions this book wouldn't have been possible. They have made vital efforts to compile up to date information on the varied aspects of this subject to make this book a valuable addition to the collection of many professionals and students.

This book was conceptualized with the vision of imparting up-to-date information and advanced data in this field. To ensure the same, a matchless editorial board was set up. Every individual on the board went through rigorous rounds of assessment to prove their worth. After which they invested a large part of their time researching and compiling the most relevant data for our readers.

The editorial board has been involved in producing this book since its inception. They have spent rigorous hours researching and exploring the diverse topics which have resulted in the successful publishing of this book. They have passed on their knowledge of decades through this book. To expedite this challenging task, the publisher supported the team at every step. A small team of assistant editors was also appointed to further simplify the editing procedure and attain best results for the readers.

Apart from the editorial board, the designing team has also invested a significant amount of their time in understanding the subject and creating the most relevant covers. They scrutinized every image to scout for the most suitable representation of the subject and create an appropriate cover for the book.

The publishing team has been an ardent support to the editorial, designing and production team. Their endless efforts to recruit the best for this project, has resulted in the accomplishment of this book. They are a veteran in the field of academics and their pool of knowledge is as vast as their experience in printing. Their expertise and guidance has proved useful at every step. Their uncompromising quality standards have made this book an exceptional effort. Their encouragement from time to time has been an inspiration for everyone.

The publisher and the editorial board hope that this book will prove to be a valuable piece of knowledge for researchers, students, practitioners and scholars across the globe.

List of Contributors

Kunimitsu Iwadare and Tatsuo Oyama
National Graduate Institute for Policy Studies (GRIPS)
7-22-1 Roppongi Minato-ku, Tokyo, Japan

R. Alan Bowman and James Lambrinos
School of Management, Union Graduate College, Schenectady, NY, USA

Thomas Ashman
Behavioral Sciences Collegium, Eckerd College, St. Petersburg, FL, USA

Elias Munapo
Graduate School of Business and Leadership, University of KwaZulu-Natal, Westville Campus, Durban, South Africa

Santosh Kumar
Department of Mathematics and Statistics, University of Melbourne, Parkville, Australia

Hongyun Wang
Department of Applied Mathematics and Statistics, Baskin School of Engineering, University of California, Santa Cruz, USA

Hong Zhou
Department of Applied Mathematics, Naval Postgraduate School, Monterey, USA

Sushil Kumar and U. S. Rajput
Department of Mathematics & Astronomy, University of Lucknow, Lucknow, India

Onkabetse A. Daman and Sulaiman Sani
Department of Mathematics, University of Botswana, Gaborone, Botswana

Sulaiman Sani
Department of Mathematics and Computer Science, Umaru Musa Yar'Adua University, Katsina, Nigeria

Shujuan Luo, Sijun Bai and Suike Li
School of Management, Northwestern Polytechnical University, Xi'an, China

Abbas Al-Refaie, Mohammad D. Al-Tahat and Ruba Najdawi
Department of Industrial Engineering, University of Jordan, Amman, Jordan

Han Shih
University of Missouri, Columbia, USA

Zhenping Li
School of Information, Beijing Wuzi University, Beijing, China

Shihua Zhang and Xiangsun Zhang
National Center for Mathematics and Interdisciplinary Sciences, Academy of Mathematics and Systems Science, Beijing, China

Fangjun Mu
Management School, Jinan University, Guangzhou, China

Bingyu Lan
College of Mathematics and Sciences, Shanghai Normal University, Shanghai, China

Jin Peng
Institute of Uncertain Systems, Huanggang Normal University, Hubei, China

Lin Chen
Institute of Systems Engineering, Tianjin University, Tianjin, China

Mayank Verma and Renduchintala Raghavendra Kumar Sharma
Department of Industrial and Management Engineering, Indian Institute of Technology Kanpur, Kanpur, India

Zhenping Li and Wenyu Li
School of Information, Beijing Wuzi University, Beijing, China

Pham Van Khanh
Institute of Economics and Corporate Group, Hanoi, Vietnam